Bioorganic Marine
Chemistry   Volume 1

Edited by Paul J. Scheuer

With Contributions by
J. W. Blunt   W. Fenical   N. Fusetani
P. Karuso   R. T. Luibrand
M. H. G. Munro   V. J. Paul

Springer-Verlag
Berlin Heidelberg New York
London Paris Tokyo

Professor Paul J. Scheuer
University of Hawaii at Manoa, Department of Chemistry
2545 The Mall, Honolulu, Hawaii 96822, USA

ISBN 3-540-17884-8 Springer-Verlag Berlin Heidelberg New York
ISBN 0-387-17884-8 Springer-Verlag New York Berlin Heidelberg

Library of Congress Cataloging in Publication Data
Bio-organic marine chemistry. 1. Marine ecology. 2. Environmental chemistry. 3. Chemical
oceanography. 4. Bioorganic chemistry. 5. Biological products. 6. Marine pharmacology.
I. Scheuer, Paul J. II. Blunt, J. W.
QH541.5.S3B57 1987   574.5′2636   87-16641
ISBN 0-387-17884-8 (U.S.: v. 1)

© Springer-Verlag Berlin Heidelberg 1987
Printed in Germany

Typesetting, printing, and binding: Brühlsche Universitätsdruckerei, Giessen
2152/3140-543210

Preface

The present series, "Bio-organic Marine Chemistry," is being launched at a time when we have the fundamental knowledge and the requisite instrumentation to probe the molecular basis of many biological phenomena.

The final volume of "Marine Natural Products–Chemical and Biological Perspectives" (Academic Press), which may be considered the precursor of this series, was published in 1983. In that series, which I edited, primary emphasis was placed on molecular structure and phyletic relationships. This focus was compatible with the major concerns of a growing research community in the field of marine natural products. Moreover, a need existed for timely reviews of a rapidly expanding and widely scattered primary literature. As I read again the Preface to Volume 1 (1978), I am amazed at the changes in direction and emphasis which have taken place during these few intervening years. Sufficient basic data are now at hand to gauge the breadth of the marine natural product spectrum and to raise questions of functions, both within and outside the marine ecosystem. Although we have few answers, the questions have become meaningful and pointed. Furthermore, the task of tracking and cataloguing the steady stream of fascinating new structures has been assumed by Faulkner's periodic surveys in *Natural Product Reports,* a bimonthly publication of the Royal Society of Chemistry.

The study of marine natural products remains firmly anchored in chemistry, i.e. in molecular integrity, yet continues to seek greater involvement in functional biology. Involvement with biology has taken place on three fronts simultaneously: a study of chemical interactions occurring in the marine habitat, i.e. marine ecology; a search for meaningful and appropriate bioassays designed to guide isolation and to predict the nature and degree of biological activity; and finally, a delineation of biological activity that might lead to application in medicine or agriculture.

The first two chapters, by Karuso and by Paul and Fenical, address the organic chemistry occurring on a coral reef. Paul and Fenical focus on green algae, their ability to survive, and how this ability might be a predictor of biological activity in general. Karuso's chapter deals with the defense mechanisms of nudibranchs, a group of soft-bodied marine invertebrates which seem to have evolved chemical as well as physical adaptation. These often beautifully colored animals provided the original stimulus to the study of chemical marine ecology some twelve years ago; they continue to be fascinating and productive research targets.

In Chapter 3 Fusetani describes his recent work in the context of a new bioassay which uses a marine invertebrate system to monitor laboratory separation, while at the same time it provides an indicator activity in a general physiological (i.e. non-marine) setting.

The final chapter by Munro and coworkers describes the efforts over the past few years toward the discovery of new useful antiviral and anticancer compounds. While traditional natural product chemistry was nurtured by folklore and serendipity, the current approach is systematic, thorough, and wide in scope.

I take great pleasure in introducing this new series which, I hope, will fill a current need and provide an incentive for future research. I encourage comments, pro and con, from the research community in the hope that "Bio-organic Marine Chemistry" will remain responsive to new insights and opportunities. Finally, I should like to thank the contributors, the colleagues who have provided manuscripts and preprints, and Springer-Verlag for their trust and cooperation.

June 1987   Paul J. Scheuer

Table of Contents

Natural Products Chemistry and Chemical Defense in Tropical Marine Algae of the Phylum Chlorophyta

Valerie J. Paul [1] and William Fenical [2]

Contents

Abstract

This article reviews the natural products chemistry of marine green algae (Chlorophyta), including investigations to December, 1986. The natural products chemistry is presented with an emphasis on chemical ecology, and in particular upon the natural biological functions of algal metabolites in the marine environment. The results of biological testing of the metabolites in laboratory assays designed to examine antimicrobial activity, cytotoxicity, larval toxicity, and ichthyotoxicity are summarized. Results of laboratory and field experiments designed to identify the deterrent effects of the metabolites toward herbivores are also reviewed. Recent studies, documenting chemical variation in the production of secondary metabolites within plants, among individual plants, and among populations of algae are discussed, with a focus on the ecological factors which may produce this phenomenon.

1. Introduction

In contrast to the more than 200 papers describing unique metabolites from marine red and brown algae [1, 2], relatively few studies have focused on metabolites from the marine green algae (Chlorophyta). Most of the research on this group of marine plants has been carried out within the last five years and it is the pur-

---

1 Marine Laboratory, University of Guam, UOG Station, Mangilao, Guam 96923 U.S.A.
2 Scripps Institution of Oceanography, University of California, San Diego, La Jolla, CA 92093 U.S.A.

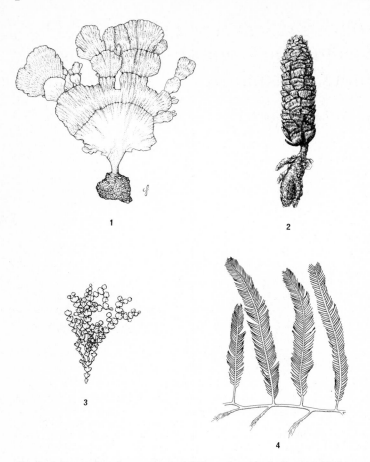

Fig. 1.–4. Examples of some tropical Chlorophyta, including calcified green algae of two families: (**1**) *Udotea flabellum* (Ellis et Solander) Lamouroux (Udoteaceae), (**2**) *Rhipocephalus phoenix* f. *breveformis* A. et E. Gepp (Udoteaceae), (**3**) *Halimeda goreauii* W. Taylor (Udoteaceae), and (**4**) *Caulerpa sertularioides* (S. Gmelin) Howe, a species of this monotypic genus of the Caulerpaceae. [Illustrations courtesy of J. Norris, Smithsonian Institution.]

pose of this chapter to review this recent work in relation to the ecological significance of the highly bioactive terpenoids which are found within this group of marine algae.

Most species of the green algae are tropical. In particular, members of the order Caulerpales are some of the most abundant and widely distributed algae on coral reefs and seagrass beds throughout the tropical oceans [3, 4]. It is this group of marine green algae that has been most thoroughly investigated both chemically and biologically. The order is subdivided into the families Caulerpaceae, containing the single large genus *Caulerpa*, and the Udoteaceae, containing the genera *Chlorodesmis, Halimeda, Avrainvillea, Penicillus, Tydemania, Udotea*, and others [5]. Many species of the Udeoteaceae have the ability to acquire deposits of calcium carbonate and some are highly calcified (between 50 to 90 percent, dry weight) [4, 6–8].

Other families of green algae which have been chemically investigated include the Dasycladaceae and Cladophoraceae. These groups have not been examined as thoroughly as the caulerpalean algae, but several nonterpenoid secondary metabolites have been isolated, and these will be reviewed here. Current research indicates that other taxonomically diverse species of green algae may warrant chemical investigation.

Studies of plant-herbivore interactions in tropical marine habitats have shown that predation on marine plants is intense, and that herbivores directly affect the community structure and biomass of benthic algae on coral reefs and other tropical marine habitats [9, 10, 12–14, 16–21]. Some excellent reviews have been published on marine herbivory [22–24]. Although many species of green algae (in particular, members of the Caulerpales) are abundant and widely distributed in reef habitats characterized by intense herbivory, they have generally been shown to be low preference components in the diets of macroherbivores such as fishes and sea urchins [9–14]. While the basis for the proliferation of these algae has not been defined, it has generally been accepted that calcification provides an effective physical defense against herbivores [9–12]. But in addition, investigations of the natural products chemistry have now shown that many of these algae produce unique secondary metabolites that appear to also play a role in the defense against herbivores.

The chemical ecology of plant-herbivore interactions has been extensively investigated in terrestrial habitats, but until recently, neglected in marine ecosystems. Information on the natural products chemistry of the green algae and preliminary information on the biological activities of these compounds provide the background necessary to test some of the predictions resulting from terrestrial research in chemical ecology. Some of the hypotheses of defense theory, based on studies of terrestrial plant-herbivore interactions [15], were examined for the first time in the marine environment by studying the green algae.

This chapter reviews the natural products chemistry of the green algae including several recent contributions. The results of biological testing of these metabolites in laboratory bioassays designed to examine antimicrobial effects, cytotoxicity, larval toxicity, and ichthyotoxicity are presented. Results of laboratory and field experiments designed to identify the deterrent effects of the metabolites toward herbivores are discussed. Finally, recent studies of chemical variation in the production of secondary metabolites within plants, among individual plants, and among populations of algae are presented. The ecological factors which may lead to the observed chemical variation are discussed.

2. Chemical Isolation and Structural Elucidation

One important factor contributing to the early paucity of chemical studies of the caulerpalean algae is that many of the secondary metabolites are very unstable and difficult to isolate. If freshly collected algae are frozen or stored in solvent for any period of time, the metabolites have been found to decompose. Recent investigators have found it necessary to extract the fresh algae and immediately purify the extracts in order to isolate the terpenoids, and to investigate the chemical and biological activities of these metabolites.

Of the many algal species from the Order Caulerpales that have been studied, virtually all have been found to produce unprecedented secondary metabolites. Recent work has focused on the families Caulerpaceae and Udoteaceae. The majority of the new compounds discovered are sesquiterpenoids and diterpenoids, often acyclic. Enol acetates and aldehydes are functional groups common to most of these metabolites. The terminal, bis-enol acetate moiety is the unifying feature of many of these metabolites, and it is a functional group uniquely found in this group of marine algae. This functionality represents an acetylated bis-enol form of a 1,4-dialdehyde constellation, to which high biological activity and chemical reactivity are generally attributed.

Secondary metabolites possessing aldehyde groups are among the most toxic and deterrent compounds that have been isolated from terrestrial plants and insects [25, 26]. Aldehyde groups, and especially unsaturated aldehydes, are potent electrophiles which can react with proteins in a number of ways to result in protein deactivation and interference with normal enzyme function. Aldehydes can react with protein primary amines to form Schiff bases (imines), or nucleophiles (e.g. alcohols, sulfhydryl groups) can add to the beta-carbon of unsaturated aldehydes in a typical "Michael Addition" reaction. Little is known or predictable, however, regarding the chemical and biological reactivity of the bis-enol acetate functionalities found in these algae. *In vivo* hydrolysis of the enol acetates could occur to yield highly reactive 1,4-dialdehydes, but addition reactions to the unsaturation may also be possible.

Algae of the Family Caulerpaceae

Species of the pantropical and subtropical genus *Caulerpa* (the only genus within the family Caulerpaceae) were the first algae of this group that were chemically investigated. Australian workers [27], studying the abundant alga *Caulerpa brownii* in Tasmanian waters, isolated the diterpenoid caulerpol (**1**) which, curiously, is the reduction product and olefin positional isomer of the important visual pigment retinal. Studies of other Australian *Caulerpa* species continued with the isolation of the sesquiterpenoid flexilin (**2**), and the diterpenoid trifarin (**3**), from the southern Australian algae *C. flexilis* and *C. trifaria*, respectively [28]. These two metabolites were the first natural products isolated that were shown to possess the 1,4-diacetoxybuta-1,3-diene (bis-enol acetate) moiety. This functional group was unknown in terrestrial or marine natural products research, and

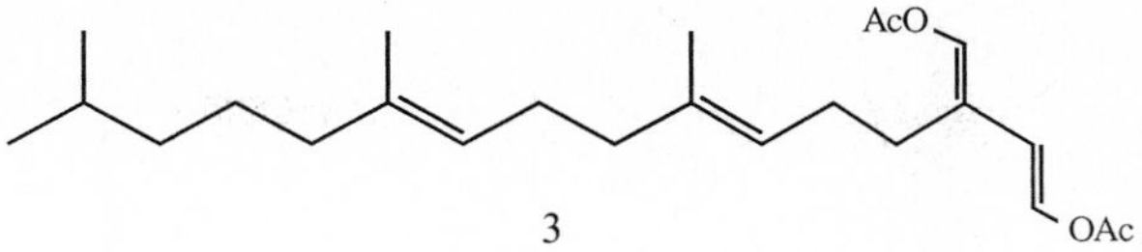

3

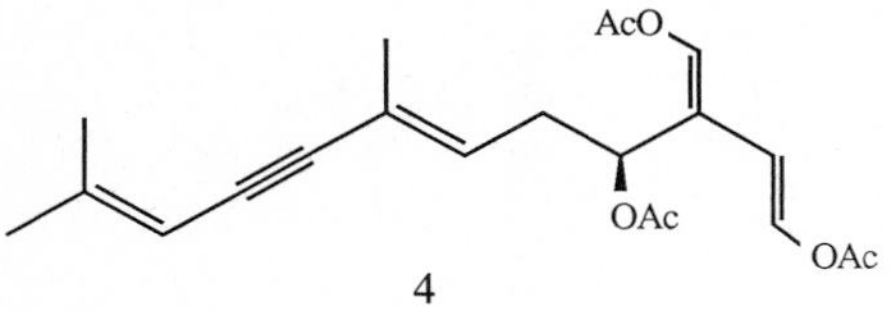

4

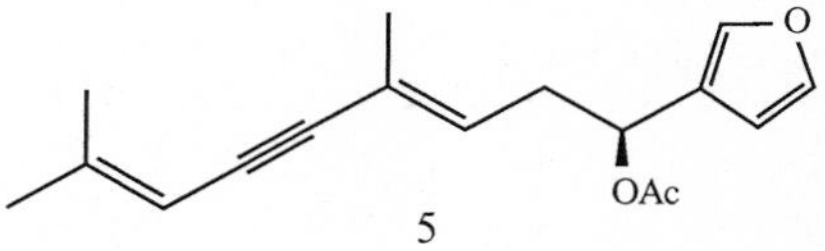

5

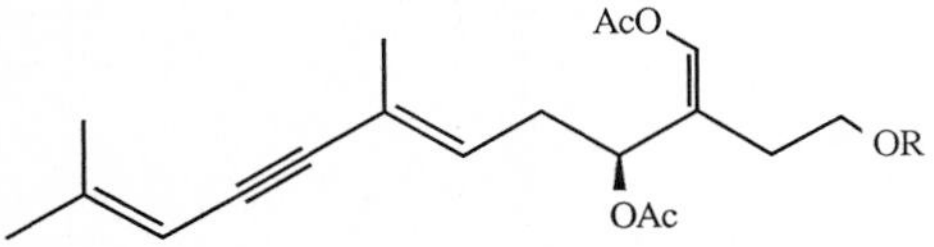

6a R=CH3(CH2)7CH=CH(CH2)7CO-
6b R=CH3(CH2)4CH=CHCH2CO-
6c R=C15H31CO-
6d R=CH3(CH2)5CH=CH(CH2)7CO-
6e R=C13H27CO-

it has subsequently been shown to be a characteristic feature of the majority of the green algal metabolites.

Caulerpenyne (**4**), a unique acetylenic sesquiterpenoid, is closely related to flexilin, and this interesting compound was first isolated from a Mediterranean collection of *C.prolifera* [29]. During subsequent investigations [30], caulerpenyne has been isolated from nine other species of the genus *Caulerpa* common in oth the tropical Pacific Ocean and the Caribbean Sea. These species include *C.prolifera* (Caribbean), *C.racemosa* (Caribbean, Pacific Mexico, western Pacific), *C.mexicana* (Florida Keys), *C.sertularioides* (Caribbean), *C.taxifolia* (Saipan), *C.paspaloides* (Bahamas), *C.lanuginosa* (Florida Keys), *C.cupressoides* (Caribbean, western Pacific), and *C.verticillata* (Caribbean).

Continued investigation of the Mediterranean *C.prolifera* led to the isolation of several additional acetylene-containing metabolites, furocaulerpin (**5**), and the fatty acid esters (**6**), in minor amounts [31, 32]. Although the fatty acid esters can be formed only after an overall reduction of caulerpenyne, furocaulerpin is easily envisioned as a dehydration product formed from the 1,4-dialdehyde derivative of caulerpenyne, obtained by hydrolysis. Acetylenes are common metabolites from terrestrial sources, but they are less common in the marine environment. These represent the only acetylenic compounds from green algae.

Several monocyclic sesquiterpenoids have been reported from species of *Caulerpa*. The major metabolites of two Indo-Pacific Caulerpa species, *C.bikinensis* (Palau) and *C.flexilis* var. *muelleri* (W. Australia), were the olefin isomers 7 and 8 [33, 34]. The minor metabolites of these two algae were different and included the dialdehyde 9 and butenolide 10 from *C.bikinensis* [33] and the acetoxy-aldehyde 11 from *C.flexilis* var. *muelleri* [34]. In each case the minor metabolites reflected partial hydrolysis of the bis-enol acetate functional group. The related monocyclic sesquiterpenoids, 12–16, have recently been reported from Florida collections of *C.ashmeadii* [35]. In relation to the chemistry of *C.prolifera*, it may be important to point out the presence of related fatty acid esters in *C.ashmedii*. In both cases, the bioactive compounds could either be stored or transported as

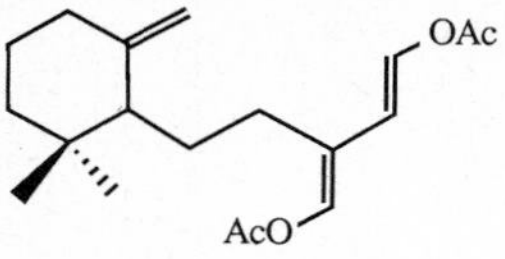

7 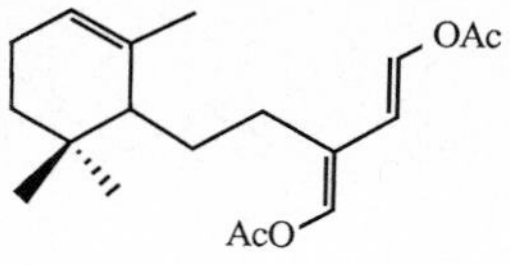   8 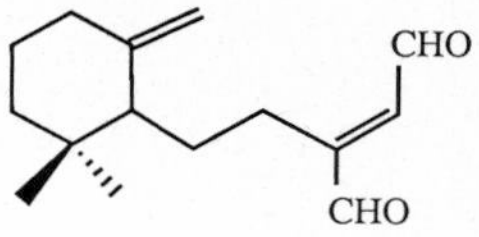   9

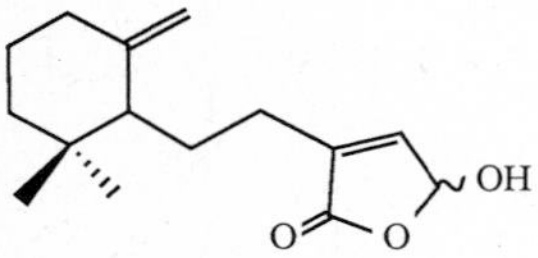

10 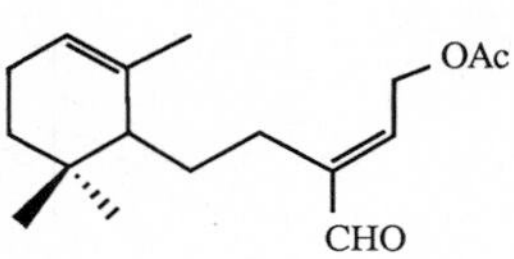   11 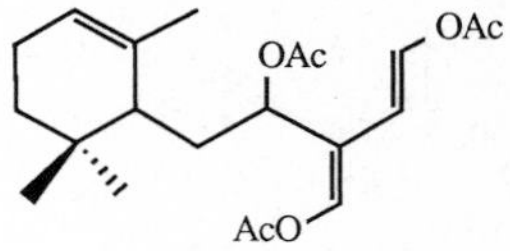   12

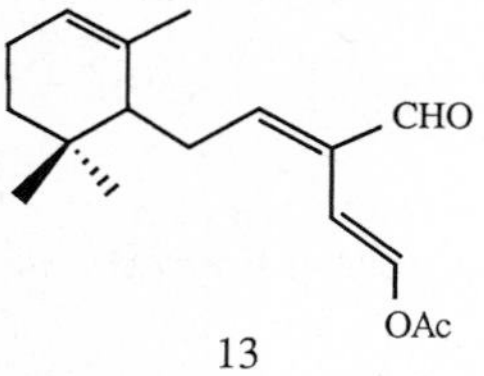

13 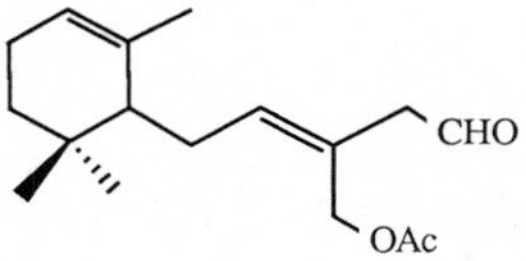   14 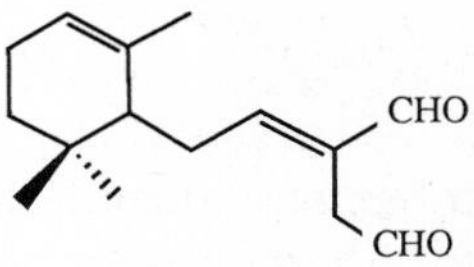   15

fatty acid ester-conjugates. This conjugation may render the compounds less reactive, hence reducing their *in vivo* toxicities.

Continued research involving Australian *Caulerpa* species has resulted in the description of two new cyclic ditperpenoids each possessing the terminal bis-enol acetate functionality. A monocyclic diterpenoid **17**, possessing a structure similar to that of caulerpol (**1**), has recently been isolated from another Australian collection of *C. brownii* [36]. One bicyclic diterpenoid **18**, which is composed of a traditional "labdane" (bicyclo[4.4.0]decane) skeleton, has also been reported from a Western Australian collection of *C. trifaria* [37].

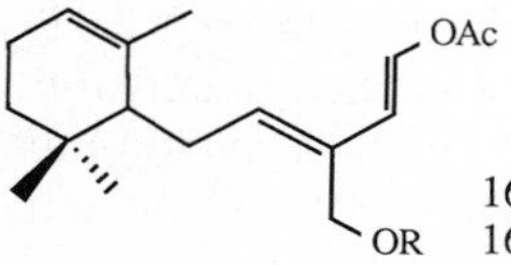

16a R=C15H31CO-
16b R=C15H29CO-

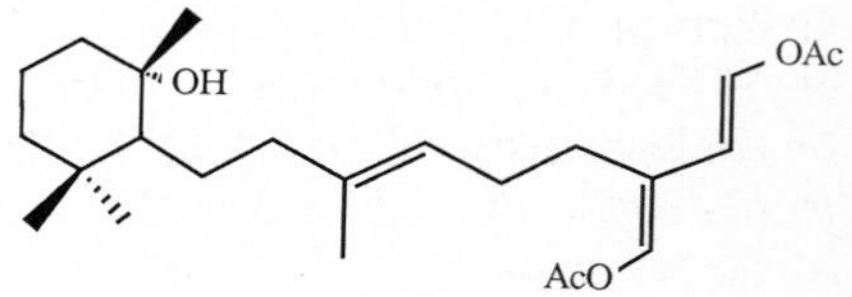

17

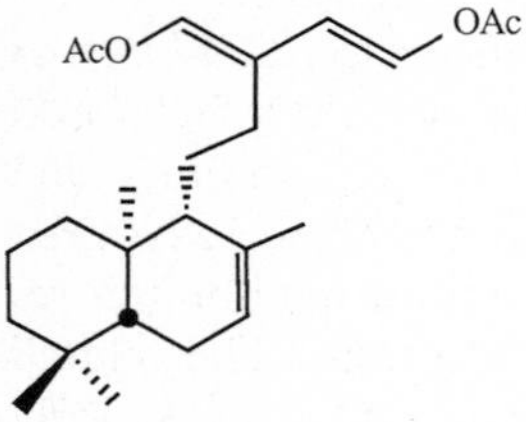

18 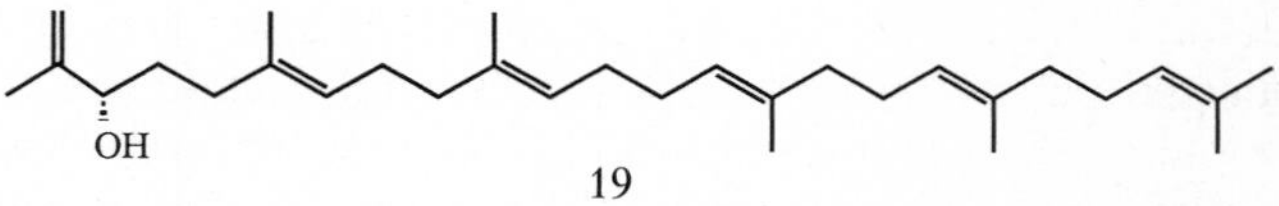   19

The mediterranean alga *C. prolifera* also contains several squalene derivatives, which are typical uncyclized isoprenoids. These include squalene, (3*S*)-squalene-

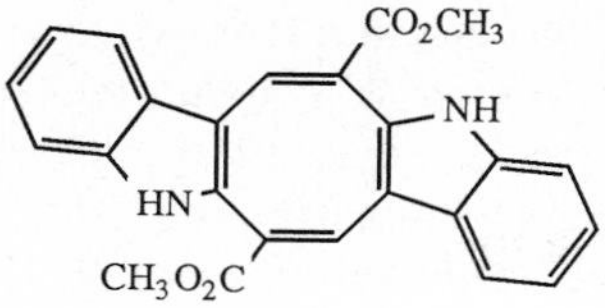

20

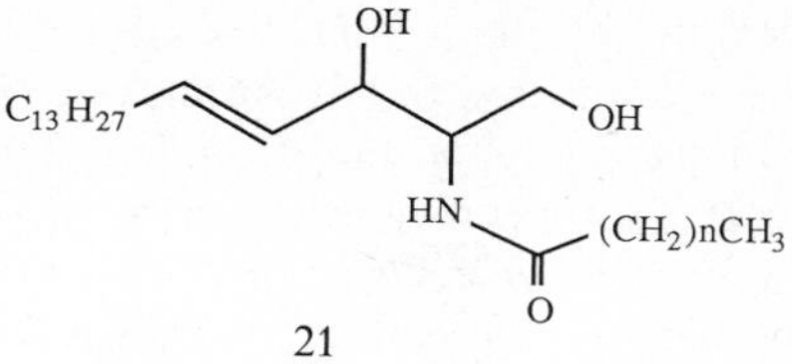

21

2,3-epoxide, (6S,7S)-squalene-6,7-epoxide, (10S,11S)-squalene-10,11-epoxide, and the alcohol **19** [38, 39].

Some of the first chemical studies of the genus *Caulerpa* involved the discovery of the pigment caulerpin (**20**), which was first isolated from the Philippine species *C. racemosa* and *C. sertularioides*. Caulerpin is a bright yellow-orange compound derived from obvious indole synthesis, and it has since been found in over 50 percent of the *Caulerpa* species investigated [40, 41]. Although caulerpin was originally described as a bioactive metabolite, recent studies show this compound to lack substantive effects [35].

Among the early metabolites reported from *Caulerpa* species is the toxin caulerpicin (**21**), a minor metabolite reported from several species [42, 43]. Unfortunately, caulerpicin does not represent a single pure compound. Caulerpicin is a mixture of N-acylsphingosines and conflicting chemical evidence indicates that no single structure represents the suggested natural product. Faulkner [2] appropriately suggests that the name caulerpicin be retired until a single pure compound is shown to possess toxicity.

Algae of the Family Udoteaceae

Chemical investigations of green algae from the family Udoteaceae have been undertaken more recently. Surprisingly, many of the metabolites were found to be closely related to, or identical with, those isolated from *Caulerpa* species. The first metabolites isolated from members of the Udoteaceae were the sesquiterpenoids rhipocephalin (**22**) and rhipocephanal (**23**). These compounds were isolated from the common Caribbean alga *Rhipocephalus phoenix* [44]. Rhipocephalin is closely related to the *Caulerpa* metabolite, caulerpenyne (**4**), but lacks the acetylenic functionality of the latter compound.

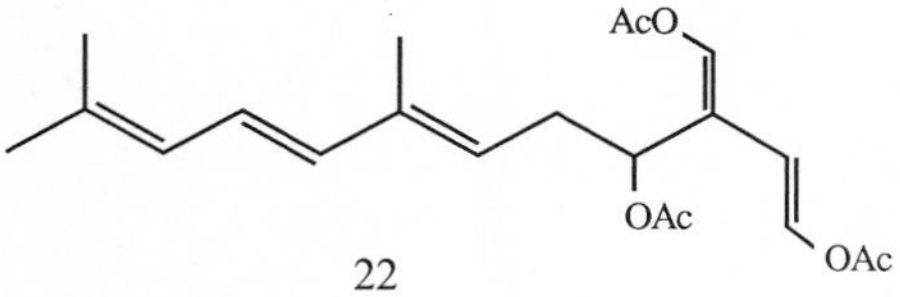

22

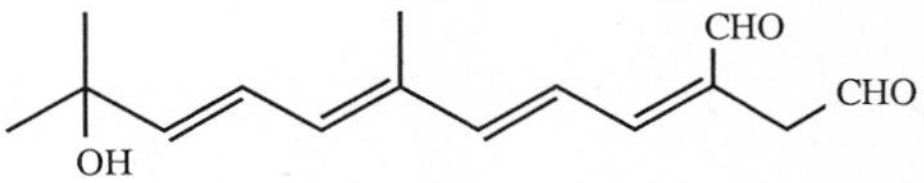

23

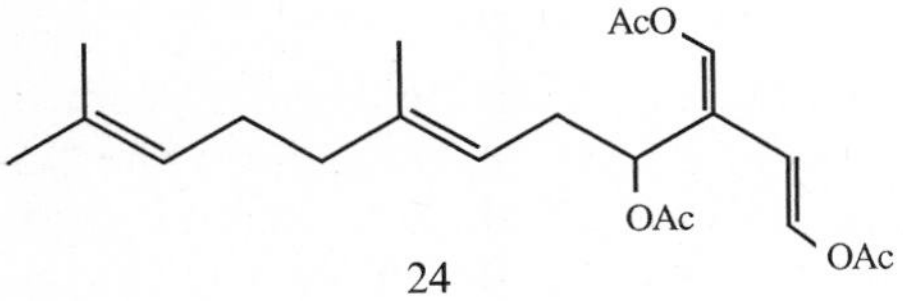

24

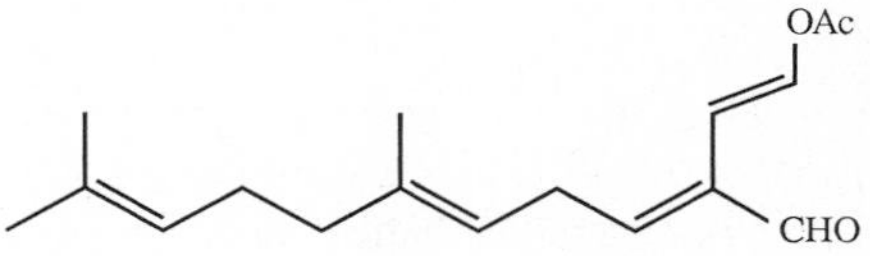

25

Flexilin (**2**), the acyclic sesquiterpenoid, bis-enol acetate from *C. flexilis* has also been identified in two species of the genus *Udotea, U. conglutinata* (Caribbean) [45] and *U. geppii* (western Pacific) [36]. Two other related sesquiterpenoids, **24** and **25**, have been identified from three different genera within the Udoteaceae, *Udotea cyathiformis, Penicillus capitatus* and *Rhipocephalus phoenix* [30, 45]. The sesquiterpenoid **26**, isolated from *Penicillus capitatus*, is unique in that the molecule contains three enol acetate functionalities [30]. Although *Penicillus* is a very

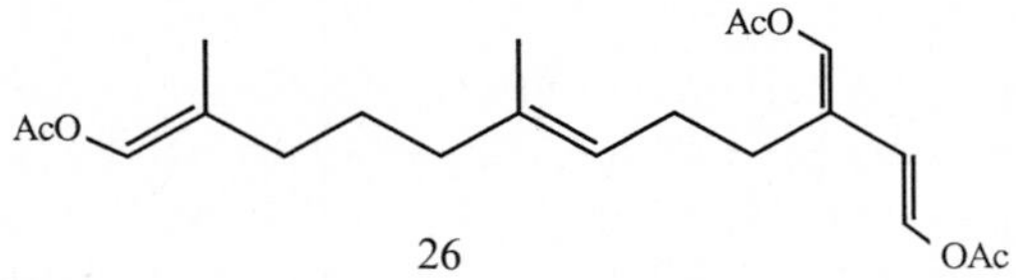

common Caribbean algal genus, natural products were not reported until 1984 [45]. This is no doubt due to the extreme instability of secondary metabolites in *Penicillus* species. It appears that *Penicillus* contains powerful oxidative enzymes which quickly degrade terpenoids on storage. If the algae are immediately extracted, however, and the metabolites liberated from all other components by chromatography, the compounds are usually readily obtained.

Diterpenoid metabolites with similar structural features have also been identified from algae within the family Udoteaceae. Studies of the noncalcified alga *Chlorodesmis fastigiata* in Australia [46], led to the isolation of the compounds didehydrotrifarin (**27**), chlorodesmin (**28**), and dihydrochlorodesmin (**29**). These compounds are unique in that they possess ketone functionalities and enol acetate groups on both ends of the linear diterpenoid skeleton. Chlorodesmin was also

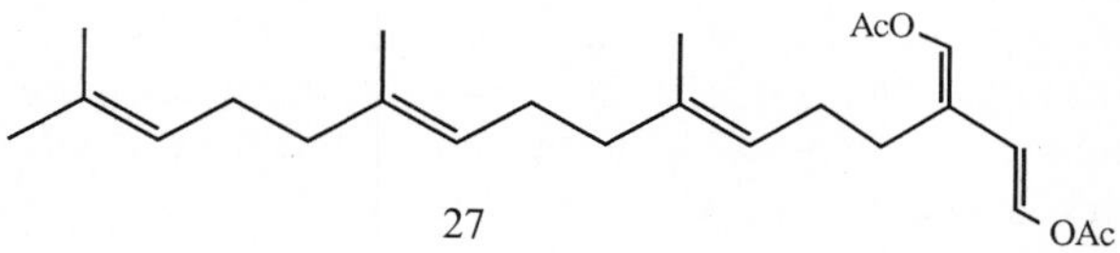

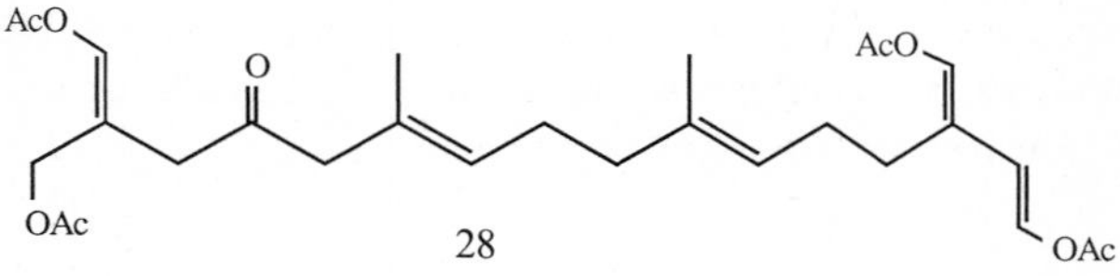

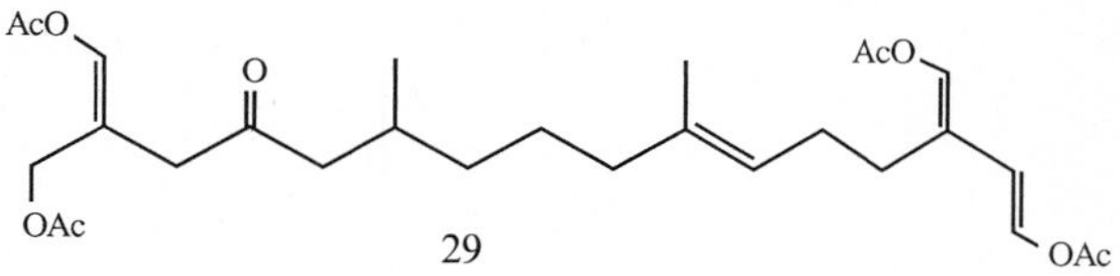

isolated from a Guam collection of *C. fastigiata* as was the hydrolysis product, aldehyde **30** [36]. A closely related diterpenoid **31**, was isolated from the calcareous tropical Pacific alga *Tydemania expeditionis* from the same family of marine algae [36].

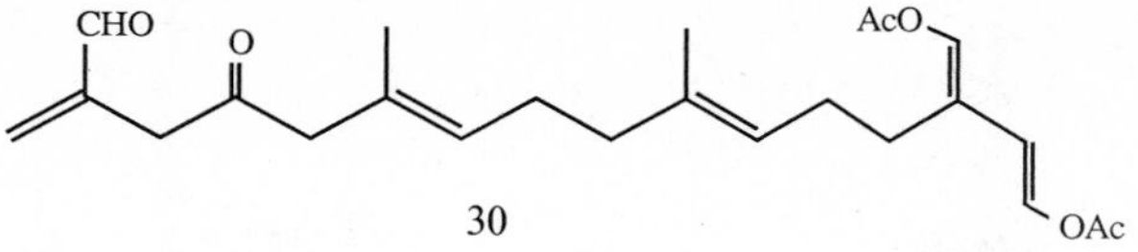

30

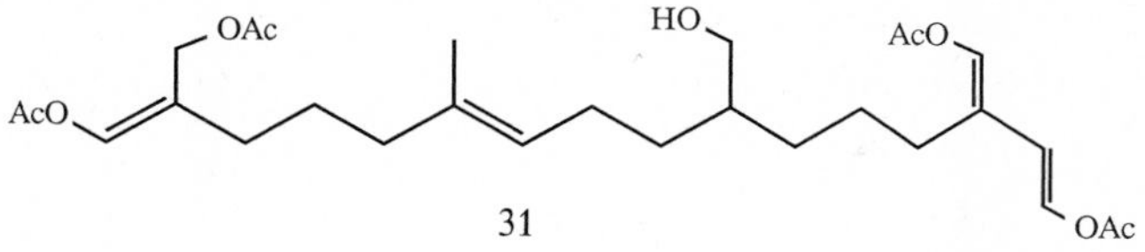

31

Udotea is one of the most common genera in the family Udoteaceae. Found world-wide, these algae usually inhabit sandy areas in warmer tropical and subtropical areas. Studies of the abundant Caribbean alga Udotea flabellum led to the isolation of a number of interesting new diterpenoids. Chemical investigations of the solvent-preserved U. flabellum led to the isolation of udoteafuran (**32**) as a minor metabolite and to a complex mixture of hydrates of udoteatrial (**33**) [47]. The stereochemistry at C-7 of udoteatrial (**33**) has recently been revised following a total synthesis [48]. On the basis of comprehensive studies of freshly collected algae, it appears that udoteatrial is generated upon prolonged storage.

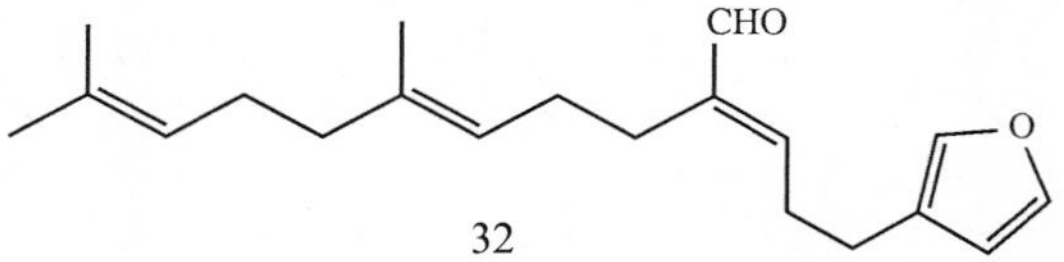

32

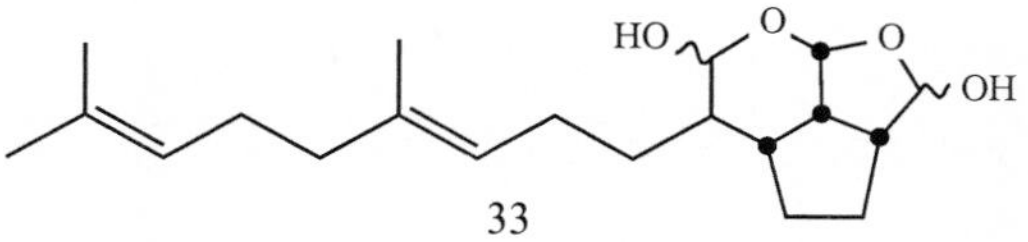

33

Udoteal (**34**) is the major metabolite in several species of freshly collected Udotea found in different biogeographical habitats. These include U. flabellum from the Caribbean Sea [49], U. argentea from the western Pacific [36], and U. petiolata from the Mediterranean Sea [50]. Minor metabolites of fresh U. flabellum include the dialdehydes petiodial (**35**) and the alcohol (**36**) [45]. Petiodial has also been reported in U. petiolata from the Mediterranean Sea [50]. Investigations of U. spinulosa from the Caribbean led to the isolation of a diterpenoid (**37**), which is isomeric with udoteal (**34**) [30]. Studies of U. argentea from Guam showed the alga to contain the α,β-epoxylactone **38** as the major metabolite [36].

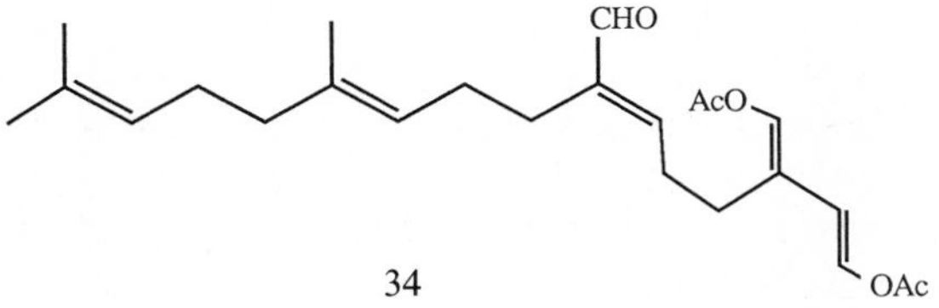

34

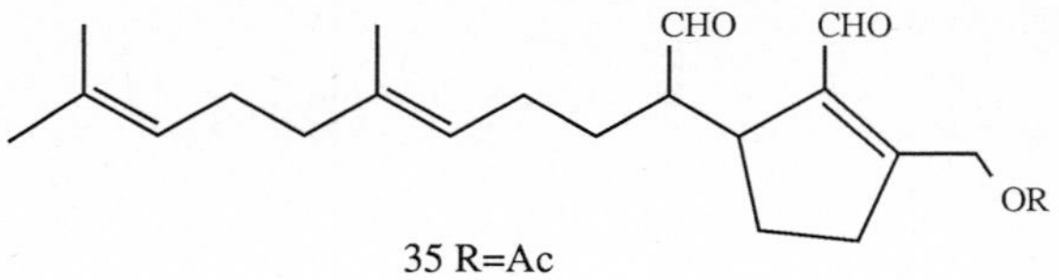

35 R=Ac
36 R=H

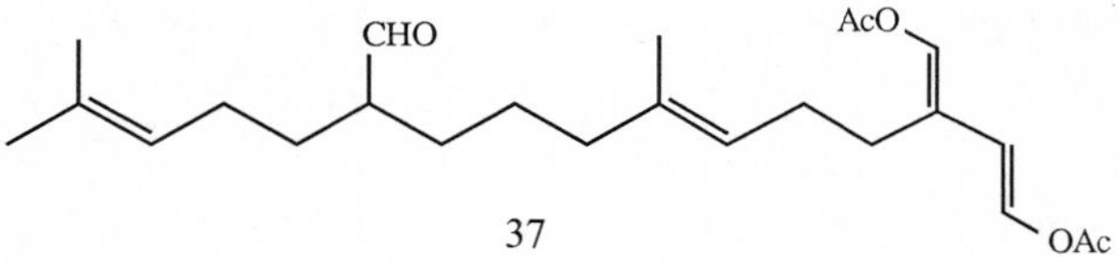

37

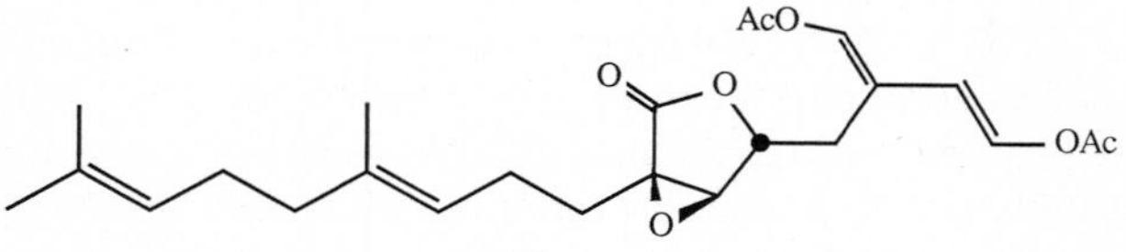

38

Several related diterpenoids have also been reported from Caribbean *Penicillus* species. Dihydroudoteal (**39**) has been isolated from *P. dumetosus* and from *P. pyriformis* [45]. The related dialdehyde **40** was also isolated from *P. pyriformis* (30), and the isomeric diterpenoids **41** and **42** (isomeric with udoteal (**34**)) were also found as minor metabolites in *P. dumetosus* [45]. In other collections of *P. dumetosus*, a different diterpenoid, triacetate **43**, was reported. This variation in secondary metabolite production will be discussed later in this chapter.

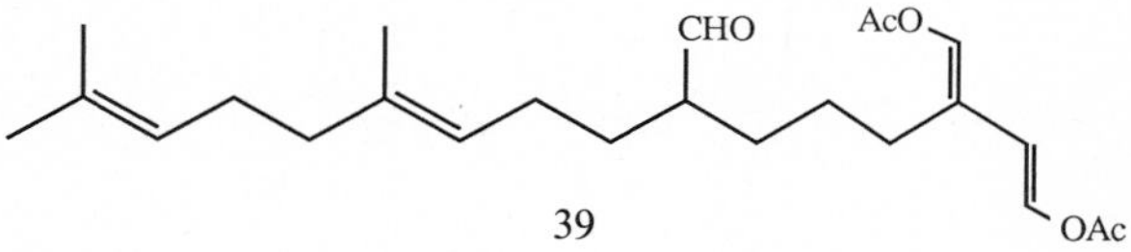

39

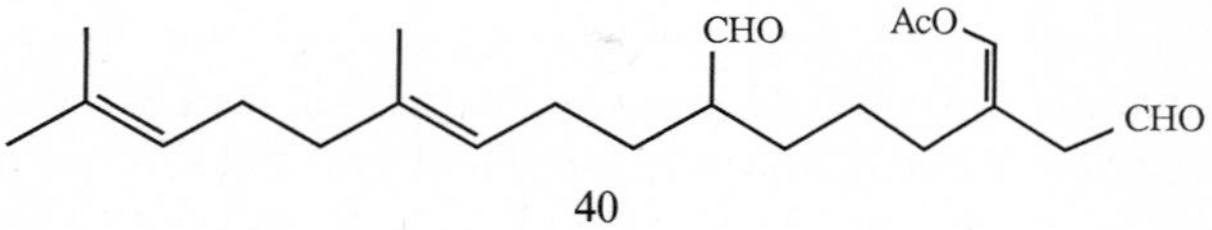

40

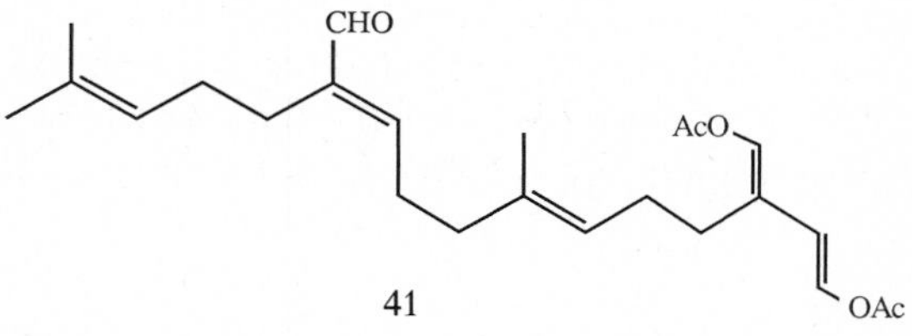

41

Algae of the genus *Halimeda* are among the most important inhabitants of tropical reefs. These algae are also the most chemically unique and biologically interesting in the family Udoteaceae. Perhaps the single most interesting compound isolated from this green algae is halimedatrial (**44**), a cyclopropane-con-

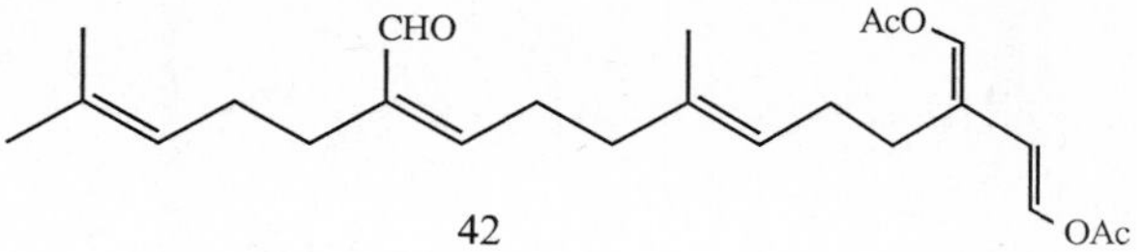

42

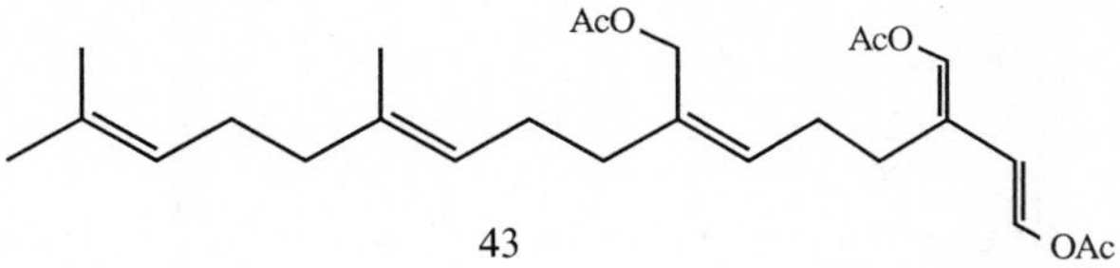

43

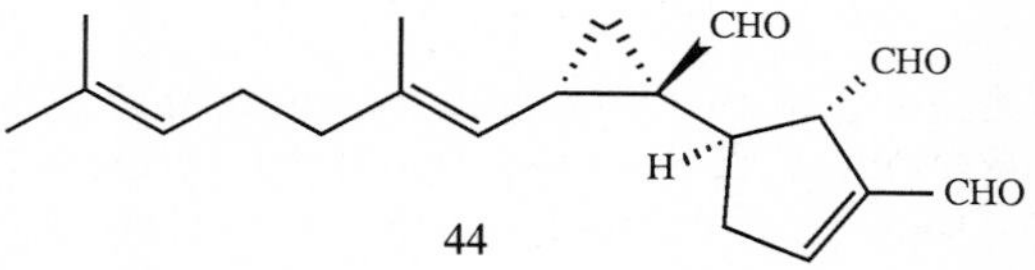

44

taining trialdehyde isolated from numerous species of this interesting genus. Hali-medatrial was first isolated from Caribbean species, but it was later identified in about 40 percent of all *Halimeda* species studied. The trialdehyde functionality of halimedatrial is unique but reminiscent of the 1,4-dialdehydes found in several terrestrial defensive compounds [51]. Several *Halimeda* species also contain epiha-limedatrial (**45**), which is the epimer at the central aldehyde functionality. Unlike halimedatrial, this compound is readily hydrated and is highly unstable, thus in-dicating the importance of stereochemistry at this center [30, 52]. The related lac-tone (**46**), a formal redox isomer of epihalimedatrial, has been isolated as a minor metabolite from several species of *Halimeda* [53].

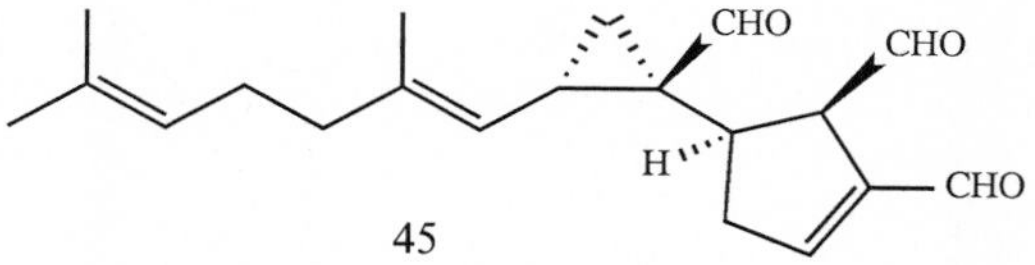

45

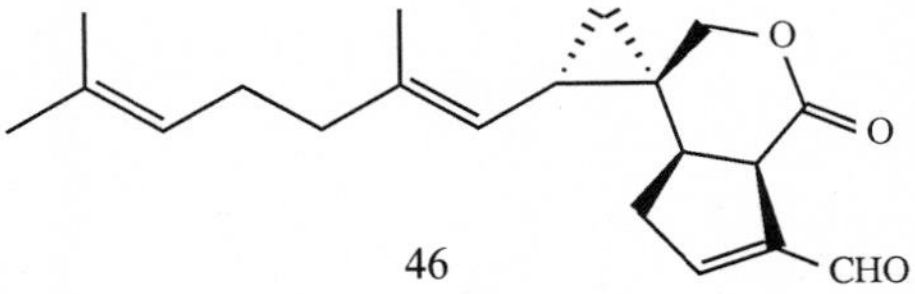

46

A major metabolite isolated from most of the *Halimeda* species is the diter-penoid tetraacetate (**47**) [53]. The metabolite was first reported from collections of *H. opuntia* from Puerto Rico [54], but subsequently was found in nine other Caribbean and tropical Pacific species of the genus *Halimeda* [53]. A structural relationship exists between the tetraacetate **47** and halimedatrial (**44**) which sug-gests that hydrolysis could convert **47** into halimedatrial. However, under no lab-oratory conditions (both acid and base catalysis) could the tetraacetate be in-duced to form halimedatrial [53]. A related tetraacetate metabolite **48** has recently

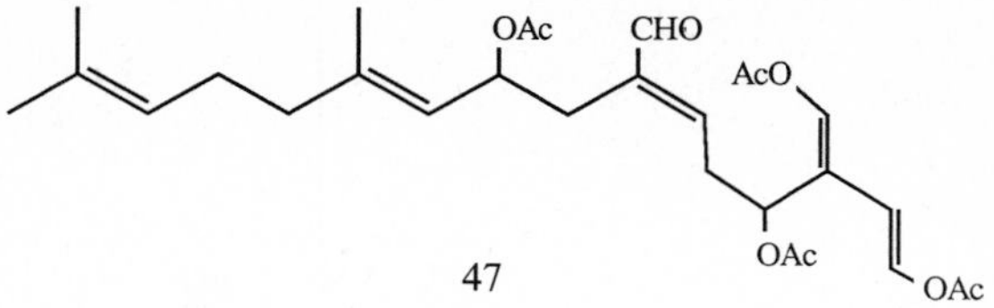

47

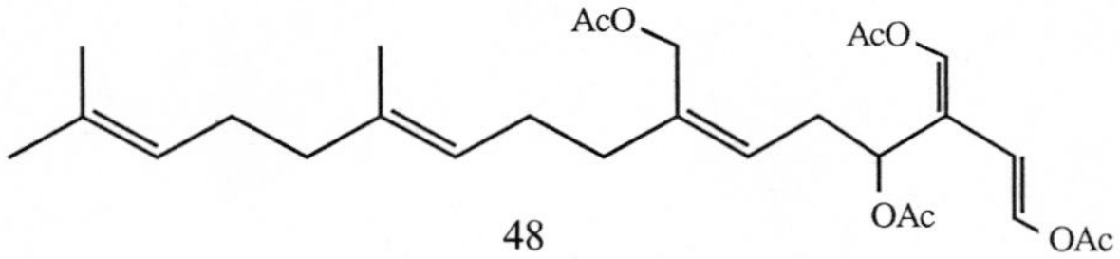

48

been isolated from *H. goreauii* [30, 52]. Lastly, the unusual bis-nor diterpenoid **49** has been reported as a minor metabolite from several species of *Halimeda* [53].

Recent studies of the alga *Pseudochlorodesmis furcellata* from Guam have led to the isolation of two new and related diterpenoids [55]. This alga, which is taxonomically related to *Chlorodesmis fastigiata*, produces an epoxylactone **50** as the major metabolite. This compound is structurally related to the epoxylactone **38**, produced by *Udotea argentea* [36]. The compound is the first δ-lactone isolated from this group. The stereochemistry of the substitutents on the lactone ring could only be suggested on the basis of NMR studies and CD experiments. The acyclic terpenoid **51** which is related to trifarin [3] was also isolated as the minor metabolite of *P. furcellata* [55].

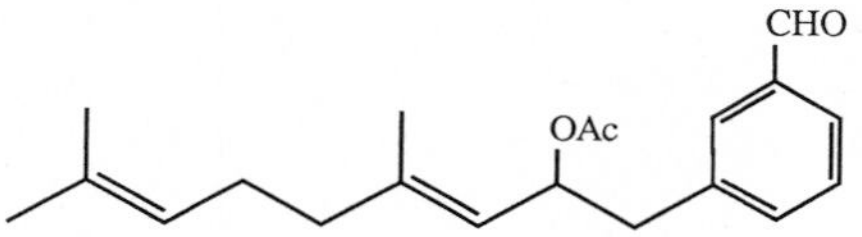

49

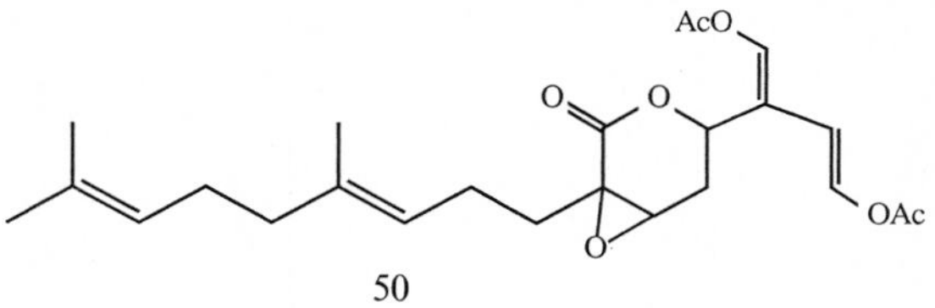

50

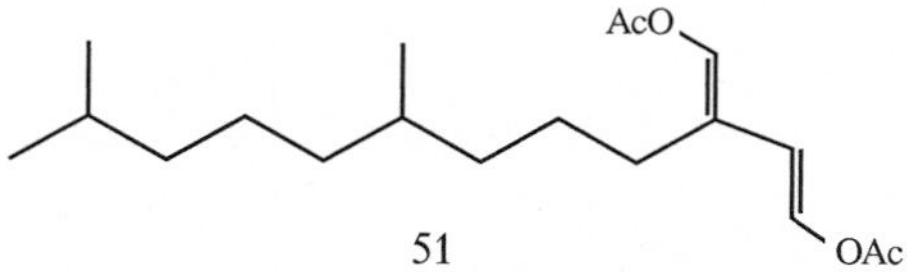

51

Studies of the Indo-Pacific alga *Tydemania expeditionis* have, on occasion, shown rather curious chemical results. Some collections from Guam and Palau were found to contain three related norcycloartane triterpenoids, **52–54** [56]. The structure of triterpenoid **53** was determined by X-ray analysis, and the derivatives were inter-converted through oxidation and hydrogenation. Some collections of *T. expeditionis* do not contain these triterpenoids, and most collections contain the more typical bis-enol acetate **31**, mentioned earlier. This result perhaps indicates that induced terpene synthesis occurs in this alga.

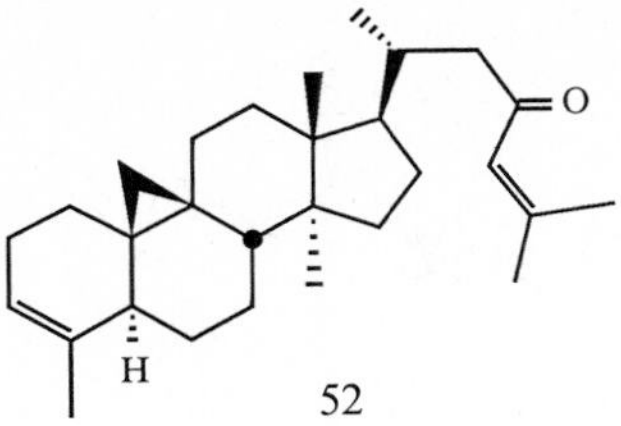

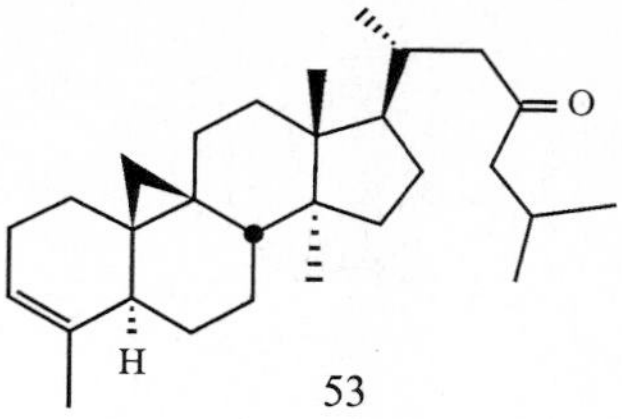

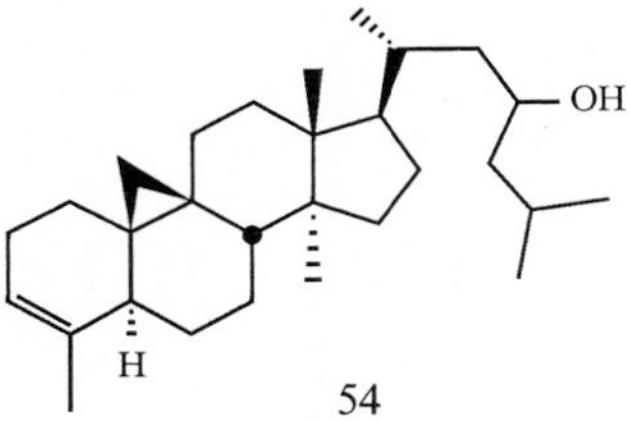

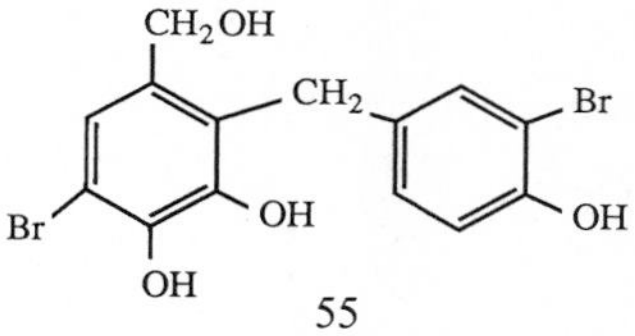

Several green algae of the genus *Avrainvillea* produce compounds which are not the terpenoids typically found from this family. Studies of Caribbean collections of *Avrainvillea longicaulis*, for example, led to the isolation of the bromine-containing diphenylmethane, avrainvilleol (**55**) [57]. This metabolite has also recently been isolated from collections of *A. rawsonii* from the Bahama Islands [30]. Avrainvilleol appears to be produced by dimerization of a p-hydroxybenzyl alcohol precursor, followed by bromination of the activated phenols. Recent chemical investigations of the *Avrainvillea*-related alga *Cladocephalus scoparius*, from the Bahama Islands, showed however that this alga produced no unusual natural products [30]. On the basis of their typical metabolites, *Avrainvillea* and *Tydemania* species should be considered unique members of the order Caulerpales. Taxonomically, it may be significant that these two genera produce compounds other than the terpenoid enol acetates and aldehydes which are characteristic of caulerpalean algae.

Other Green Algae

Although the majority of the natural products from the green algae are structurally similar sesqui- and diterpenoids, several different metabolites are produced by some members of this group. Several species of algae classified within the order Dasycladales, family Dasycladaceae, have been chemically investigated including *Cymopolia barbata* and *Dasycladus vermicularis*. The cymopols, a group of prenylated bromohydroquinones from the calcareous alga *C. barbata* were first reported in 1976 from collections made in Bermuda [58]. The major metabolite was the simple geranyl hydroquinone derivative, cymopol (**56**). Minor metabolites included cymopol monomethyl ether (**57**), cyclocymopol (**58**) and its monomethyl ether (**59**), cymopolone (**60**), isocymopolone (**61**), and cymopochromenol (**62**). Cyclocymopol and its methyl ether **58**, **59** are interesting monocyclic

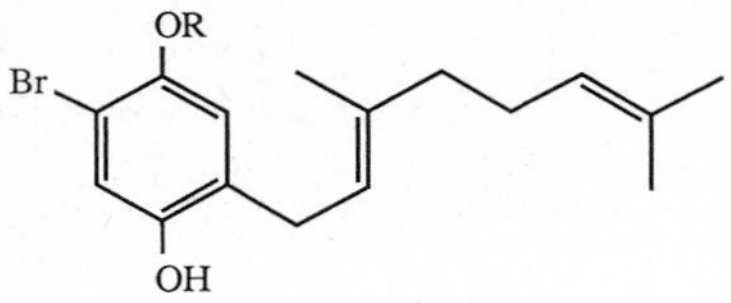

56 R=H
57 R=CH3

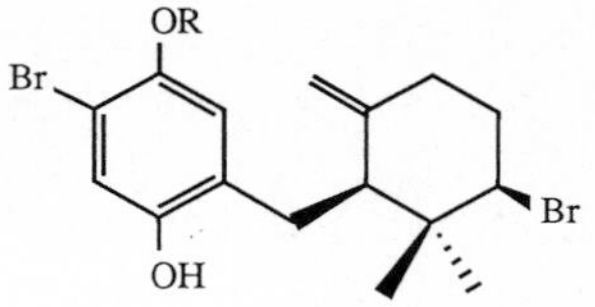

58 R=H
59 R=CH3

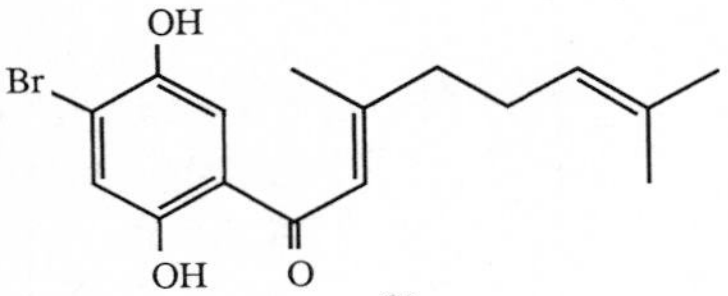

60

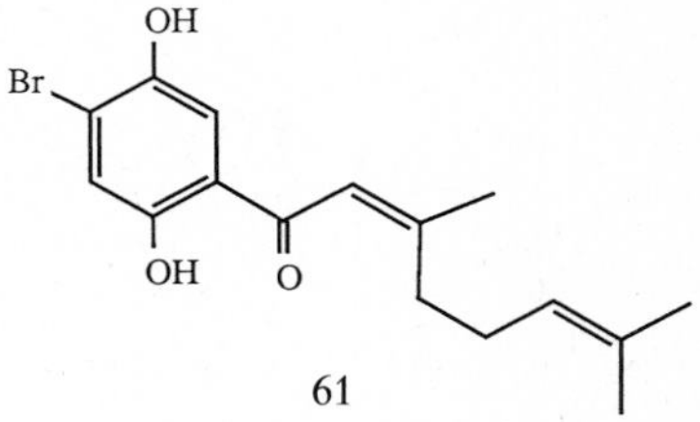

61

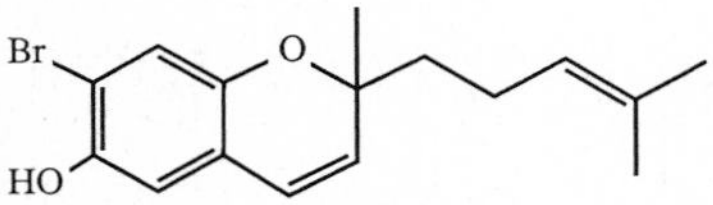

62

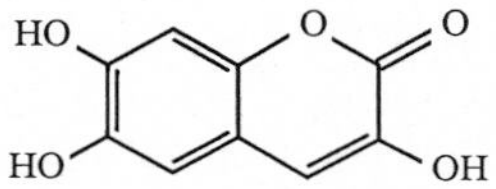

63

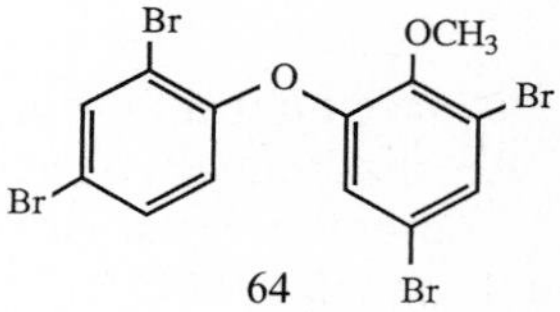

64

products apparently produced by bromocyclization in a process more commonly found in the red seaweeds (Rhodophyta). In 1982, diastereomers of cyclocymopol and cyclocymopol monomethyl ether were reported from *C. barbata* collected in the Florida Keys [59].

Chemical investigation of the Mediterranean alga *Dasycladus vermicularis* led to the isolation of 3,6,7-trihydroxycoumarin (**63**) [60] and two related coumarins. These yellow, water-soluble compounds had been observed by earlier investigators to be exuded from wounded plants, leading to yellow discoloration of surrounding seawater. Similar compounds may be produced by other species of *Dasycladus, Cymopolia, Batophora*, and related algae [60].

A biologically active brominated diphenyl ether **64** was recently reported from the green alga *Cladophora fascicularis* from Okinawa [61]. The compound was also isolated from the sea hare *Aplysia dactylomela* which was feeding on the alga at the time of collection. Brominated diphenyl ethers have not been isolated from other algal sources, but they have been isolated from sponges. A mixture of free fatty acids showing ichthyotoxic activity was isolated from an alga of the family Cladophoraceae, *Chaetomorpha minima* [62].

3. Biological Activities

Much research in the chemical ecology of marine algae has focused on intense herbivory in tropical waters as a major factor selecting for chemical defense in tropical algae. The effects of macroherbivores on algal community structure have been well documented in tropical waters (see Introduction). Certainly, intense herbivory could lead to the selection of chemical deterrents. However, other factors may also play a role in the evolution of chemical deterrents in marine algae.

Hypothesized factors selecting for chemical defenses include:

1) the need to inhibit heterotrophic or pathogenic microorganisms such as bacteria and fungi;

2) the need to inhibit fouling organisms such as algal spores and settling larval invertebrates;

3) interspecific competition for space;

4) the inhibition of herbivorous micrograzers; and

5) inhibition of grazing by macroherbivores. The effects of most of these selective pressures on algal growth and survivorship have not been documented for tropical algae. Algal secondary metabolites may be active against a number of biological factors simultaneously, thereby increasing the adaptive benefit of secondary metabolite production [63].

Consideration of these selective factors led to the development of bioassays used to assess the toxic properties of the compounds of caulerpalean algae toward microorganisms, fertilized sea urchin eggs, sea urchin sperm, urchin pluteus larvae, and tropical fish. Laboratory bioassays used natural predatory and pathogenic organisms to strengthen the biological significance of the results [30, 52, 64]. Most of the biological testing on the caulerpalean algae was the result of our work at Scripps Institution of Oceanography. A broad spectrum of physiological effects were demonstrated for these metabolites and are presented with a brief description of the methods used in each assay.

Antimicrobial assays were performed by a standard agar plate-disk method. Petri plates containing suitable growth media were inoculated with test microorganisms. Known strains of marine and terrestrial bacteria and fungi were used, as well as several marine isolates of bacteria and fungi obtained from the surfaces of tropical marine brown algae. Although terrestrial pathogenic microorganisms are generally used for antimicrobial screening, the use of marine microorganisms provides more information about the possible antimicrobial roles for secondary metabolites in nature.

Compounds were tested at concentrations of 0.10 mg/disk. After the microorganisms had grown (1–2 days), the zones of inhibition (clear zones) were measured around each filter disk. The results of the assays are presented in Table 1 (antibacterial activities) and Table 2 (antifungal activities). Similar results of antimicrobial assays have been previously reported for some of these metabolites (30, 64). Extracts of *Cymopolia barbata* were reported to possess antibiotic activity which has been attributed to the cymopols [58]. The coumarins from *Dasycladus vermicularis* showed antibiotic activity toward several terrestrial and marine bacteria [60]. The diphenyl ether from *Cladophora fascicularis* exhibited antimicrobial activity toward *E. coli, B. subtilis*, and *S. aureus* [61].

Cytotoxicity, measured as the inhibition of cell division in the sea urchin egg cleavage, is a bioassay now often used in screening the bioactivities of marine natural products. The procedure we used for testing the green algal metabolites was that of Jacobs *et al.* [65, 66]. The results of the cytotoxicity tests of the green algae metabolites are summarized in Table 3 [30, 64]. The results of cytotoxicity testing for other algal secondary metabolites have been presented [67].

Table 1. Results of antibacterial assay with green algal metabolites

Bacteria	Compounds tested											
	2	4	7	12	13	14	15	16	17	22	24	25
Staphylococcus aureus	+	+							+	+	+	−
Bacillus subtilis	−	+							+	+	+	−
Serratia marinoruba	−										+	+
Vibrio splendida	+										+	+
V. harveyi	−	+	+						+	+	+	+
V. leiognathi	−	+	+	−	+	+	+			+	+	+
Vibrio sp.	+			−	+	+	+	−	+			+
VJP Cal 8101	−	+									+	+
VJP Cal 8102	+	+	−							+	+	+
VJP Cal 8103	+	+										+

+ = Inhibition of bacterial growth
− = No inhibition, normal growth
Results obtained using standard agar plate-assay disk methods at 0.10 mg per disk

We designed a bioassay as a general toxicity screen toward sea urchin sperm obtained from the California urchin *Lytechinus pictus*. Sperm were in contact with known concentrations of the compounds dissolved in seawater for 30 min, and solvent and seawater controls were run simultaneously. Toxicity was measured as the complete loss of flagellar motility. Results of these assays are presented in Table 3 [30]. Assays examining larval toxicity may have ecological significance if the compounds are present naturally in seawater, or upon algal surfaces, at the concentrations employed in the tests. Relevant assays of this type must also involve assay organisms known to be potential predators or fouling species.

We also measured toxicity against pluteus larvae (36 hours after fertilization) of the sea urchin *Lytechinus pictus*. Both acute toxicity (1 hour) and toxicity after 24 hours were assessed. The results were viewed microscopically and toxicity was defined as 100% inhibition of larval swimming and ciliary motion. Results of these assays are also summarized in Table 3 [30].

Ichthyotoxic effects of the green algal metabolites were determined in our laboratory by assaying for death of tropical damselfishes within one hour. The Pacific species, *Pomacentrus coeruleus* and *Dascyllus aruanus*, were used in these assays. Details of this assay have been reported [51, 68]. The results for the green algal metabolites are presented in Table 4 [30]. The values reported are the lowest effective concentrations. At least three fish were tested at these minimum effective concentrations.

As the results in Tables 1–4 show, these metabolites show physiological activity in a variety of assays. Very few marine natural products have been assayed as extensively as this group of green algal compounds. Few other algal metabolites show the wide spectrum of bioactivities that these compounds show.

In particular, the compounds possessing one or more aldehyde functional groups were the most biologically active in these assays. In other studies of ter-

Table 1 (continued)

Compounds tested

26	28	30	31	33	34	35	36	37	38	39	40	43	44	46	47	49	55
+	+	+	+	+	+	−	−		+	+		+	+	+	+	−	−
+	+	+	+		+	−	−		+		−	+	+		+	+	+
−					+	+				+		−	+	+	−		−
+	+	+	+		+	+	+	+	−	+	+	−	+	−	+	+	−
+	−	+	+		+	+	+	+		+	−	+	+	+	−		−
+	+						+			+		+	+	+	+	−	+
−						−	+			−		−	+	−	+		
−	−	−	+		−	+	+			+	−	−	+	+	+	−	
+							+			+		+	+	+	+		

restrial plants and insects, secondary metabolites possessing aldehyde groups were also toxic and feeding deterrent compounds. Terpenoid metabolites such as warburganal, polygodial, isovelleral and the iridoid aldehydes are potent toxins produced by terrestrial plants and insects [25, 26, 69–71].

The antimicrobial activities of extracts and natural products from marine algae have been investigated previously [72–76]. Little is known about the ecological role of antimicrobial activity in algae and the selective pressures pathogenic microorganisms in tropical waters may exert. Bacterial films are known to play an important role in inducing larval settlement [77–80], and algae may produce antimicrobial defenses for these reasons.

The antifouling activities of some algal phenolic metabolites toward settling algae and bryozoans have been reported [81–83]. The terpenoids produced by caulerpalean algae are toxic to larvae of some marine invertebrates. However, more studies specifically involving fouling organisms should be conducted before conclusions regarding the antifouling role of these metabolites can be made.

Many of the green algal metabolites also showed ichthyotoxic effects. These results support the proposal of herbivore deterrent functions for the secondary metabolites. Further discussions of the role of the secondary metabolites in caulerpalean algae as defenses against herbivory are presented in the next section of this chapter.

4. Herbivore Deterrent Role

Herbivory in both marine and terrestrial communities can be very intense, reducing the biomass and survival of individual plants and thereby influencing interspecific competition and community structure [9–18, 22–24, 84–93]. The importance of plant defensive mechanisms, especially the role of secondary metabolites

Table 2. Results of antifungal assays with green algal metabolites

Fungi	Compounds tested																			
	2	4	7	22	24	25	26	28	30	31	33	34	35	36	39	40	43	44	46	47
Candida albicans	−	−	−	−	−	−	+	−	−	−	+	−	−	−	−	−	+	+	−	−
Leptosphaeria sp.	−	+	−	+	+	+		−	−	−		−	−	−	−	−	+	−	+	−
Lulworthia sp.	−	+	−	−	+	+	−	−	−	−		−	−	+	+	−	−	+	+	+
Alternaria sp.	−	+	−	+	+	+	+	−	−	+		+	+	+	+	−	−	+	+	+
Dreschleria haloides	+	+	+	−	−	−	−		+	−		−	−	+	+	−	−	+	+	−
Lindra thallasiae	+	+	+	−	+			+	+	+		−	+		+	−	−	+		−
VJP Cal 8104	+	+	+	+	+	+	−	+	+	+		−	−	+	−	+	−	+	+	+
VJP Cal 8105	−	+	+	+	+	+	−	−	−	−		−	−	−	+	−	+	+	+	−

+ = Inhibition of fungal growth
− = No inhibition, normal growth
Results obtained during standard agar plate disk methods at 0.10 mg per disk

Table 3. Results of bioassays with sea urchin fertilized eggs, sperm and larvae (*Lytechinus pictus*, *Echinometra mathaei*)

Bioassay	Compounds tested (µg/ml)																						
	2	4	7	9	10	22	24	25	26	28	30	31	34	35	36	37	38	39	40	43	44	46	47
Lytechinus pictus Fertilized egg cytotoxicities ED$_{100}$	16	8	4	2	16	8	2	2	16	8	8	8	8	1	1	16	8	16	8	16	1	16	8
Echinometra mathaei Fertilized egg cytotoxicities	+	+	+				+						+						+		+	+	+
L. pictus Sperm toxicity (30 min) ED$_{100}$	−	8	4	4	16	8	8	2		8	4	−	8	1	1	8	8	8	4	16	1	8	8
L. pictus Larval toxicity (1 h) ED$_{100}$	8	8	4	4	−	4	2	2		4	4	8	8		0.5	8	16	4	2	16	1	8	8
L. pictus Larval toxicity (24 h) ED$_{100}$	4	2	1	2	4	2	1		2	2	4	8	0.2		8	8	2	2	8	0.2	4	4	16
E. mathaei Larval toxicity (1 h)	+	+	+				+						+					+		+	+	+	

Active concentrations in µg/ml
+ = Active at ED$_{100}$ = 16 µg/ml
− = Compound not active at 16 g/ml
Blanks = Not tested

Table 4. Results of fish toxicity

Bioassay													
Compounds testet	2	4	7	9	10	12	13	14	15	16	22	24	25	26
Fish toxicity 1 h ($n=3$)	–	20	10	5	–	10	2.5	5	2.5	–	10	5	5	–
Copmounds testet	28	30	31	34	35	36	37	39	40	43	44	46	47	49
Fish toxicity 1 h ($n=3$)	10	5	–	–	5	5	–	–	10	–	5	20	–	–

Fish toxicity ED_{100} concentrations in µg/ml

as protection against herbivores, has been extensively investigated in terrestrial communities [94–99]. The diversity and ubiquity of these secondary metabolites has brought up many questions regarding their costs, benefits, and the selective forces influencing their evolution.

In contrast, much less is known about plant-herbivore interactions in the marine environment, especially in regard to the role of algal defenses against herbivores. Chemical adaptation has been hypothesized to play an important role in marine algae growing in areas of high herbivory [10, 12–14, 91, 93], but this function of secondary metabolites in marine algae has rarely been investigated [51, 100–104].

Rhoades [15] proposes several methods of testing the antagonistic functions of secondary metabolites against herbivores. To examine deterrent effects of metabolites he suggests:
1) the food preferences of herbivores should be related to the secondary metabolite composition of plants;
2) the chemical substances should be incorporated into artificial diets;
3) within-plant and between plant comparisons of metabolite concentrations and food preference should be made. In addition, he suggests that the physiological effects of the metabolites on the survival, growth, and fecundity of herbivores should be examined. Detailed studies in this area have been conducted for terrestrial herbivores, especially plant-insect interactions [98–99], but these studies are lacking in marine chemical ecology.

Several studies have investigated the susceptibility of marine algae to grazing by tropical herbivorous fishes in field experiments [9, 12–14, 91–93]. Some of these investigators have made an effort to correlate the results of feeding preference experiments with the natural products of marine algae reported in the chemical literature [93]. Only recently have the feeding preferences of herbivorous fish and the composition of secondary metabolites from the algae actually used in the preference experiments been studied simultaneously. This allows the determination of the correlation between feeding preferences and chemical defenses [104–105]. Most species of the Udoteaceae have been found to be low preference algae, and species of the noncalcified genus *Caulerpa* are medium to low preference dietary items [14, 35, 91, 93, 103–104].

Lobel and Ogden [11] showed that the parrotfish, *Sparisoma radians*, demonstrated very low survivorship (lower than starvation) when fed on diets of *Caulerpa mexicana, Halimeda incrassata*, and *Penicillus pyriformis*. They attributed this mortality to toxins in *Caulerpa* and calcification in *Halimeda* and *Penicillus*. However, it is likely that the toxic terpenoids present in all three of these species were responsible for the high mortality observed. Although many of the caulerpalean algae are highly calcified, many reef herbivores, including the parrotfishes, are well adapted for the consumption of calcareous plants [51, 104].

We used a laboratory assay to investigate the feeding deterrent effects of many of the green algal metabolites toward tropical fishes. This assay was a modification of feeding deterrent assays reported previously [44, 68]. Natural concentrations of the metabolites were prepared in diethyl ether and applied volumetrically to pellets of fish food. Control pellets were treated with the same volume of ether which was allowed to evaporate at room temperature. Results of the assay were measured by counting the number of bites taken of treated and control pellets by individual fish in a school of 6–8 fish. (Tropical damselfish, *Pomacentrus coeruleus* and *Eupomacentrus leucostictus*, were used in separate experiments.) The results were analyzed statistically with the Mann-Whitney Test and are summarized in Table 5 [30].

Feeding deterrence experiments were also conducted by one of us (VJP) in large outdoor aquaria with mixed populations of reef fishes from Guam [52]. A diet of a preferred green alga, *Enteromorpha* sp., mixed with agar was treated with either the metabolite halimedatrial (**44**) at natural concentrations or with solvent only (diethyl ether). Treated and control foods were offered to the fishes in seven trials, each on separate days for 12–16 hours. The results showed that halimedatrial significantly reduced feeding by 79% [52].

The deterrent effects of several of the green algal metabolites toward the parrotfish, *Sparisoma radians*, on St. Croix and the rabbitfish, *Siganus spinus*, on Guam have also been tested in aquarium assays. Caulerpenyne (**4**) and halimedatetraacetate (**47**) were effective feeding deterrents toward the parrotfish [106]. These compounds were not effective toward the rabbitfish, however chlorodesmin (**28**) and extracts containing the *Tydemania* diterpenoid **31** were effective feeding deterrents toward these herbivorous fishes [107].

Table 5. Results of feeding deterrence assays

	Compounds tested											
	4	7	9	10	22	24	25	34	35	36	44	47
Feeding deterrence	+	+	+	−	+	+	+	+	+	+	+	+

+ = Active feeding deterrent at natural concentrations (2000–5000 ppm)
− = Not active deterrent at 5000 ppm

The deterrent role of some extracts of *Caulerpa*, the sesquiterpenoid caulerpenyne, caulerpin, extracts of *Cymopolia barbata*, and cymopol toward the sea urchin *Lytechinus variegatus* have also been reported [101]. Caulerpenyne and cymopol appeared to reduce levels of grazing, however results were not statistically significant at $p = 0.05$.

Avoidance behavior of the marsh periwinkle snail, *Littorina irrorata* was used as an assay for the presence of noxious secondary metabolites in crude algal extracts. The avoidance behavior was also examined for several pure metabolites including caulerpenyne, cymopol, and a diterpene alcohol from *Penicillus dumetosus*. All were a actively avoided by snails. Most of the green algal extracts that were tested were also avoided by the periwinkle snails [108].

The effects of several green algal metabolites, incorporated into the diets of Caribbean juvenile conch (*Strombus costatus*), on the survival of these herbivores were measured in laboratory assays [30]. The *Strombus* were fed diets of *Enteromorpha* coated with caulerpenyne (**4**), udoteal (**34**), halimedatrial (**44**), and ether only (control) for two weeks. At the end of this time, all of the conch on the control diets survived, 55% of the conch on the caulerpenyne diet had died, 44% of the conch on the udoteal diet died, and all of the conch on the halimedatrial diet had died within six days. Longterm feeding on these metabolites was clearly detrimental toward these herbivorous mollusks.

Field assays examining the deterrent effects of algal metabolites on natural populations of herbivores have been successfully utilized for some of the green algal metabolites. These are some of the most promising experiments for examining the deterrent role of secondary metabolites under natural foraging conditions. The first metabolite tested was the diterpenoid halimedatrial (**44**) on reefs in St. Croix [109]. The compound was coated on a preferred alga, *Acanthophora specifera*, and the seagrass, *Thalassia testidinum*, at concentrations found in *Halimeda*. The coated and control (ether only) plants were loaded onto polypropylene ropes and replicate samples were placed on the reef for 6–7 hours. The compound was significantly deterrent to foraging herbivores and reduced grazing by about 50% relative to controls. Similar studies have been conducted on Guam for the metabolites halimeda tetraacetate (**47**), chlorodesmin (**28**), and the epoxylactone from *Pseudochlorodesmis* **50**, which were also effective deterrents under field conditions, when coated on palatable algae at natural concentrations (1% of algal dry mass) [110]. Udoteal (**34**) and extracts from three common species of *Caulerpa* were not effective feeding deterrents under these same conditions on Guam [110].

Mark Hay and coworkers recently used similar field studies in the Caribbean to examine the feeding deterrent effects of cymopol (**56**) and several natural products from red and brown algae [111]. Cymopol significantly reduced grazing by herbivorous fishes but significantly increased grazing by the sea urchin *Diadema antillarum* when coated on the seagrass *Thalassia testidinum* at concentrations of 1% of the algal dry mass [111]. Other field experiments examining the effects of secondary metabolites toward different species of herbivores will be useful in determining the deterrent roles of other algal secondary metabolites under natural foraging conditions.

Chemical Variation

Studies of terrestrial plant-herbivore interactions have shown that a great deal of variation occurs in secondary metabolite production. This may be attributed to both biological and physical environmental factors. Seasonal fluctuations in the amount, type, and localization of deterrent chemicals within plants is known to occur [112]. An increase in secondary metabolites may occur in response to varying herbivore pressure due to fluctuations in herbivore populations and activity [15, 99, 113–117]. Similarly, phytoalexins are plant defensive compounds not normally found in healthy plants but induced by the attack of microorganisms, usually fungi [96, 118–119].

Many predictions involving concentrations and variations in secondary metabolite production have resulted from terrestrial studies of plant chemical adaptations and plant-herbivore biochemical evolution [15, 63, 86, 87, 112, 120]. However, many of these hypotheses have not been rigorously tested in terrestrial habitats and have not been examined in the marine environment.

In our research at the Scripps Institution of Oceanography, we examined chemical variation in the production of secondary metabolites by caulerpalean algae. Our findings regarding chemical variation in different plant parts, between individual plants in a population, and between populations of the same species of algae are presented here. In addition, recent work by M. Hay, V. Paul, and coworkers is also reviewed in this section.

The chemical composition of the growing tips, whole blades, stipes, rhizoidal holdfasts, and reproductive portions of the algae were compared by thin layer chromatography (TLC) analysis. High performance silica gel liquid chromatography (HPLC) was also used for quantitative analyses. No significant difference in the secondary metabolite composition was detected in the blades, stipes, and holdfasts of several species of algae. *Penicillus dumetosus*, *P. capitatus*, *Rhipocephalus phoenix*, *Caulerpa paspaloides*, *Udotea flabellum*, and *Halimeda monile* from the Caribbean were tested for chemical variation in plant parts [30, 121]. *Halimeda scabra* was examined for the secondary metabolite composition in the newly produced, uncalcified growing terminal segments. The metabolites halimedatrial (**44**) and halimeda tetraacetate (**47**) which were present in the whole plant, were also found in the growing segments (total 2% of the algal dry weight). No calcium carbonate or chlorophyll were present in these new tissues. The reproductive structures (gametangia) in *Halimeda tuna* were separated from the rest of the plant and were shown to contain halimeda tetraacetate (**47**), as 2% of the dry weight of the gametangia. These dry weight concentrations were approximately four times greater than in the calcified plant tissues [30, 121].

M. Hay and coworkers conducted further investigations of the chemical composition of the newly produced segments of *Halimeda* in a recent study in the saturation diving facility, Hydrolab, on St. Croix (NOAA, National Undersea Research Program) [109]. They found that the concentrations of halimedatrial (**44**) and epihalimedatrial (**45**) were approximately an order of magnitude greater [5–10% of the ash free dry weight (AFDW)] in the newly produced, uncalcified segments than in the mature segments (about 0.5% of the AFDW) of *H. incrassata*. The newly produced segments were grazed by herbivores significantly more than older segments (new lost 31%, old lost 22%). The dry mass of new segments was only 36–50% ash while the dry mass of old segments was 83% ash, and nitrogen content was over 300% greater in new segments, so it was surprising that the difference in susceptibility to grazing between old and new segments was not greater. They concluded that the *Halimeda* metabolites (especially halimedatrial, **44**) function as deterrents toward grazers, and high concentrations of the compounds were critical defenses in the young, uncalcified tissues [109].

Examinations of individual plants within a population were conducted with *Halimeda tuna* and *Halimeda incrassata* in the Florida Keys and *H. incrassata* and *Rhipocephalus phoenix* in the Bahama Islands. Approximately 20 individual plants were collected from each population, extracted separately and analyzed by proton nuclear magnetic resonance (NMR). For *H. incrassata* and *R. phoenix*, a known amount of chloroform was added to the NMR solvent (carbon tetrachloride) as a standard. The chloroform peak was integrated with downfield absorptions of the secondary metabolites to more accurately estimate the concentration of terpenoids. A striking difference in the concentrations of the secondary metabolites was observed for all four species examined. Even though plants were collected within meters of one another, concentrations of metabolites varied from 0 to over 2% of the algal dry weight [30].

These observations led to a further study by V. Paul and M. Hay on chemical variation and its relationship to growth rates of the algae [122]. Several terrestrial studies have shown that secondary metabolite production and growth rates of plants are inversely correlated. It has often been proposed that a metabolic cost is involved in the production of secondary metabolites by plants. If secondary metabolite concentration is high, individuals or populations will grow more slowly, yet be more resistant to grazers, than plants with lower secondary metabolite concentrations. To test this prediction, 57 individual *Halimeda incrassata* plants were caged (to minimize grazing effects), the production of new segments was measured for nine days, and each plant was individually analyzed for its secondary metabolite composition. Although a great deal of variation in both growth rates and the concentration of the *Halimeda* diterpenoids occurred, no correlation between these variables was observed [122]. TLC and HPLC comparisons were used to examine chemical variation in different populations of algae. Extracts of the same species of algae collected in different habitats were analyzed by HPLC to quantitatively compare the concentrations and types of metabolites present. Proton NMR was used to identify all metabolites after HPLC analyses. A large amount of variation was observed in populations of the same species of algae collected in various habitats [30, 121]. Reef collections of *Udotea cyathiformis* and *Rhipocephalus phoenix* showed 3–6 times greater concentrations of the major metabolites (sesquiterpenoids **24**, **25**) than grassbed collections of the same plants. The diversity of minor metabolites, including a variety of sesquiterpenoid aldehydes, was also much greater in the reef collections. Reef and grassbed collections of *P. dumetosus* were compared and showed that different diterpenoids were produced (as mentioned in Sect. II of this review). All reef collections of this alga contained the metabolite dihydroudoteal (**39**) while the grassbed plants produced the triacetate **43**. *P. capitatus* also showed variation in the metabolites produced from different habitats [30, 121].

Spatial variation in herbivory is well known in tropical habitats [12, 13, 92, 93, 123]. In particular, reef habitats usually have much greater levels of herbivory than seagrass beds, and shallow reef areas have higher levels than deep reef areas [12, 13, 92, 123]. The chemical variation observed in different populations of these green algae may reflect a differential commitment to chemical defense under varying levels of herbivory. In all comparisons, the reef populations of algae either showed different secondary metabolites, higher concentrations of major metabolites, or a greater variety of metabolites produced than grassbed populations [30].

Recent studies in the western Pacific [124] have also shown that populations of *Caulerpa racemosa* and other species of caulerpalean algae show variation in secondary metabolite production. *Caulerpa racemosa* growing on reef slopes (high herbivory habitats) on Guam and Pohnpei produces different secondary metabolites than the reef flat (low herbivory) populations. In addition, the extract from a reef slope collection from Pohnpei was significantly avoided by herbivorous fishes compared to the extract from a reef flat collection [124]. These observations on the variation in secondary metabolite production by algae exposed to different levels of herbivory suggest that chemical production might be induced in response to increased grazing pressure.

To further examine the role of herbivory on the induction of secondary metabolite production by marine algae, artificial grazing experiments were conducted. Individual plants were clipped with scissors in their natural habitat. The

Table 6. Summary of artificial grazing experiment results

Algae	Extract/ dry wt.	Major metabolites	Concentration	
			% extract	% dry wt.
Phipocephalus phoenix				
Control	5.5%	22	1.7	0.10
Clipped	5.8%	22	4.5	0.26
Udotea cyathiformis				
Control	1.9%	24	8	0.15
Clipped	1.9%	24	15	0.28
Penicillus dumetosus				
Control	0.2%	43	20	0.04
Clipped	0.44%	43	40	0.18
Halimeda tuna				
Control	1.85%	47	5	0.10
Clipped	2%	47	20	0.40
Halimeda incrassata, Florida				
Control	1.4%	47	12	0.20
Clipped	1.3%	47	20	0.26
Halimeda incrassata, Bahamas				
Control	3.5%	47	25	0.8
Clipped	3.5%	47	30	1.0

algae examined in these studies were *Penicillus dumetosus, Halimeda tuna,* and *H. incrassata* in the Florida Keys and *H. incrassata, Rhipocephalus phoenix,* and *Udotea cyathiformis* in the Bahama Islands. Approximately 20 plants were marked as controls and were not clipped, 20 nearby plants were marked and clipped at least daily. At the end of 3–7 days the control and clipped plants were collected and extracted. Proton NMR was used to estimate the concentrations of secondary metabolites present in the extracts [30]. The results of these experiments are summarized in Table 6. The results of these preliminary studies indicate that terpenoid production may be induced in these green algae. The concentrations of the major metabolites were higher in the clipped algae for each species. Further studies with long term artificial grazing experiments and transplants of algae between habitats could yield important information on the induction of terpenoid production in response to levels of herbivory in marine algae.

6. Conclusion

Many species of caulerpalean algae have been chemically investigated and almost all species produce terpenoid metabolites possessing enol acetate and aldehyde functional groups. These structural features which are chemically reactive may be responsible for the wide spectrum of biological activity these metabolites demonstrate toward microorganisms, invertebrate eggs, sperm, and larvae, marine fishes, and mollusks. Recent studies have focused on the feeding deterrent effects of many of these metabolites toward herbivores in biologically relevant aquarium and field assays.

Future studies should include chemical investigation of green algae in other families such as the Dasycladaceae and Cladophoraceae. Preliminary studies of these algae have led to the identification of several new bioactive metabolites, and many other species have not been chemically examined.

Biologically relevant assays should be used to investigate the toxic and feeding deterrent properties of the isolated natural products. Field assays have now been developed which test the antiherbivore properties of algal metabolites under natural foraging conditions. In addition, aquarium assays can be utilized to examine the deterrent effects of extracts and isolated metabolites toward different species of marine herbivores. These studies are essential to investigate the chemical ecology of the marine plant-herbivore interactions that are especially important in tropical marine communities.

Understanding the complexities involved in the chemical defenses of marine algae will be a particular challenge to natural products chemists and chemical ecologists. Are some herbivores resistant to the chemical defenses of marine algae? What environmental factors influence the chemical variability observed in secondary metabolite production? Is the production of defensive compounds induced when marine algae are subjected to attack by predators or pathogens? What roles may the green algal metabolites play in defense against competitors, fouling organisms, and pathogens in the marine environment? These are just some of the questions that can only be answered by the close cooperation of chemists and biologists and an interdisciplinary approach to marine chemical ecology.

Acknowledgements

The research conducted at Scripps Institution of Oceanography was the result of financial support from the National Science Foundation both as a graduate fellowship to VJP and as research grants to WF (CHE81-11907 and CHE83-15546). Unpublished research conducted in Guam was also the result of financial support from the National Science Foundation (OCE 8600998). The authors thank M. Hay and S. Nelson for their helpful comments on earlier drafts of this chapter, and James N. Norris, National Museum of Natural History (Smithsonian Institution) for providing the figures illustrating tropical marine algae. These figures are the property of the Smithsonian Institution and they have been reproduced here with their permission. This is contribution No. 242 of the University of Guam Marine Laboratory.

References

1. Scheuer PJ (ed.) (1978–1983) Marine natural products: chemical and biological perspectives, Vols 1–5, New York, Academic Press
2. Faulkner DJ (1984) Nat Prod Rep 1:251
3. Taylor WF (1960) Marine algae of the eastern tropical and subtropical coasts of the Americas, Ann Arbor, Univ. of Michigan Press
4. Hillis-Colinvaux L (1980) Adv Mar Biol 17:1
5. Round FE (1971) Brit Phycol J 6:235
6. Bold HC, Wynne MJ (1978) Introduction to the algae, New Jersey, Prentice-Hall
7. Littler ME (1976) Micronesica 12:27
8. Borowitzka MA (1977) Oceanogr Mar Biol Ann Rev 15:189
9. Ogden JC (1976) Aquat Bot 2:103
10. Ogden JC, Lobel PS (1978) Env Biol Fish 3:49
11. Lobel PS, Ogden JC (1981) Mar Biol 64:173
12. Hay ME (1981) Aquat Bot 11:97
13. Hay ME, Colburn T, Downing D (1983) Oecologia 58:299
14. Littler MM, Taylor PR, Littler DS (1983) Coral Reefs 2:111
15. Rhoades DF (1979) Evolution of plant chemical defense against herbivores, in: Herbivores: their interaction with secondary plant metabolites (ed.) Rosenthal GA, Janzen DH, p 1, New York, Academic Press
16. Randall JE (1961) Ecology 42:812
17. Randall JE (1965) Ecology 46:255
18. Randall JE (1974) The effect of fishes on coral reefs, in: Proc 2nd Int Coral Reef Symp., p 159, Brisbane, Great Barrier Reef Committee
19. Earle SA (1972) The influence of herbivores on the marine plants of Great Lameshur Bay, with an annotated list of plants, in: Results of the Tektite program: ecology of coral reef fishes, (eds.) Collette BB, Earle SA Natl Hist Mus LA County Sci Bull 14:17
20. Wanders JBC (1977) Aquat Bot 3:357
21. Brock RE (1979) Mar Biol 5:381
22. Lubchenco J, Gaines SD (1981) Ann Rev Ecol Syst 12:405
23. Gaines SD, Lubchenco J (1982) Ann Rev Ecol Syst 13:111
24. Hawkins SJ, Hartnoll RG (1983) Oceanogr Mar Biol Ann Rev 21:195
25. Cimino G, De Rosa S, De Stefano S, Sodano G, Villani G (1983) Science 219:1237

26. Sterner O (1985) The Russulaceae sesquiterpenes, Ph D Dissertation, Lund Institute of Technology, 123 pp
27. Blackman AJ, Wells RF (1976) Tetrahedron Lett 2729
28. Blackman AJ, Wells RF (1978) Tetrahedron Lett 3063
29. Amico V, Oriente G, Piattelli M, Tringali C, Fattorusso E, Magno S, Mayol L (1978) Tetrahedron Lett 3593
30. Paul VJ (1985) The natural products chemistry and chemical ecology of tropical green algae of the order Caulerpales, PhD Dissertation, U.C. San Diego, 217 pp
31. De Napoli L, Fattorusso E, Magno S, Mayol L (1981) Experientia 37:1132
32. De Napoli L, Magno S, Mayol L, Novellino E (1983) Experientia 39:141
33. Paul VJ, Fenical W (1982) Tetrahedron Lett 23:5017
34. Capon RJ, Ghisalberti EL, Jeffries PR (1981) Aust J Chem 34:1775
35. Paul VJ, Littler MM, Littler DS, Fenical W (1987) J Chem Ecol 13:1171
36. Paul VJ, Fenical W (1985) Phytochem 24:2239
37. Capon RJ, Ghisalberti EL, Jeffries PR (1983) Phytochem 22:1465
38. De Napoli L, Fattorusso E, Magno S, Mayol L (1980) Tetrahedron Lett 21:2917
39. De Napoli L, Fattorusso E, Magno S, Mayol L (1982) Phytochem 21:782
40. Maiti BC, Thomson RH, Mahendran M (1978) J Chem Res (S) 126
41. Vest SE, Dawes CJ, Romeo JT (1983) Bot Mar 26:313
42. Doty MS, Aguilar-Santos G (1966) Nature 211:990
43. Mahendran M, Somasundaram S, Thomson RH (1979) Phytochem 18:1885
44. Sun HH, Fenical W (1979) Tetrahedron Lett 658
45. Paul VJ, Fenical W (1984) Tetrahedron 40:2913
46. Wells RJ, Barrow KD (1979) Experientia 35:1544
47. Nakatsu T, Ravi BN, Faulkner DJ (1981) J Org Chem 46:2435
48. Whitesell JK, Fisher M, Da Silva Jardine P (1983) J Org Chem 48:1557
49. Paul VJ, Sun HH, Fenical W (1982) Phytochem 21:468
50. Fattorusso E, Magno S, Mayol L, Novellino E (1983) Experientia 39:1275
51. Paul VJ, Fenical W (1983) Science 221:747
52. Paul VJ (1985) Chemical adaptation in pantropical green algae of the genus *Halimeda*, in: Proc Fifth Inter Coral Reef Symp., Tahiti Vol 5:39
53. Paul VJ, Fenical W (1984) Tetrahedron 40:3053
54. Tillekeratne LMV, Schmitz FJ (1984) Phytochem 23:1331
55. Paul VJ, Ciminello P, Fenical W (1987) in review
56. Paul VJ, Fenical W, Rafii S, Clardy J (1982) Tetrahedron Lett 23:5017
57. Sun HH, Paul VJ, Fenical W (1983) Phytochem 22:743
58. Hogberg H-E, Thomson RH, King T (1976) J Chem Soc Perkin Trans I, 1696
59. McConnell OJ, Hughes PA, Targett NM (1982) Phytochem 21:2139
60. Menzel D, Kazlauskas R, Reichelt J (1983) Bot Mar 26:23
61. Kuniyoshi M, Yamada K, Higa T (1985) Experientia 41:523
62. Fusetani N, Ozawa C, Hashimoto Y (1976) Bull Jap Soc Sci Fish 42:941
63. Feeny P (1975) Biochemical evolution between plants and their insect herbivores, in: Coevolution of animals and plants (eds.) Gilbert LE, Raven PH, Austin, Texas, Univ. of Texas Press
64. Fenical W, Paul VJ (1984) Hydrobiologia 116/117:135
65. Jacobs RS, White S, Wilson L (1981) Fed Proc Am Soc Exp Biol 40:26
66. Cornman I (1950) J Nat Cancer Inst 10:1112
67. Norris JN, Fenical W (1982) Chemical defense in tropical marine algae, in: The Atlantic Barrier Reef ecosystem at Carrie Bow Cay, Belize 1: structure and communities (eds.) Rutzler K, Macintyre IG, Smithsonian Contr Mar Sci 12:417
68. Paul VJ, McConnell OJ, Fenical W (1980) J Org Chem 45:3401
69. Kubo I, Lee Y-W, Pettei M, Pilkiewicz F, Nakanishi K (1976) Chem Commun, 1013
70. Cavill GWK, Hinterberger H (1960) Aust J Chem 13:514
71. Camazine SM, Resch JF, Eisner T, Meinwald J (1983) J Chem Ecol 9:1439
72. Almodovar LR (1964) Bot Mar 6:143
73. Hornsey IS, Hide D (1974) Br Phycol J 9:353
74. Henriquez P, Candia A, Norambuena R, Silva M, Zemelman R (1979) Bot Mar 22:451

75. Caccamese S, Azzolina R, Furnari G, Cormaci M, Grasso S (1980) Bot Mar 23:285
76. Hodgson LM (1984) Bot Mar 27:387
77. ZoBell CE, Allen EC (1935) J Bact 29:239
78. Woollacott RM (1981) Mar Biol 65:155
79. Mihm JW, Banta WC, Loeb GI (1981) J Exp Mar Biol Ecol 54:167
80. Brancato MS, Woollacott RM (1982) Mar Biol 71:551
81. McLachlan J, Craigie JS (1966) J Phycol 2:133
82. Sieburth JM, Conover JT (1965) Nature 208:52
83. Al-Ogily SM, Knight-Jones EW (1977) Nature 265:728
84. Morrow PA, LaMarch VC (1978) Science 201:1244
85. Rauscher MD, Feeny PD (1980) Ecology 61:905
86. Coley PD (1983) Ecol Monogr 53:209
87. Coley PD (1985) Science 230:895
88. Vadas RL (1977) Ecol Monogr 47:337
89. Lubchenco J (1978) Am Nat 112:23
90. Lubchenco J (1980) Ecology 61:333
91. Lewis SM (1985) Oecologia 65:370
92. Lewis SM (1986) Ecol Monogr 56:183
93. Hay ME (1984) Oecologia 64:396
94. Sondheimer E, Simeone JB (1970) Chemical ecology, New York, Academic Press
95. Whittaker RH, Feeny P (1971) Science 171:757
96. Levin DA (1976) Ann Rev Ecol Syst 7:121
97. Harborne JD (1977) Introduction to ecological biochemistry, New York, Academic Press
98. Harborne JD (1978) Biochemical aspects of plant and animal coevolution, New York, Academic Press
99. Rosenthal GA, Janzen DH (1979) Herbivores: their interaction with secondary plant metabolites, New York, Academic Press
100. Geiselman JA, McConnell OJ (1981) J Chem Ecol 7:1115
101. McConnell OJ, Hughes PA, Targett NM, Daley J (1982) J Chem Ecol 8:1437
102. Steinberg PD (1984) Science 223:405
103. Steinberg PD (1985) Ecol Monogr 55:333
104. Paul VJ, Hay ME (1986) Mar Ecol Prog Ser 33:255
105. Paul VJ, Potter TS, manuscript in preparation
106. Targett NM, Targett TE, Vrolijk NH, Ogden JC (1986) Mar Biol 92:141
107. Paul VJ, Nelson SG, in review
108. Targett NM, McConnell OJ (1982) J Chem Ecol 8:115
109. Hay ME, Paul VJ, Lewis SM, Gustafson K, Tucker J, Trindell R (1987) manuscript in preparation
110. Paul VJ (1987) Bull Mar Sci, in press
111. Hay ME, Gustafson K, Fenical W (1987) Ecology, in press
112. Feeny P (1976) Rec Adv Phytochem 10:1
113. Freeland WJ, Janzen DH (1974) Amer Nat 108:269
114. Whitham TG, Slobodchikoff CN (1981) Oecologia 49:287
115. Haukioja E, Hanhimaki S (1985) Oecologia 65:223
116. Denno RF, McClure MS (1983) Variable plants and herbivores in natural and managed systems, New York, Academic Press
117. Whitham TG, Williams AG, Robinson AM (1984) The variation principle: individual plants as temporal and spatial mosaics of resistance of rapidly evolving pests, in: A new ecology (eds.) Price PW, Slobodchikoff CN, Gaud WS, p 15, New York, Wiley
118. Deverall BJ (1972) Proc R Soc Lond 181:233
119. Cline K, Wade M, Albersheim P (1978) Plant Physiol 62:918
120. Fox LR (1981) Amer Zool 21:853
121. Paul VJ, Fenical W (1986) Mar Ecol Prog Ser, 34:157
122. Paul VJ, Hay ME (1987) J Exp Mar Biol Ecol, in preparation
123. Hay ME (1985) Spatial patterns of herbivore impact and their importance in maintaining algal species richness, in: Proc Fifth Inter Coral Reef Symp, Tahiti, Vol 4:29
124. Paul VJ (1987) work in progress

Chemical Ecology of the Nudibranchs

Peter Karuso [1]

Contents

Abstract

The fact that nudibranchs often contain interesting natural products has been known for over a decade. These natural products were quickly recognized as defensive agents usually sequestered from their particular prey. Recently more fundamental questions, requiring the close collaboration between chemists and biologists, have been addressed. These include the questions of chemoreception and chemosignalling of nudibranchs. The activities in this field are reviewed for the first time with particular emphasis on the defense mechanisms of nudibranchs and the chemical ecology of nudibranchs in all stages of their life cycle.

1 Texas A & M University, Department of Chemistry, College Station, Texas, 77843-3255, U.S.A.

Bioorganic Marine Chemistry, Volume 1
© Springer-Verlag Berlin Heidelberg 1987

1. Introduction

Nudibranchs are exclusively marine, slug-like invertebrates and are the major representatives of the molluscan subclass Opisthobranchia (Fig. 1). They have undergone complete detorsion of the visceral mass and totally lost the restrictive shell of other gastropods. This has allowed an immense diversity of body form and habitat to develop. However, the loss of the protective shell made necessary the development of more dynamic biological and chemical defenses to replace the rather static shell. The evolutionary implications of this shell loss has been reviewed by Faulkner and Ghiselin [1] who concluded that development of chemical defenses was a preadaption, enabling the animal to dispense with the shell. This view was intimated some time earlier by Thompson [2] with the finding that the partially shelled prosobranch *Velutina velutina* was rejected as food by fish. All other nudibranchs tested similarly were also rejected suggesting that the opisthobranchs in the evolutinary process of losing their shell already possess alternative defenses.

The defense mechanisms of nudibranchs have been reviewed from a biological viewpoint by Thompson [3] and Ros [4], and their general ecology by Todd [5]. These authors concentrated on the morphological and behavioral defenses as they were, at the time, largely unaware of the chemical defenses.

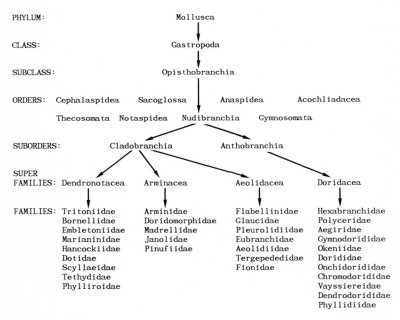

Fig. 1. Classification of the Nudibranchs after Ref. [8]. Recently there has been a reaction by nudibranch systematists against the large number of family names proposed by earlier workers (particularly for the Doridacea and Aeolidacea). This list therefore embodies a shortened list of families within the Doridacea and Aeolidacea. Similar reductions can be expected within the Dendronotacea

Nudibranch defensive mechanisms fall into three basic categories: behavioral, morphological and chemical. These defenses may be employed in a hierarchical fashion. In essence, a nudibranch's primary defense is to avoid detection by potential predators. This is largely achieved by a reclusive habit; many nudibranchs conceal themselves under rocks or in sediment and appear only under cover of darkness. Alternatively, they may spend their entire lives in caves or under ledges. Those that do not hide are generally cryptic (camouflaged). This type of defense can be further subdivided as follows.

1) Homochromy, in which the animals' body color matches that of the background. For instance the bright red dorid *Rostranga pulchra* feeds exclusively on sponges of the genera *Microciona, Ophlitospongia, Esperiopsis,* and *Plocamia* all of which are red also [6, 7]. In fact, carotenoids from the sponges are accumulated by the nudibranch giving it an identical color. This is an example of a morphological defense and of a passive chemical defense as this nudibranch uses pigments from its food. Many other nudibranchs display similar adaptations [6]. In addition, *R.pulchra* has a rough texture that closely matches that of the sponge and a flattened body barely raised above the sponge surface to maximize its cryptic coloration.

2) Countershading is also employed by some nudibranchs. In this strategy the outline of the animal is minimized so as to disguise its shape. For example the pelagic aeolid, *Glaucus*, floats on the surface of the ocean; its upper surface is dark blue and the underside is white [5]. The tropical chromodorid *Chromodoris coi* [8] is countershaded in such a way as to render its outline indistinct and the markings on its dorsum disguise its shape.

3) Disruptive coloration – in which the coloration and pattern break up the outline of the body – is also common among the nudibranchs. For instance, *Peltodoris atromaculata* has large brown spots which effectively camouflage it against its food source, the brown sponge, *Petrosia ficiformis* [4].

Many nudibranchs have shapes that closely match those of their food. For instance, the spongiverous chromodorid *Verconia verconis* has a mantle edge raised into conules that mimic those of its food, *Aplysilla rosea* [9]. The aeolid *Cuthona kuiteri* feeds on the hydroid *Zyzzyzus spongicola*. The shape and color of its cerata are remarkably similar to those of the tubulariid polyps. Also, the eggs are laid under the protecting stinging tentacles of the hydroid [10]. Other nudibranchs mimic rocks or seaweed (e.g. *Melibe* spp. [8]). The red *Aldisa* sp. from Hawaiian waters [11] does not camouflage itself against its background; rather it mimics a sponge to the extent of having three false exhalent canals in a row on its dorsum.

Many nudibranchs, notably the chromodorids and polycerids, are not cryptic and make no attempt to conceal themselves. Many authors have taken this to show aposomatic or "warning" coloration [12, 13]. So far there has been no experimental justification for this and the trend in recent thought is that most strikingly colored nudibranchs are cryptic due to disruptive coloration which breaks up the animal's shape [3, 14]. However, there is no doubt that some nudibranchs are aposomatic, e.g. *Phyllidia* spp. which are brightly colored and make no attempt to conceal themselves. They are highly toxic to fish and crustaceans [15]. In addition, there is some evidence of aposomatic circles of nudibranchs in certain areas. For example, Ros [4] describes five blue species of *Glossodoris* from the Iberian coast that he believes to represent a Müllerian mimicry circle. According to this concept [16] all the nudibranchs are distasteful and all look alike; thus, a predator that has learned not to eat one species will also not eat the others. This phenomenon may be widespread as Rudman [17, 18, 19] has described a number of color groups of chromodorids from the east coast of Australia.

If a potential predator can locate the potential prey nudibranch either visually, tactilely or chemically, then secondary defenses may be brought into operation. Many nudibranchs can swim for several minutes, e.g. *Tritonia gilberti* will swim if contact is made with the predatory starfish *Pycnopodia helianthus* [20]. *Hexabranchus sanguineus* from Hawaii is a mottled sandy red color when at rest. However, if the animal is poked, it will suddenly unfold its mantle to reveal bright red and white coloration which according to Edmunds [21] is a startle defense. Continued harassment causes the nudibranch to swim [22], which can last for several hours according to Eliot [23], which has given rise to its common name "Spanish dancer".

Should the predator get to the stage of mouthing the nudibranch, further defenses may be employed. Many dorids have a mantle reinforced with spiney spicules that makes eating them difficult. The predatory bullomorph, *Navanax inermis*, has been reported to reject spiculated dorids [24]. However, *Navanax* will consume shelled bullomorphs [14] which may be just as difficult to eat. This indicates that other defenses could be involved in the rejection of dorids [5].

Autonomy is also a common defense in the large discodorids; e.g. *Peltodoris atromaculata* will discard its mantle edge in response to attack [25]. All aeolids will readily autonomize their cerata on persistent irritation. The sacoglossans are not nudibranchs, although many have no shells or a very reduced shell. They do, however, resemble nudibranchs and have similar defensive needs. Autonomy is well developed in this group, e.g. *Oxynoe panamensis* will discard its tail which will continue to wriggle as the animal attempts to escape [26].

The most elaborately defended nudibranchs are the coelenterate feeding aeolids. These animals possess some of the defenses mentioned above and in addition accumulate intact nematocysts (stinging cells) from their prey. These stinging cells are transported through the gut to cnidosacs at the cerata tips. There they are stored and released automatically in response to touch, firing on contact with water. Many aeolids have demonstrated the ability to selectively accumulate specific nematocysts from their prey(s) [27, 28]. These nematocysts are invariably the most potent. The mechanisms by which the nudibranchs prevent the nematocysts from firing while feeding, and how the nematocysts are selectively accumulated is unknown.

As mentioned earlier, many nudibranchs are not cryptic and do not make use of the defenses mentioned above. These animals often rely on the third category of defense, namely chemical.

2. Chemical Defense

It has long been recognized that nudibranchs are infrequently preyed upon. Herd-mann and Chubb in 1892 [29] discovered that nudibranchs are not accepted by aquarium fish as food. Crossland in 1911 [39], noted that live chromodorids were rejected by fish as food when thrown to them in shallow water. Admittedly, these experiments were crude, inconclusive and subjective. In 1960, Thompson [3] tested 24 nudibranchs and showed that all animals were rejected by fish as food. Even cryptic nudibranchs were inedible, presumably using camouflage only as a primary defense. Thompson found that many dorids secreted acid in response to aggravation. This was first noted by Garstang in 1890 [12]. Since then this defense has been noted in many dorid nudibranchs, especially those that feed on acidic tunicates. In all cases the pH is low (1–2) and the acid invariably sulfuric [31]. Thompson also recognized glands in the non-acidic dorids that secreted a fluid that tasted bitter or had no taste to the human palate. He postulated that these glands release defensive chemicals, but did not speculate on their nature.

The first chemical investigation of an opisthobranch was carried out in the early 1960's by Yamamura and Hirata [32] who isolated brominated terpenoids from the sea hare *Aplysia kurodai*. These metabolites have subsequently been shown to originate from their diet, the red alga *Laurencia* sp. [33]. Since then, many other aplysiomorphs have been studied and all were shown to contain me-tabolites from their algal diet. These relationships have been reviewed [34].

The transfer of toxic metabolites from prey to predator opisthobranch was recognized in 1970 by Doty and Anguilar-Santos [35] in the herbivorous sacoglos-san *Oxynoe panamensis*, which accumulates caulerpicin and caulerpin from its food, the green algae *Caulerpa* spp. The authors found that these metabolites were not transferred to other invertebrates that lived in the same habitat.

2.1 Superfamily Doridacea

2.1.1 Family Phylidiidae

In 1963 Johannes [15] observed that lobsters kept in an aquarium with a specimen of *Phyllidia varicosa* died within 30 minutes. He showed that the nudibranch se-creted a poisonous mucus lethal to fish and crustaceans. Johannes characterized the toxin as a non-proteinaceous, volatile small molecule with an unusual smell. He correlated this observation with earlier work by Risbec [36] who noted that all species of *Phyllidia* which he had studied possessed an identical unusual odor. Johannes's work prompted Scheuer and co-workers to investigate the chemical defense of *Phyllidia varicosa* from Hawaii [37, 38]. They characterized the pun-gent toxin as a sesquiterpene isocyanide (ν_{max} 2120 cm^{-1}), but were unable to as-sign a structure from the small sample. By chance, they observed a specimen of *P. varicosa* feeding on an off-white sponge (*Ciocalypta* sp.) and noted that the sponge had the same odor as the nudibranch. Extraction of the sponge revealed a rich mixture of lipids of which the toxin was a prominent component. The structure was elucidated as 9-isocyanopupukeanane (1), which was difficult to separate from its 2-isomer (2). This work represents the first chemical investigation of a nudibranch and the first report of a nudibranch selectively accumulating a chemical defense agent from its diet. In subsequent work using gas chromatography/ mass spectroscopy (GCMS), Roll [39] found *P. varicosa* to contain a mixture of sesquiterpenoids. *Phyllidia loricata, P. rosans* and an unidentified *Phyllidia* sp. also contained the same compounds. In addition to these compounds, *P. varicosa quadrilineata* and *P. pustulosa* contained an isothiocyanate identified as the isothiocyanoaromadendrane (3). The source of this compound was the sponge *Densa* sp. on which the nudibranchs feed. Both sponges contained many other sesquiterpenoids, but these were apparently not detected in the nudibranchs.

Similarly, *P. pulitzeri* from the Mediterranean contained an isocyano terpene, axisonitrile-1 (4), as its major metabolite, as well as a mixture of minor axisonitriles 6–9 [40], all of which had previously been isolated from the sponge *Axinella cannabina* [41, 42, 43] upon which the nudibranch was known to feed. The major metabolite (4) was ineffective as an antifeedant but showed pronounced ichthyotoxicity. An unidentified species of *Phyllidea* from Sri Lanka has recently been found to contain 3-isocyanotheonellin (5) [44]. In addition, some 20 species of *Phyllidia* from Guam, Saipan, Pohnpei, and Papua New Guinea and one species of *Fryeria* from Guam invariably contained sesquiterpene isocyanides derived from their sponge diet [22].

2.1.2 Family Chromodorididae

The only other nudibranch found to contain isonitriles was the primitive chromodorid *Cadlina luteomarginata* [45]. Several collections from southern California of this dorid were made and each contained different compounds, reflecting its

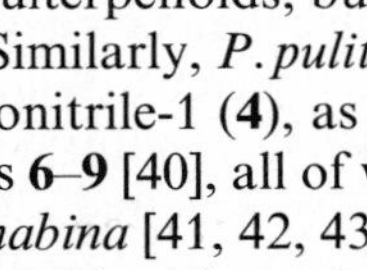

R = NC   R' = H   (1)
R = H   R' = NC   (2)

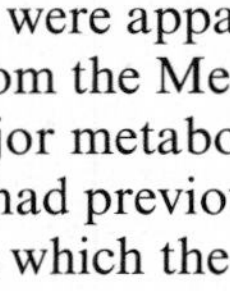

(3)

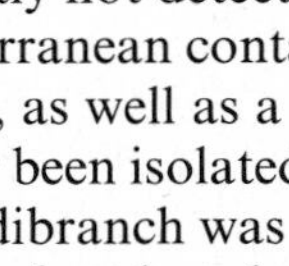

R = NC   (4)
R = NCS/NHCHO (6)

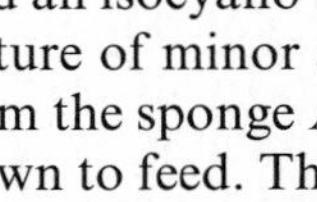

R = NC/
NCS/NHCHO (7)

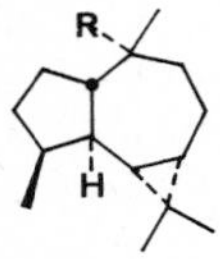

R = NC   (8)
R = NCS   (9)

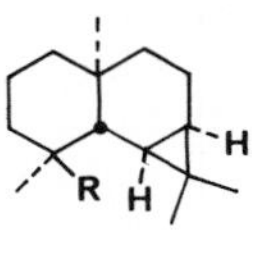

(10)
(11)

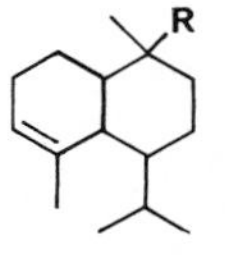

(12)
(13)

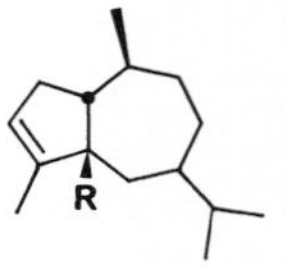

(14)
(15)

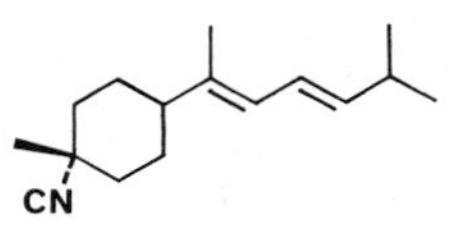

(5)

variable diet. A July 1977 collection, from Scripps Canyon, La Jolla, contained isonitrile **10** as the major metabolite, as well as the corresponding isothiocyanate **11** and the isonitrile **8** all derived from the sponge *Axinella* sp. which was found to contain four sets of isonitriles, isothiocyanates **10–15** and the corresponding formamides [46]. The same collection also contained pallescensin-A (**16**) previously isolated from the Mediterranean sponge *Dysidea pallescens* [47] and subsequently from the southern Californian sponges *Dysidea amblia* and *Euryspongia* sp. [48]. In addition, dihydropallescensin-2 (**17**) closely related to pallescensin-2 (**18**) from *D. pallescens* [49], was isolated from this collection of *C. luteomarginata*. A July 1980 collection, from Pt. Loma, again yielded **10** as the major metabolite along with **11, 8, 9** and another isonitrile/isothiocyanate pair, **12/13** or **14/15**. This collection also yielded dihydropallescensin-2 and furodysinin (**19**). The latter had been isolated from two Australian species of *Dysidea* [50, 51] and southern Californian *D. amblia* [46]. Pleraplysillin-1 (**20**) was detected as a minor metabolite; it had previously been isolated from the Mediterranean sponge *Pleraplysilla spinifera* [52]. An earlier collection from Scripps canyon in January 1977 contained no isonitriles, only furodysinin (**19**), pallescensin-A (**16**), pleraplysillin (**20**) and dendrolasin (**21**). The latter has been isolated from the sponge *Oligoceras hemorrhages* [53]. A minor component of this collection was idiadione (**22**), previously isolated from the Californian sponge *Leiosella idia* [54]. The same nudibranch collected from two locations in Canada by Hellou *et al.* [55, 56] again contained different compounds. Collections from Barkley Sound, British Columbia, demonstrated varying amounts of the known sponge metabolites furodysin (**23**) and furodysinin (**19**). The latter is a common constituent of the California collections of *C. luteomarginata*. An exceptional collection from Barkley Sound contained only microcionin-2 (**24**) previously known from the sponge *Microciona toxystilla* [57]. The major component of all Howe Sound and Barkley Sound collections was a new compound, albicanol acetate (**25**) along with minor quantities of albicanol (**26**), related to the *Dysidea* metabolites possessing the drimane skel-

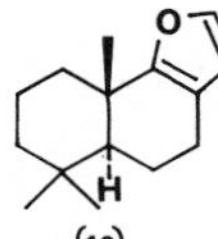

(16) 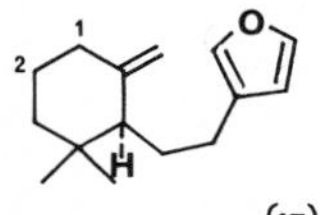   (17)   (19)   (20)
1,2 – dehydro (18)

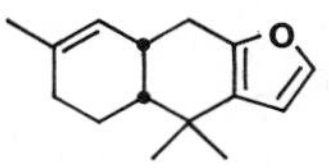

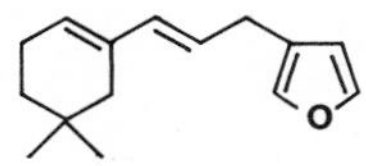

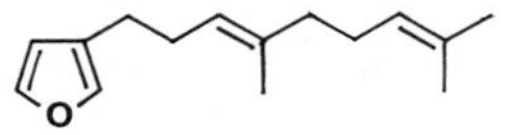

(21) 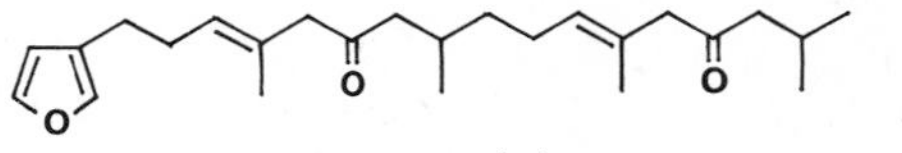   (22)

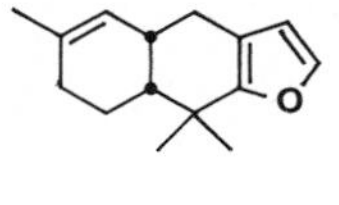

(23) 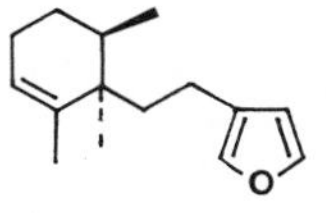   (24) 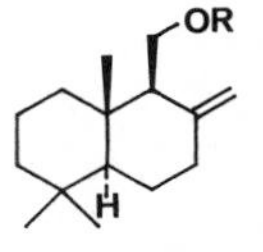   R = Ac (25)   (27)
R = H (26)

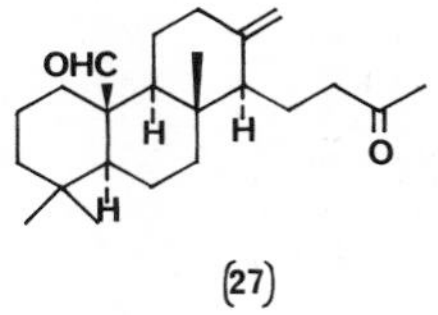

eton. The odoriferous metabolite responsible for the sweet fragrance of British Columbian specimens of *C. luteomarginata* was present in only trace amounts. It was isolated as its DNP derivative and identified as luteone (**27**), a degraded sesterterpene. Although C-21 degraded sesterterpenes are common in sponges [58], this was the first example of a C-23 terpenoid. Collections of the same nudibranch from Queen Charlotte Island, British Columbia, contained the novel diterpene marginatafuran (**28**) [58] that bears a close enough similarity to the sponge metabolite spongia-13(16),14-dien-19-oic acid (**29**) isolated from a Canary Island specimen of *Spongia officinalis* [60] to postulate a sponge origin for **28**.

The related *C. limbaughorum* and *C. flavomaculata*, like *C. luteomarginata*, feed on *Dysidea amblia*, *Leiosella idia* and *Axinella* sp. All three nudibranchs were shown to selectively accumulate one or two metabolites from each sponge [1, 46] suggesting their ability to fractionate the sponge metabolites and accumulate only those with the most potent ecological advantages, metabolizing the remainder.

The selective accumulation of terpenoid furans (masked 1,4-dialdehydes) and derivatives from their food sponges has come to typify the Chromodoridae. Scheuer and co-workers [61] isolated nakafuran-8 (**30**) and nakafuran-9 (**31**) as the major metabolites of the sponge *Dysidea fragilis* and the nudibranchs *Hypselodoris* (≡ *Chromodoris*) *maridadilus* and *Risbecia* (≡ *Hypselodoris*) *godeffroyana*. *Hypselodoris infucata* also contained the same two compounds [62]. All three nudibranchs feed exclusively on *D. fragilis* but live in different habitats thus avoiding competition. Both compounds are potent antifeedants against common reef fish. Nakafuran-8 (**30**) has also been isolated from *Hypselodoris californiensis* [48] along with dendrolasin (**21**) while *H. ghiselini* contained nakafuran-9 (**31**) along with dendrolasin (**21**) and the nakafuran-9-butenolide (**32**) – a compound closely related to nakafuran-9 [48]. The latter nidbranch also contained the unique metabolite, ghiselinin (**33**) which is related to ambliol-B (**34**) isolated from

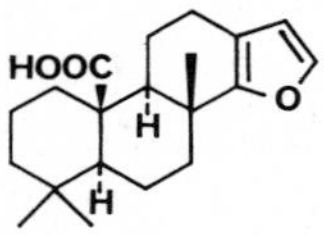

(28) 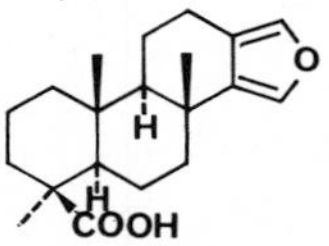   (29)

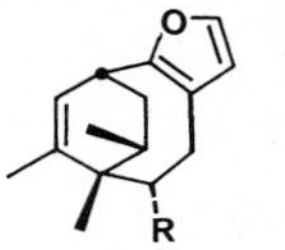

R = H   (30)
R = OH   (36)
R = OAc (37)

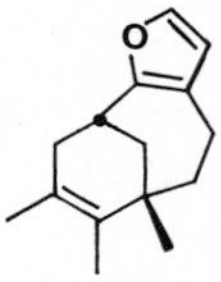

(31)

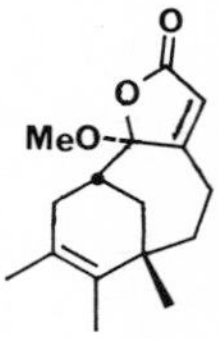

(32)

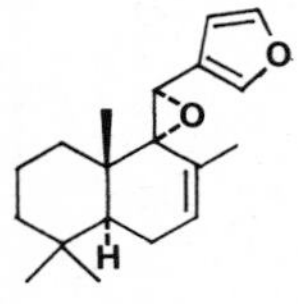

(33) 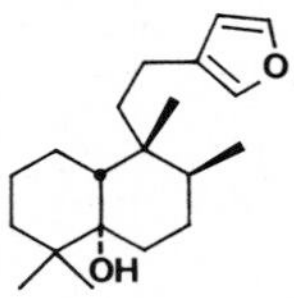   (34) 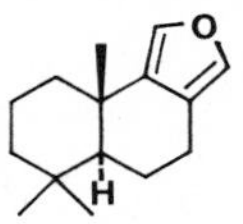   (35)

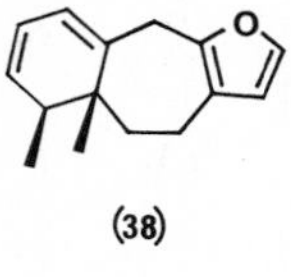

(38) 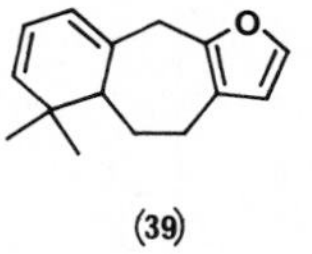   (39) 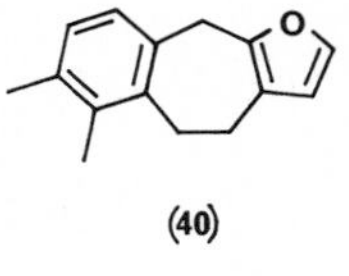   (40) 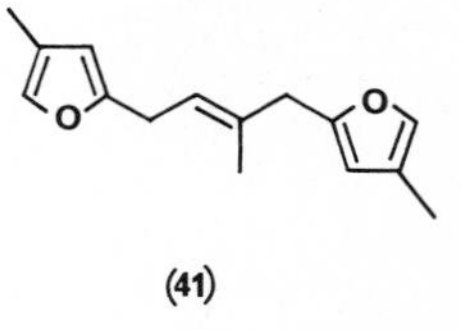   (41)

Dysidea amblia [63]. *Hypselodoris zebra* was found to concentrate furodysinin (**19**) from *Dysidea etheria* on which it feeds. The nudibranch also contained eury-furan (**35**), 5-hydroxynakafuran-8 (**36**) and 5-acetoxynakafuran-8 (**37**) probably obtained from other Dysideidae sponges. The chromodorid *Mexichromis* (≡ *Hypselodoris*) *porterae* [48] collected exclusively while feeding on *D. amblia* contained furodysinin (**19**) and euryfuran (**35**), both minor components of the food sponge [46]. *Hypselodoris agassizi* yielded only one compound agassizin (**38**) [48], which is closely related to pallescensin-G (**39**) from *D. pallescens*. The closely related chromodorid, *Thorunna* (≡ *Hypseldoris*) *daniellae* [64] yielded spiniferin-2 (**40**) previously isolated from *Pleraplysilla spinifera* [65]. The same sponge also contains longifolin (**41**) which was isolated from the nudibranchs *H.* (≡ *Glos-sodoris*) *gracilus* [66] and *H.* (≡ *Glossodoris*) *valenciennesi* [41]. Faulkner and co-workers isolated five closely related sesquiterpene esters from *Chromodoris marislae* [67]. The major metabolite, marislin (**42**), on normal handling, rearranged to the known sponge metabolite pleraplysillin-2 (**43**) (from *Pleraplysilla spinifera* [68]). A series of four minor sesquiterpene esters (**44**)–(**47**) was also isolated. *Chro-modoris albonotata* [64] and a species of *Chromodoris* referred to as "snowflake" [69, 70] both contained pu'ulenal (**48**) another drimane masked 1,4-dialdehyde found to be a potent antifeedant towards fish.

Some chromodorids have been reported to contain sesterterpenes derived from their sponge diet. *Hypselodoris* (≡ *Glossodoris*) *tricolor* [41] a case in point, was found to contain furoscalarol (**49**) and deoxoscalarin (**50**). The nudibranch was observed to feed on a sponge thought to be *Cacospongia mollior*. Extraction of the sponge yielded not only **49** and **50** but also scalaradial (**51**) all of which had previoulsy been isolated from *C. mollior* [71] and *C. scalaris* [72], thereby again de-monstrating selective incorporation of sponge metabolites. Biological testing re-vealed all three compounds to have antifeedant properties. Faulkner and co-

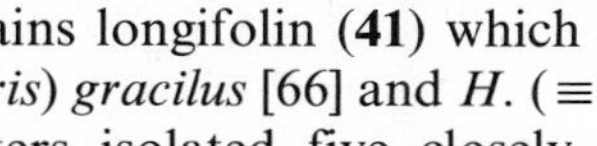

R = a (**42**) R = b (**43**)

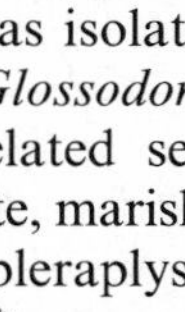

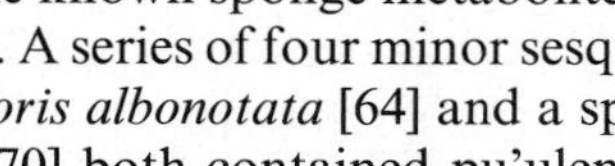

R = a (**44**) R = b (**45**)

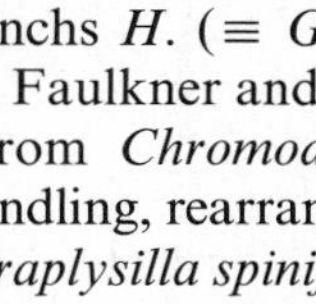

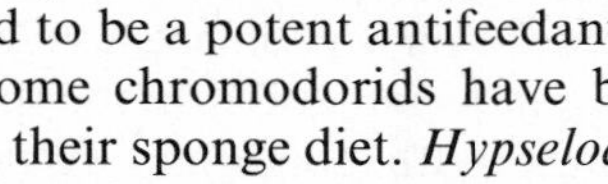

R = a (**46**) R = b (**47**)

a

b

(**48**)

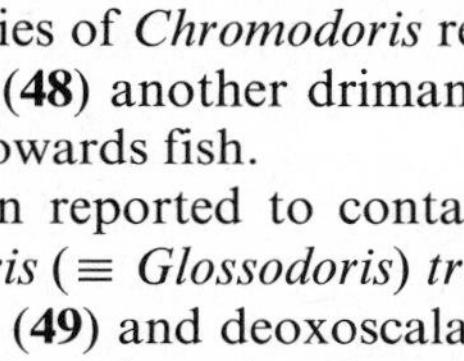

(**49**)

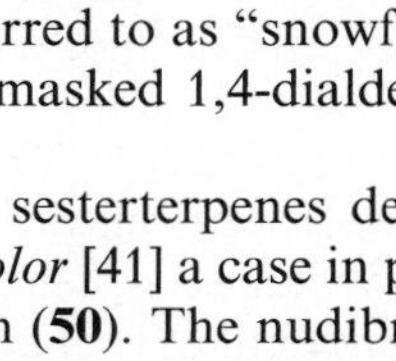

(**50**)

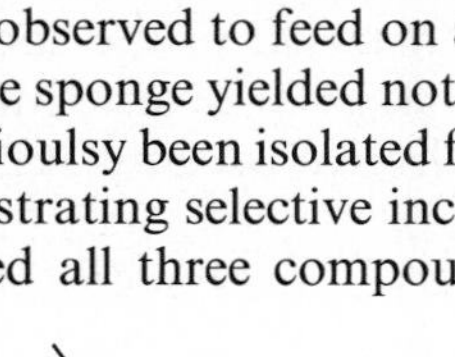

R = H   (**51**)
R = Ac (**61**)

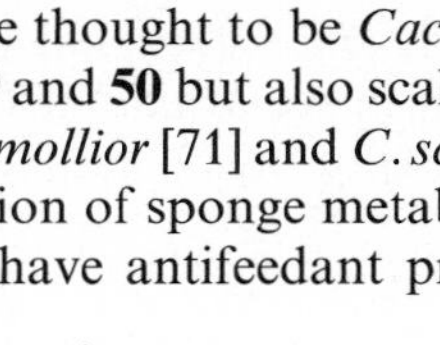

α OH, R = β H (**52**)
β OH, R = α Ac (**53**)

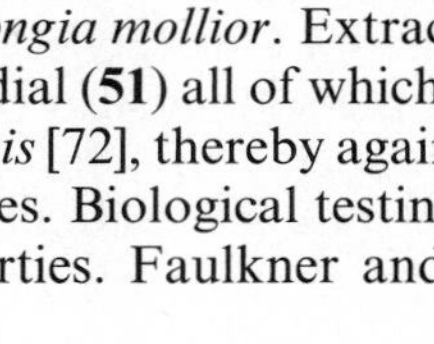

(**54**)

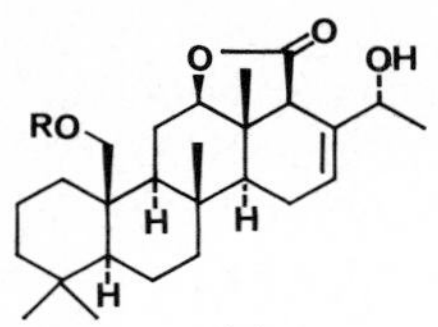

R = H (55)
R = Ac (56)

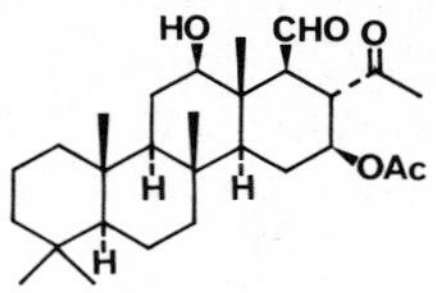

(57)

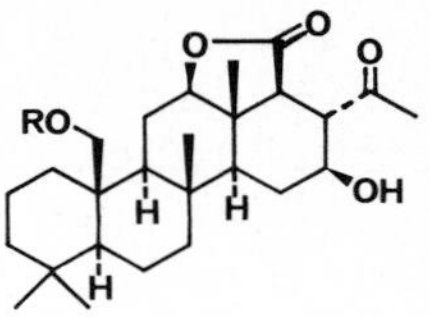

R = H (58)
R = Ac (59)

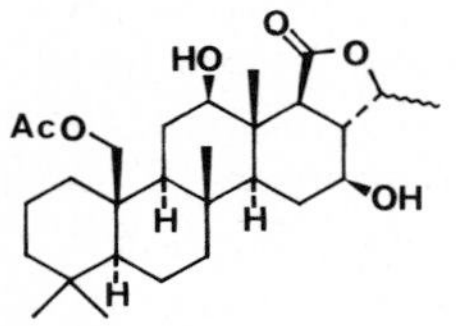

(60)

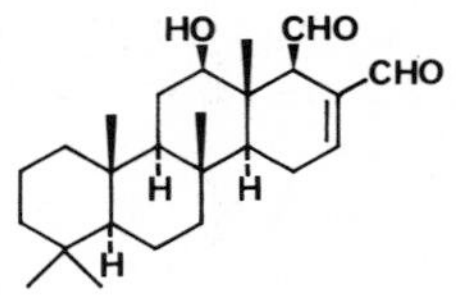

(62)

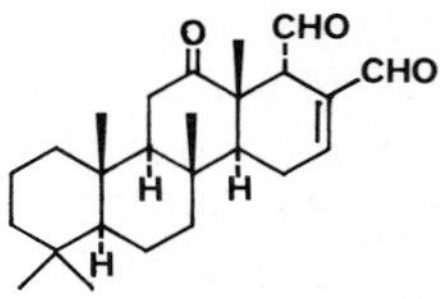

(63)

workers [73] isolated five 24-methylscalarane derivatives (52)–(56) from *Glossodoris* (≡ *Chromodoris*) *sedna*. The food source of this nudibranch was not located but these compounds bear definite similarities to 12,16-di*epi*scalarherbacin-A (57) isolated from *Carteriospongia* sp. [74] and the 24-methylscalaranes (58–60) isolated from *Lendenfeldia* sp. [75]. *Chromodoris youngbleuthi* feeds on the sponge *Spongia oceania*. The nudibranch and the sponge both contained 12-deacetylscalaradial (61). The nudibranch also contained 12-deacetyl-12-*epi*scalaradial (62) and 12-deacetyl-18-*epi*-12-oxoscalaradial (63) not detected in the sponge, while the sponge also contained scalaradial (51), which was not detected in the nudibranch [76]. This association again implies the selective accumulation of minor sponge metabolites.

Diterpenoids of sponge origin have also been isolated from nudibranchs. *Chromodoris norrisi* was found to contain one major compound, norrisolide (64), identified by X-ray crystallography [77]. The same compound was latter isolated from a Palauan sponge referred to as *Dendrilla* sp. [77] and New Zealand collections of *Chelonaplysilla violacea* [78]. In 1979 Kazlauskas *et al.* [79] reported spongiadiol (65) and spongiatriol (66) and the peracetates (67, 68) as major components of an Australian "*Spongia*" sp.[2] some time later, Scheuer and co-workers [80] found that the nudibranch *Glossodoris* (≡ *Casella*) *atromarginata* collected from Sri Lanka contained 67 and 68 along with spongiadiol-19-acetate (69) and spongiatriol-17,19-diacetate (70). Two further spongian compounds (71, 72) featuring a more highly oxidized A-ring were also found. This would indicate that the nudibranch fed on a different, though closely related, "*Spongia*" sp. (there are 20–30 species of *Spongia* in the western Indo-Pacific). *Chromodoris macfarlandi* contained four spongian diterpenes macfarlandin-A-D (73)–(76) [81, 82]. Macfarlandin-A (73) and -B (74) are closely related to aplysulphurin (77) previously isolated from the sponge *Darwinella oxeata* (≡ *Aplysilla sulfurea*) [83], which suggests that the nudibranch is feeding on an aplysillid sponge. The east australian nudibranch, *Chromodoris tasmaniensis* was found to contain high concentrations

---

2 Currently being reclassified into a new genus within the family Spongidae by Prof. P. R. Bergquist.

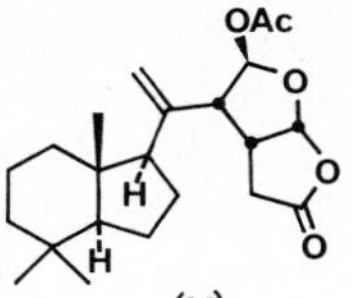

(64)

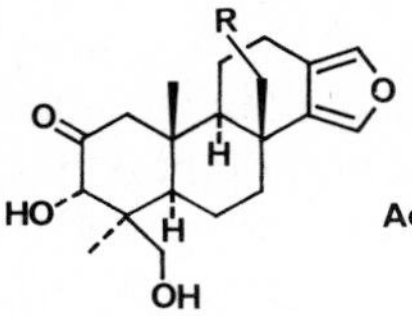

R = H (65)
R = OH (66)

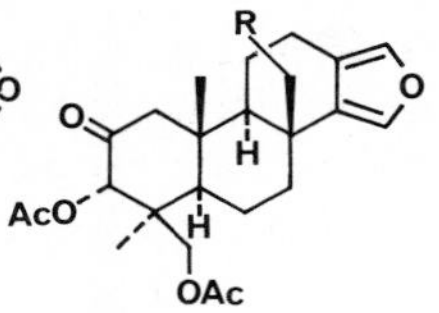

R = H (67)
R = OAc (68)

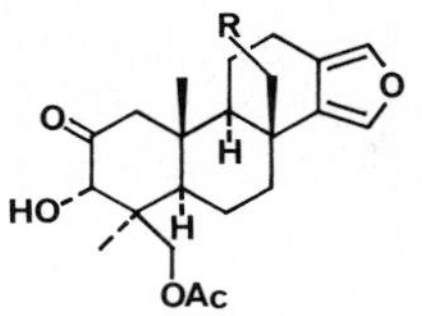

R = H (69)
R = OAc (70)

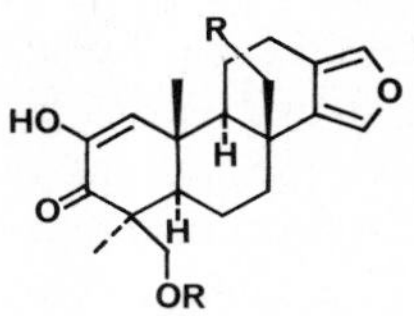

R = H (71)
R = Ac (72)

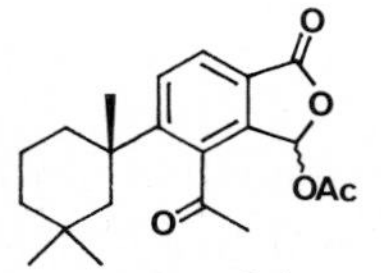

α OAc (73)
β OA (74)

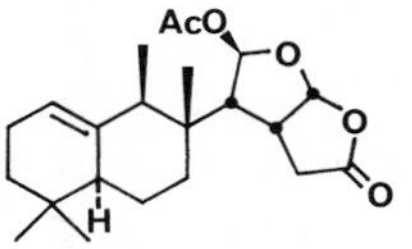

(75)

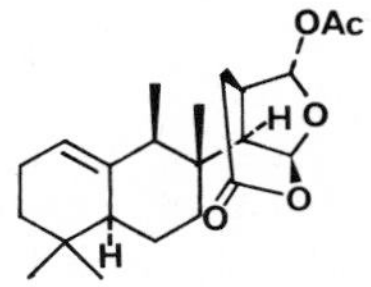

(76)

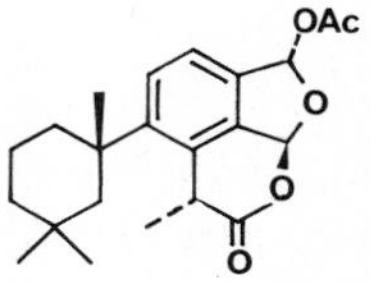

(77)

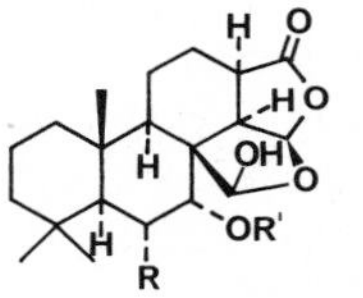

R = H , R' = COCH$_2$CH$_2$CH$_2$CH$_3$ (78)
R = H , R' = Ac (79)
R = OAc , R' = COCH$_2$CH$_2$CH$_2$CH$_3$ (80)
R = OH , R' = COCH$_2$CH$_2$CH$_2$CH$_3$ (81)
R = OCOCH$_2$CH$_2$CH$_2$CH$_3$, R' = OH (82)
R = OCOCH$_2$CH$_2$CH$_2$CH$_3$, R' = OAc (83)

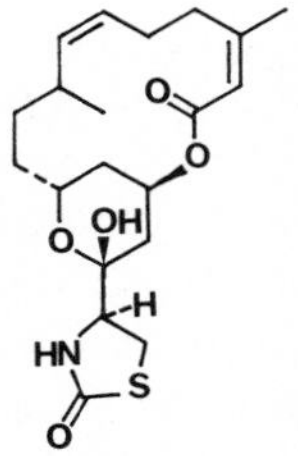

(84)

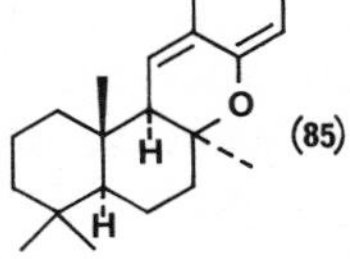

(85)

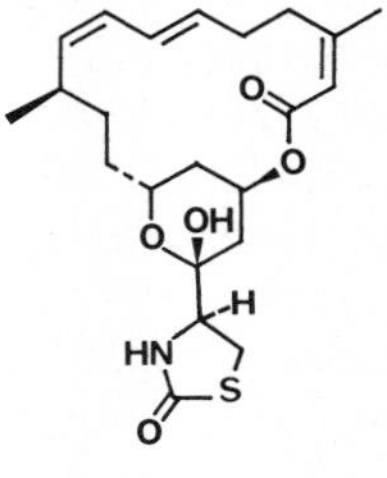

(86)

of six aplyroseol diterpenoids (78)–(83) [84] which were major components of its prey sponge, *Dendrilla* sp. (≡ *Aplysilla rosea*) [85]. The sponge contained four other components not isolated from the nudibranch while the nudibranch contained three further compounds not detected in the sponge.

Glossodoris quadricolor from the Red Sea offers one of the clearest examples of a toxic aposomatic chromodorid. The yellow, white and blue stripes of this nudibranch contrast markedly to the bright red of its food sponge, *Latrunculia magnifica*, upon which it is invariably found. Gut analysis confirmed that the nudibranch feeds on *L. magnifica* and TLC showed both the sponge and nudibranch extracts contained latrunculin-B (84) [86] a compound known to be highly ichthyotoxic [87]. Another metabolite from *L. magnifica*, latrunculin-A (85) [88], has been isolated from *Chromodoris elisabethina* collected from Guam and Enewetak by Okuda and Scheuer [89]. The nudibranch was observed feeding on a brown species of *Dysidea* (misidentified as *Heteronema*). This sponge contained puupehenone (86) and no trace of latrunculin-A. The authors postulated that the nudibranch accumulated the toxin from an occasional food source (presumably related to *L. magnifica*). Interestingly, both *G. quadricolor* and *C. elisabethina* appear almost identical and contain related compounds despite their large geographic separation.

2.1.3 Family Dendrodorididae

Two genera from this family have been studied: Cimino *et al.* [66] reported that the digestive gland of *Dendrodoris grandiflora* yields fasciculatin (87), a metabolite previously isolated from the sponge *Ircinia fasciculata* [90], as the sole metabolite apart from sterols and fatty acids. A subsequent collection [91] again yielded fasciculatin but also microcionin-1-4 (24), (88)–(90) previously isolated from a sponge referred to as *Microciona toxystilla* [57, 92]. The new C-21 furanoterpenes (91, 92) were also isolated. These are closely related to C-21 furanoterpenes isolated from *Spongia officinalis* [93, 94, 95] and other *Spongia* spp. [96]. A mixture of chromanols (93), closely related to prenylated perhydroquinones (94) isolated from the sponge *Sarcotragus* spp. [97, 98, 99], were also isolated from the digestive gland of *D. grandiflora*. All these compounds, or at least very closely related ones, occur in sponges commonly found in the habitats of the nudibranch. This strongly suggests that they are accumulated from a dietary source or are minor metabolic elaborations of dietary compounds. The digestive gland also contained a mixture of drimane fatty acid esters (95), reminiscent of many *Dysidea* metabolites. Thermolysis of these esters led to a quantitative yield of the known sponge metabolite, euryfuran (35). The skin (dorsum) extract yielded no esters, rather polygodial (96) a known antifeedant first described from the African plant *Warburgia stuhlmanni* [100]. It is interesting to note that the C-9 epimer was inactive as an antifeedant [101], as were the esters (95). The dorsum also contained 6β-acetoxyolepupuane (97); both 97 and polygodial were found to be fish antifeedants. As no dietary origin for polygodial or its derivatives was known, Cimino *et al.* performed some biosynthetic experiments to determine whether *D. grandiflora* was capable of *de novo* terpene production. Injection of [2-^{14}C]mevalonate into the digestive gland of several animals, followed by extraction after 24 h yielded radioactively labelled esters (95), polygodial (96) and 6β-acetoxyolepupuane (97), but unlabelled microcionins (24), (88)–(90), fasciculatin (87), C-21 furanoterpenes (91, 92), and chromanols (93), thus confirming the dietary origin of these terpenoids [91].

Dendrodoris limbata also contained polygodial and 6β-acetoxyolepupuane in the dorsum and the esters (95) in the digestive gland [40, 102, 103]. Biosynthetic

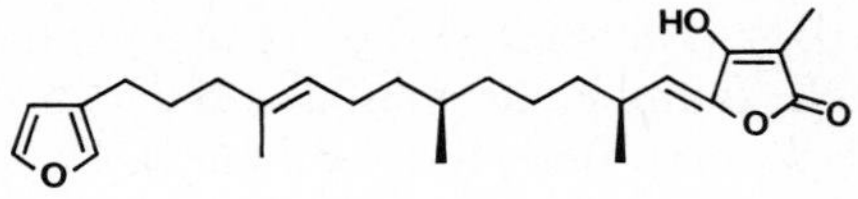

(87)

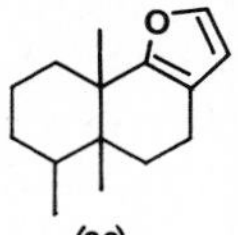

(88)

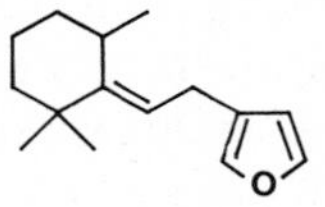

(89)

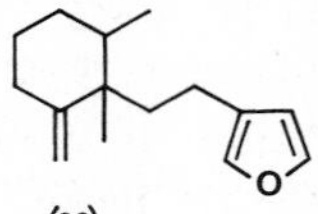

(90)

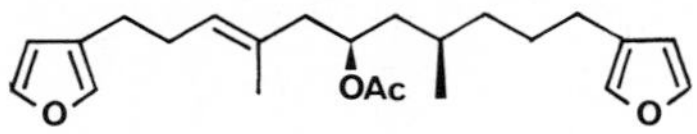

(91)

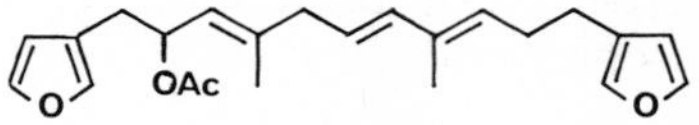

(92)

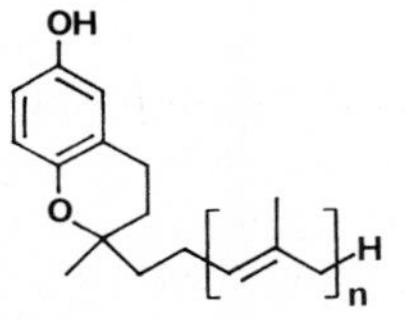

n = 1–6   (93)

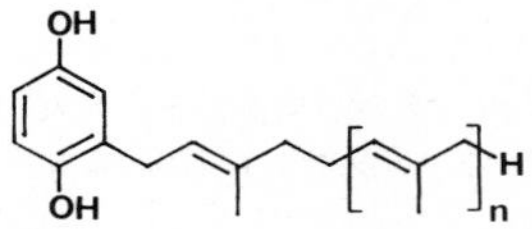

n = 1–6   (94)

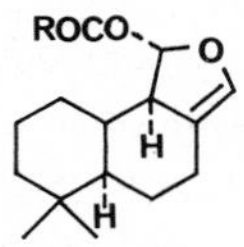

R = C$_{18}$–C$_{20}$

(95)

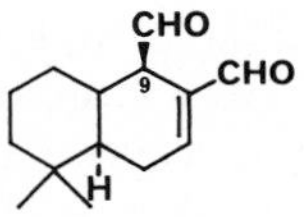

(96)

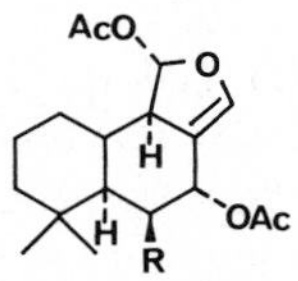

R = H   (98)
R = OAc (97)

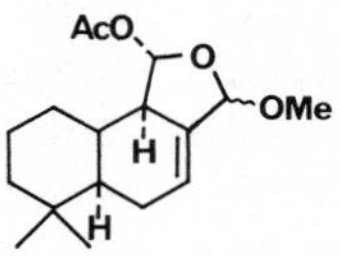

(99)

experiments, again with mevalonate [103, 104] showed that polygodial, 6β-ace-
toxyolepupuane and the fatty acid esters were biosynthesized. Experiments to determine whether the esters were precursors of polygodial or detoxification products of the latter [105] gave contradictory results suggesting that both metabolites
are biosynthesized by independent mechanisms. The authors' experiments contained too many variables for any firm conclusions on the mechanism of polygodial production, but they did show that *D. limbata* and *D. grandiflora* are capable
of *de novo* biosynthesis of drimane sesquiterpenoids.

Polygodial has also been isolated from the Pacific dendrodorids *Dendrodoris
nigra, D. tuberculosa* and *D. krebsii* [106] along with the related compound, olepupuane (**98**). The masked dialdehyde, olepupuane (**98**), and the fatty acid esters
(**95**) were isolated from the related dendrodorids *Doriopsilla albopunctata,
D. janaina*, and an undescribed doriopsillid [106]. A second collection of *D. albopunctata* contained only olepupuane and a third collection only the drimane
methoxy acetal (**99**). The same paper details the extraction of two more undescribed dendrodorids which contained only the fatty acid esters (**95**).

Dendrodoris nigra is known to feed on sponges in the genus *Suberites* but these sponges could not be found in the habitat of the nudibranch. Several Californian Dendrodoridae have been observed feeding on *Pseudosuberites pseudos* which however does not contain any drimane sesquiterpenes. Lack of evidence for a food-linked origin of these common nudibranch metabolites lends weight to the postulate that at least the family Dendrodorididae have developed the capability to biosynthesis their defense allomones. This adaptation would have had important evolutionary repercussions, allowing the animals to become independent of foods which contain antifeedant chemicals. As mentioned earlier, the chromodorid *Chromodoris albonotata* contains pu'ulenal (**48**), the enol acetate of polygodial. If one considers the capability of the dendrodorids to biosynthesize this type of compound, one must allow the possibility that similar biosynthesis occurrs in this chromodorid.

2.1.4 Family Dorididae

A somewhat diverse family of nudibranchs, the Dorididae contain very different metabolites from those of the families already discussed. Andersen and Sum [107] isolated three farnesic acid glycerides from *Archidoris odhneri*. These were shown to be the farnesate glyceride (**100**) and two monoacetates, (**101**) and (**102**). Minor metabolites [108] included the monocyclofarnesic acid glyceride (**103**), and drimane acid glyceride (**104**) plus the tricyclogeranylgeranoic acid glyceride (**105**). Of these only the drimane (**104**) was active as an antifeedant. *Archidoris montereyensis* also contained **103**, **104** and **105** [109] as well as the glyceride ether **106** [108]. The latter compound was also isolated as a minor component of the sponge *Halichondria panicea* on which the nudibranch feeds and is responsible for the antibiotic activity of the sponge and nudibranch extracts. The presence of a biogenetic series (farnesate, monocyclofarnesate, bicyclofarnesate) of terpenoid glycerides and the lack of a dietary link for these metabolites suggested the possible biosynthesis to Andersen and Sum. This was supported by ^{14}C incorporation experiments with *A. odhneri* and *A. montereyensis* [108, 109]. [2-^{14}C]Mevalonate was injected into the digestive glands of several animals. Radio-labelled compounds were isolated, **105** and **104** for *A. montereyensis* and **100**–**102** from *A. odheneri*, thus demonstrating that these nudibranchs are capable of terpene synthesis.

Cimino *et al.* [110] isolated a series of high molecular weight polyacetylenes **107**–**111** from the sponge *Petrosia ficiformis* and its predator, the nudibranch, *Peltodoris atromaculata*. Similar halogenated polyacetylenes **112**–**120** have been isolated by Walker and Faulkner [111] from *Diaulula sandiegensis* collected at La Jolla. The same compounds were detected in the crude extract of a sponge collected many years prior to the *Diaulula* work [1]. Halogenated polyacetylenes are also reasonably common in certain halosclerid and nepheliospongid sponges, e.g. *Siphonochalina* [112], *Reniera fulva* [113] and *Xestospongia testudinaria* [114]. *Diaulula sandiegensis* collected from Monterey by Fuhrman *et al.* [115] was reported to contain isoguanosine (**121**). The same authors found 1-methylisoguanosine (**122**) in *Anisodoris nobilis* [116]. Both compounds were isolated on the basis of potent cardiovascular activity, so that non-active metabolites may not have

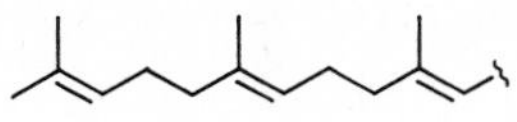

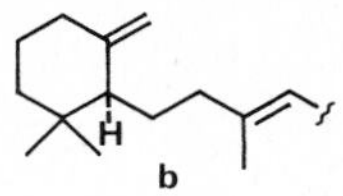

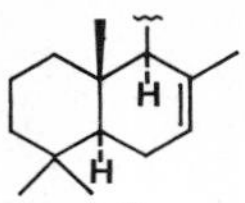

R = a   (100)
R = a , 1–Ac (101)
R = a , 2–Ac (102)
R = b   (103)
R = c   (104)
R = d   (105)
R = e   (106)

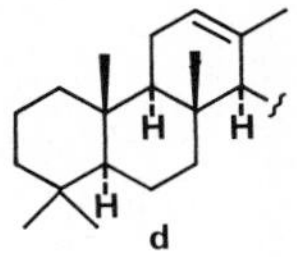

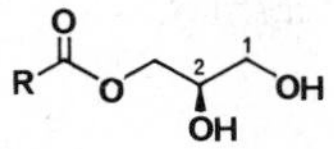

n–C$_{16}$H$_{33}$–{
e

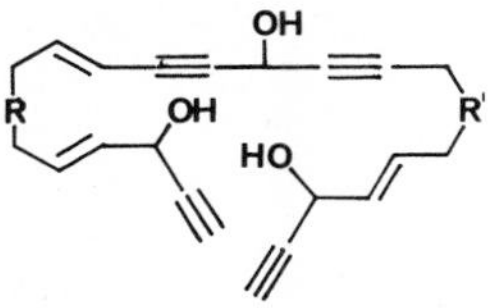

R + R' = C$_{25}$H$_{44}$ (107)
R + R' = C$_{28}$H$_{50}$ (108)
R + R' = C$_{28}$H$_{52}$ (109)
R + R' = C$_{31}$H$_{58}$ (110)
R + R' = C$_{34}$H$_{64}$ (111)

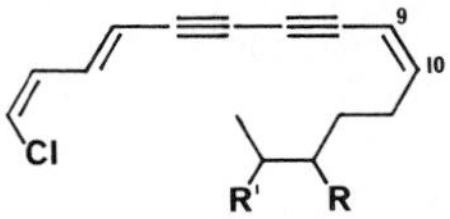

R = H , R' = OH   (112)
R = H , R' = =O   (113)
R = O , R' = H   (114)
R = OH , R' = H, 9,10–dehydro (115)

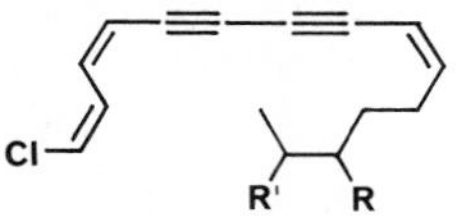

R = H , R' = OH (116)
R = H , R' = O= (117)
R = OH, R' = H   (118)

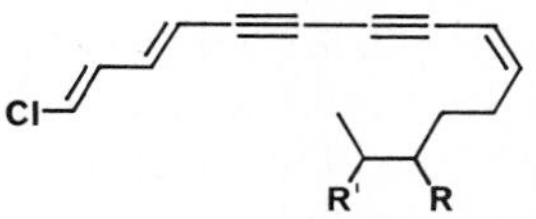

R = H , R' = H   (119)
R = OH , R' = OH (120)

been detected. 1-Methylisoguanisine has also been isolated from the sponge *Tedania digitata* [117] and adenosine (**123**) from the skin of the predatory bullomorph *Aglaja depicta* [102]. The odoriferous substance from *Anisodoris nobilis* has been isolated by Andersen and co-workers [108] and shown to be the truncated sesquiterpene 1-*nor*farnesal (**124**).

Most sponges contain high concentrations of sterols and some nudibranchs accumulate these. *Aldisa cooperi* (≡ *A. sanguinea cooperi* [118] contains two steroidal ketones, **125** and **126** [119]. The food source of the nudibranch, the sponge *Anthoarcuata graceae*, did not contain these bile acids but rather the cholestenone **127** as the major component. The bile acids were found to be potent fish antifeedants, whereas **127** was ineffective. The authors suggest this as the first demonstration of a nudibranch modifying a dietary component to produce and effective antifeedant. However, because sponges often contain up to 100 steroids in trace amounts, it remains possible that the nudibranch was actively accumulating *minor* constituents and sequestering them into its own chemical defense on the basis of biological activity.

The cosmopolitan dorid, *Jorunna funebris*, collected from Sri Lanka was found to contain a series of quinones **128**–**131** and the dihydroquinone **132** [120].

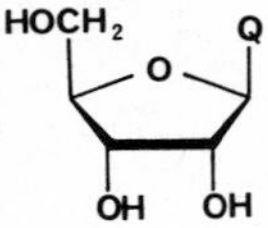

Q = a (121)

Q = b (122)

Q = c (123)

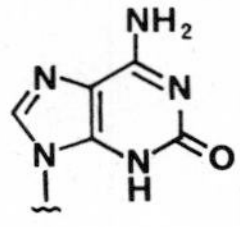

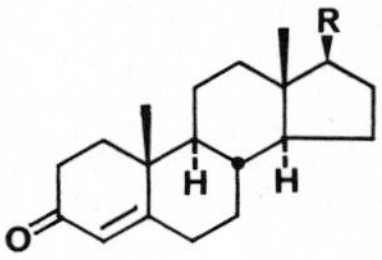

R = 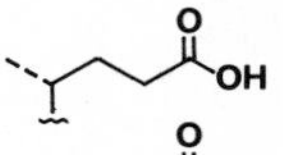   (125)

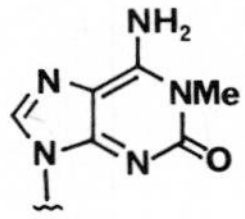

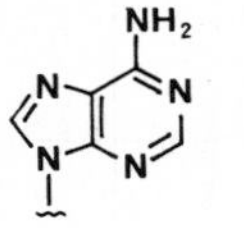

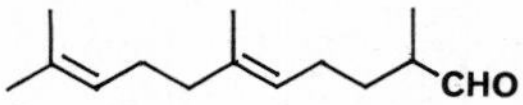

R = 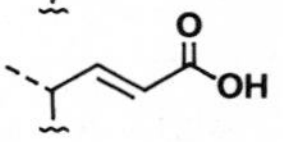   (126)

R = 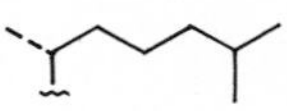   (127)

Its prey, the blue sponge *Xestospongia* sp., contained **128** and **130** and other related compounds not isolated from the nudibranch.

2.1.5 Family Onchidorididae

Only two members of this family have been studied. An unidentified species of *Adalaria* [121] was found to contain a group of sterol peroxides of which **133** is an example. The food source of the nudibranch also contained an identical sterol peroxide mixture. The original publication [121] states that the nudibranch fed on, and was collected on, a sea-pen (soft coral). A later review [108] states that

R = Ac   (128)
R = COEt (129)

R = Ac   (130)
R = COEt (131)

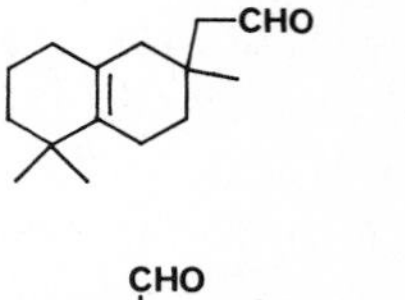

(132)

(134)

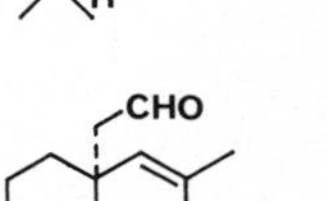

(135)

(136)

(133)

the organism feeds on an encrusting bryozoan, *Membranipora membranacea.* Since all known species of *Adalaria* feed on bryozoans, the latter report is most likely correct. The same authors [122, 123] report the isolation of three odoriferous aldehydes nanaimoal (**134**), acanthodoral (**135**) and isoacanthodoral (**136**) from *Acanthodoris nanaimoensis.*

2.1.6 Family Polyceridae

This is a large tropically centered family that feed almost entirely on bryozoans and is characterized by spectacular coloration. *Triopha catallinae* [124] and *Polycera tricolor* [108] both from British Columbia, contain the novel diacylguanidine **137**. A total synthesis established the geometry of the alkene and confirmed the overall structure [125].

  Tambja abdere and *T. diora* feed exclusively on a bryozoan, *Sessibugula translucens* [125]. Carte and Faulkner found both the bryozoan and the nudibranch to contain tambjamines A-D (**138**)–(**141**) and the hydrolysis products; the bipyrrole aldehydes **142**–**144** [127]. All seven compounds were found to be antifeedants; tambjamine-C and -D are the most potent (1–5 µg/mg food pellet) and the aldehydes are the least effective (>20 µg/mg). The same authors recently reported [128] that both *T. abdere* and *T. eliora* are attracted by water that had passed over their preferred food, *S. translucens*, but the animals were unaffected by water passing over the related bryozoan, *Bugula neritina*. Water containing tambjamine A and B at concentrations of 10^{-10}M also attracted the nudibranchs, thereby indicating that these compounds were at least partly responsible for the observed chemotaxis. The specialized carnivorous polycerid *Robastra tigris* feeds preferentially on *Tambja* spp. [126] and locates its prey by following fresh slime trails. The slime trails of *T. abdere* were found to contain 17.7 µg of tambjamines per 50 cm and it was postulated that *R. tigris* may use these compounds to identify the slime trail of *Tambja* spp. [128]. Extraction of *R. tigris* revealed that it too contained all the tambjamines and aldehydes [127]. *Tambja abdere* exudes a copious yellow mucus when attacked by *R. tigris*. This secretion contains approximately 3 mg of tambjamines (one third of the total stored) and is usually sufficient to deter *R. tigris*. The smaller *T. eliora* contains less than 2 mg per animal of tamb-

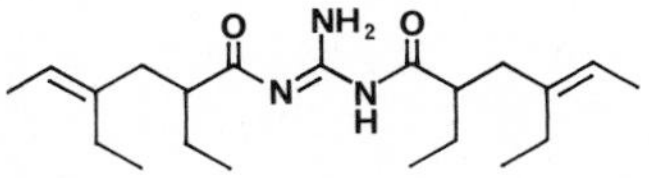

(137)

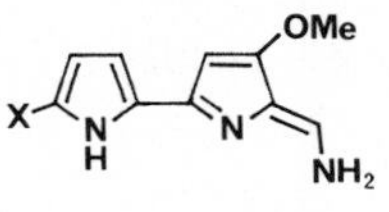

X = H   (138)
X = Br   (139)

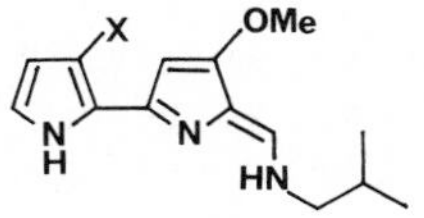

X = H   (140)
X = Br   (141)

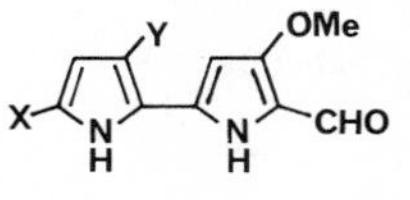

X = H , Y = H (142)
X = H , Y = Br (143)
X = Br , Y = H (144)

jamines and apparently prefers to escape from *R. tigris* by swimming. This study represents an excellent example of how the same set of compounds perform varied functions for four associated organisms.

2.2 Superfamily Aeolidacea

This group of nudibranchs consists mostly of coelenterate feeders that have the ability to store undischarged nematocysts from their prey. Those species that feed on organisms that do not possess powerful nematocysts often have their cnidosacs modified to hold glandular tissue which may contain defensive chemicals derived from the prey [129].

2.2.1 Family Tergepedidae

Phestilla melanobrachia is obligatorily associated with hard corals of the genus *Tubastrea*. The more common gold form preys on *T. coccinea* and the dark green form eats *T. diaphana*. The orange form of *P. melanobrachia* contains a series of indole alkaloids **145–149** and hokopurine (**150**) [130, 62]. The coral contains all of these compounds, in addition to the dehydroaplysinopsins **151, 152**. Two of these compounds **147** and **148** had previously been isolated from the sponge *Dercitus* sp. [131] and are closely related to the antitumor compound aplysinopsin (**153**), isolated from the sponges *Fascaplysinopsis reticulata* [132], *Smenospongia* spp. [131, 133] and *Thorectandra* sp. [134]. 6-Bromoaplysinopsin (**149**) has recently been isolated from *Smenospongia aurea* [135]. Brominated indole alkaloids and purine bases commonly occur in marine organisms [136].

  Phestilla lugubris (≡ *P. sibogae*) feeds on the hard coral *Porites compressa*. The nudibranch has non-functional cnidosacs and accumulates zooxanthellae from *P. compressa* that continue to photosynthesize in the cerate of the nudibranch [137]. Extraction of more than 1000 animals [138] led to the isolation of sterols and the ubiquitous marine metabolite, homarine (**154**) [139].

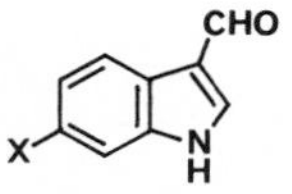

X = H (145)
X = Br (146)

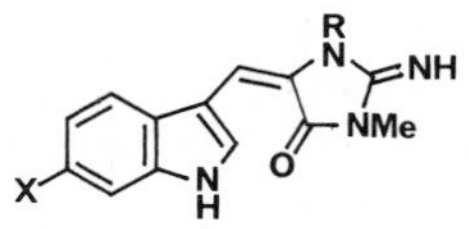

R = H , X = H (147)
R = H , X = Br (148)
R = Me, X = Br (149)
R = Me, X = H (153)

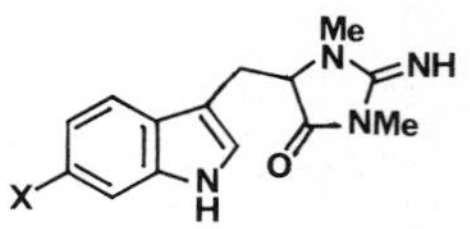

X = H (151)
X = Br (152)

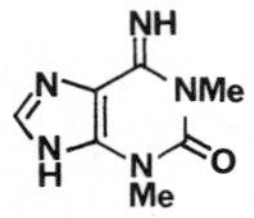

(150)

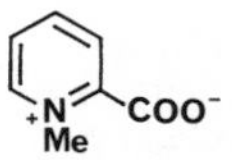

(154)

2.2.2 Family Glaucidae

Only one member of this family has been studied. *Phyllodesmium longicirra* feeds exclusively on the alcyonarian coral *Sarcophyton trocheliophorum* and also accumulates zooxanthellae from its food [129]. The nudibranch and the coral both contain two thunbergol cembranes (155), (156) and trocheliophoral (157) [140]. Selective extraction of the cerata and body wall indicates that the cembranes are located almost exclusively in the cerata, thus implying a defensive role for these terpenoids.

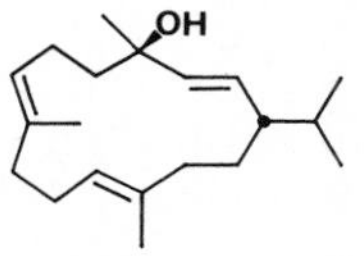

(155) 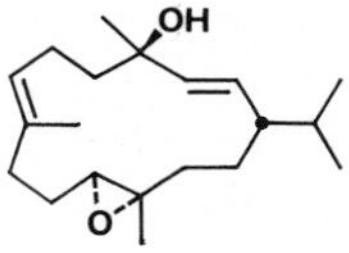   (156) 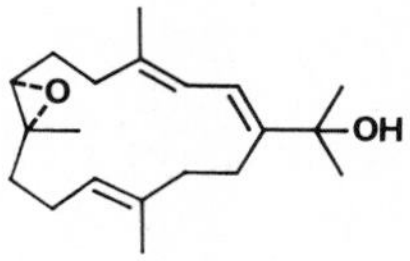   (157)

2.2.3 Family Flabellinidae

Three aeolids, *Hervia peregrina*, *Flabellina affinis* and *Coryphella lineata*, which were studied by Italian workers, were found to contain a mixture of polyhydroxylated steroids [66, 141]. The major component was identified as 158; the minor component was a mixture of several hydroxy and acetoxysterols with basic structure 159. The same sterol mixture was traced to three related hydroids, *Eudendrium rameum*, *E. racemosum* and *E. ramosum*, which comprise the nudibranchs' diet.

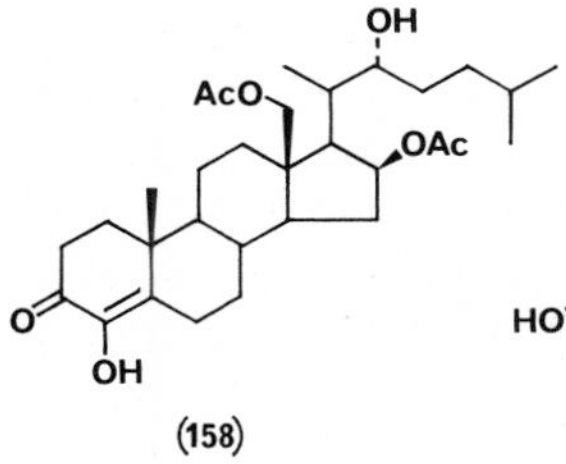

(158) 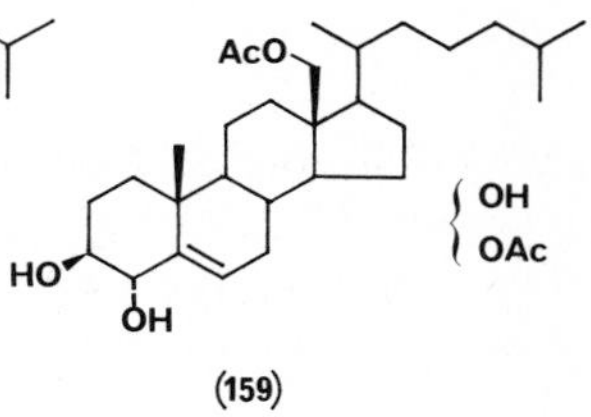   (159)

2.3 Superfamily Dendronotacea

2.3.1 Family Tethydidae

Only one genus from this family has been studied. Andersen and Ayer [142] found *Melibe leonina* to contain two unusual truncated monoterpenes, 2,6-dimethyl-4-heptenal (160) and 2,6-dimethyl-4-heptenoic acid (161). The authors found that the aldehyde 160 was responsible for the characteristic odor of the nudibranch, which had previously been reported as repugnant to potential predators [143]. An

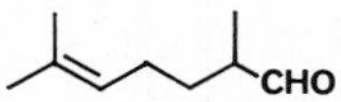

(160) 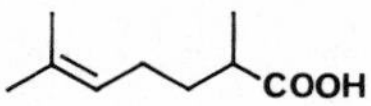   (161)

extract of *Melibe pillosa* has also been shown to contain the same two compounds by GCMS [144]. The origin of these metabolites is obscure because the nudibranchs live in kelp forests and use their unique cephalic hood to filter small crustaceans from the water.

2.3.2 Family Tritoniidae

A single specimen of a *Tritonia* sp. from Papua New Guinea [145] contained a mixture of punaglandins e.g. **162** as judged by a ^{1}H NMR spectrum of the crude extract. A similar mixture was previously isolated from the octocoral *Telesto riisei*

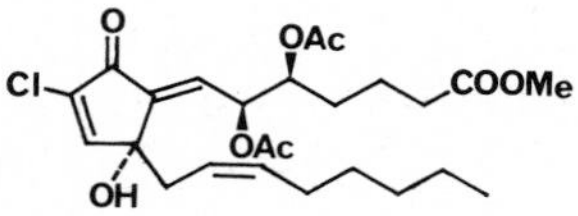

(162)

[146] on which the nudibranch was feeding when it was collected. The punaglandins are eicosanoids and possess a broad spectrum of biological activity [146, 147].

2.4 Superfamily Arminacea

2.4.1 Family Janolidae

Sodano and Spinella [148] isolated 129 mg of janolusimide (**163**) from 300 specimens of *Janolus cristatus*. This novel alkaloid incorporates an imide function and is toxic to mice (LD_{50} 5 mg/kg). Janolid nudibranchs generally feed on bryozoans but no closely related compounds have as yet been isolated from any source to indicate a dietary link.

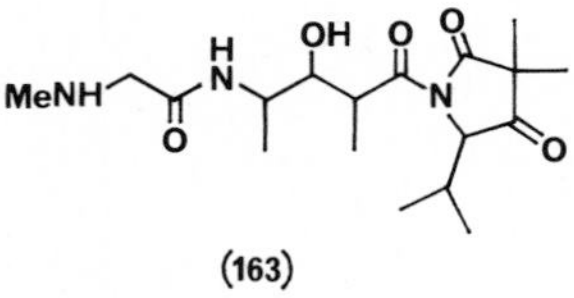

(163)

There is no doubt that most nudibranchs studied to date obtain their secondary metabolites directly from their diet. It has often been observed that accumulation of minor metabolites occur at the expense of the more common ones, which suggests that the utilized metabolites are actively accumulated at some metabolic cost. In virtually all cases, where the compounds have been tested, some antifeedant or toxic properties against potential predators has been demonstrated, thereby strongly suggesting that the metabolites are sequestered for the specific purpose of chemical defense [1].

No convincing example of a nudibranch which modifies metabolites has been shown, although the possibility exists. For instance, *Chromodoris youngbleuthi* contains **63**, a compound that may be produced by the nudibranch from the sponge metabolite scalaradial (**51**) by oxidation at C-12 and epimerization at C-18. Another case in point is the dorid *Aldisa cooperi*, which contains two bile acids (**125, 126**) possibly produced from the cholestenones (**127**) of the sponge on which it feeds. This may occur by enzymic cleavage of the side chain to convert inactive compounds to effective antifeedants. The sea hare *Stylocheilus longicauda* contains aplysiatoxin and debromoaplysiatoxin, whereas its food source, the blue-green alga *Lyngbya majuscula*, contains only debromoaplysiatoxin [149]. It remains possible that the sea hare brominates the algal metabolites thereby producing a more toxic defense allomone [150].

The family Dendrodorididae contain drimane based sesquiterpenes related to the active antifeedant polygodial. There is sufficient evidence to suggest that this family biosynthesizes its defense allomones. The family Dorididae also contains some members that biosynthesize and others that accumulate dietary chemicals. These findings are in accord with studies of other opisthobranchs. As mentioned earlier, the anaspideans (sea hares) generally accumulate metabolites from their algal diet, particularly from the chemically noxious red and brown seaweeds [34]. The Cephalaspidea (bullomorphs) seem to be capable of biosynthesis. For instance, *Navanax inermis* contains navanones A–C (**165**)–(**167**), which were shown by Fenical and co-workers [151, 152] to be biosynthesized metabolites. A similar metabolite, pulo'upone (**168**) has been isolated from another cephalaspidean, *Philinopsis speciosa* [153]. The sacoglossans feed on siphonous green algae and accumulate functional chloroplasts from their food. Some more primitive members

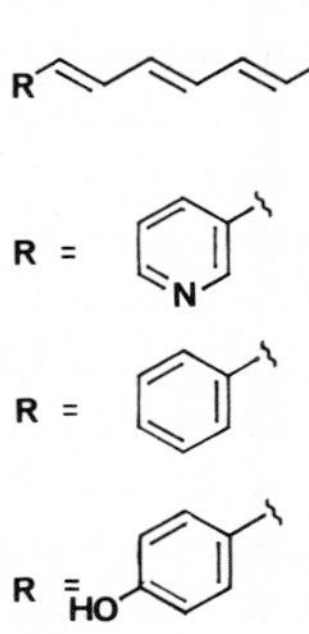

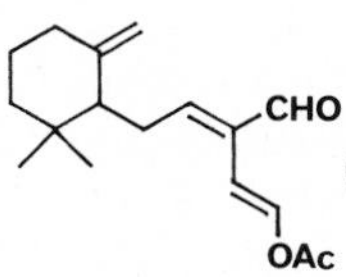

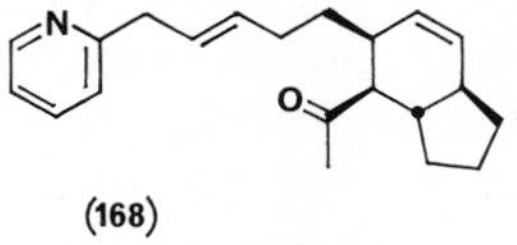

accumulate chemicals from their diet. *Oxynoe panamensis* contains the toxic algal substances caulerpicin and caulepin [35], while the more evolved shell-less forms contain polyketide metabolites unlike any algal compounds, thus suggesting biosynthesis. Similarly, some marine pulmonates contain typical algal metabolites. *Onchidella binneyi* contains onchidal (**164**) [154], which is closely related to metabolites isolated from *Caulerpa* spp., while most other marine pulmonates studied so far have yielded polyketides [34].

3. Chemotaxonomy

There is general agreement amoung taxonomists that opisthobranch classification at the family and higher levels is relatively artificial [8, 55]. For many years
nudibranchs were described and new genera created with little attempt at organiz-
ing existing genera. Predictably, this has led to many synonomous species described from different locations (e.g. *Hexabranchus sanguineus* has no fewer than
20 synonyms). Nudibranchs were traditionally classified by external appearance
and radula morphology. This rather narrow base has led to much confusion and
to the inability to define the natural limits of any particular group. The opisthobranchs appear to be polyphylectic with much parallel evolution; this has further
served to obscure relationships [156, 157]. Much work has been done recently toward reduction of the number of families and synonomizing of many species. The
classification given in Fig. 1 and in the text is that according to Willan [8] and
Rudman [9]. Research into the taxonomy of nudibranchs is still dynamic; thus the
higher order classification can be expected to change. This fluid situation makes
it difficult to state anything definitive about the chemotaxonomy of the nudibranchs. Also, as the chemicals derived from nudibranchs are usually diet-derived
and accumulated on a functional basis (toxic or antifeedant), they may not be
good taxonomic markers. Nonetheless, some trends have become apparent and
chemicals from nudibranchs may in the future prove useful in systematics. All
members of the family Phyllidiidae that have been studied contain sesquiterpene
isonitriles derived from sponges which constitute their diet. The most extensively
studied family, the Chromodorididae, have been found to contain mainly terpenoid furans, 1,4-dialdehydes or other furan precursors. This seems to suggest
that the 1,4-dialdehydes and masked dialdehydes are important marine defense
agents. These terpenoids are invariably obtained from sponges of the orders Dic-
tyoceratida and Dendroceratida (particularly the genus *Dysidea*). Both of these
orders are made up of sponges that are devoid of spicules. It would be difficult
to establish whether the nudibranchs feed on these sponges specifically to obtain
their metabolites or whether it is for some other reason (e.g. palatability, abun-
dance etc.) and the chemicals obtained are only of secondary importance. The
family Dendrodorididae generally contain drimane-based sesquiterpenes related
to the active antifeedant polygodial. These compounds are biosynthesized via
mevalonic acid. The family Dorididae contain members that biosynthesize terpene glycerides and others that sequester sterols, polyacetylenes, alkaloids or nucleosides from their sponge diet. The family Polyceridae accumulate antifeedants
from their bryozoan diet. Of the other families, which have been discussed, too
few members have been studied to allow any but the most perfunctory generalization. Suffice it to say they probably obtain their secondary metabolites from their
diet.

The storage site of the chemicals extracted from nudibranchs has been the subject of much speculation without much evidence. At various times the digestive
gland or skin glands have been implicated. Repugnatorial glands, a term coined
by Crozier [158] in 1917, is applied to a row of large glands on the projecting
mantle edge of many nudibranchs. On sufficient irritation an oily secretion is
emitted from these glands. Food particles smeared with this secretion are rejected

by fish. Nearly all opisthobranchs but particularly the nudibranchs are endowed with these curious non-mucous glands that Thompson [3] concludes must be defensive in nature to explain their position. The first evidence to support this theory came from Thompson *et al.* [45], who found that the antifeedant terpenoids from *Cadlina luteomarginata* were primarily located in the dorsum. Karuso [84] studied the nudibranch *Chromodoris tasmaniensis* and by excising the glandular tissue from over 30 specimens showed qualitatively and quantitatively that the terpenoids isolated from the nudibranch were concentrated in the mantle glands. Subsequently, Coll *et al.* [140] showed that the aeolid *Phyllodesmium longicirra*, which has a non-functioning cnidosac, stores the terpenes from its alconacian diet in the cerata, where they can best be utilized against potential predators.

All nudibranchs are simultaneous hermaphrodites and fertilization is internal following reciprocal copulation. Some time later, encapsulated eggs embedded in gelatinous ribbons are laid. The bulk of the matrix is mucopolysaccharide in nature that has generally been assumed to provide protection of the embryos from mechanical damage, infection and possibly predation [5]. No instance of parental protection of spawn masses has been recorded. Nonetheless the eggmasses are generally free of predation. This led Roesener and Scheuer [159] to investigate the chemical constituents of the eggmasses of *Hexabranchus sanguineus* from Hawaii. They isolated two macrocycles, ulapualide-A (**169**) and -B (**170**), which were potent antibacterial agents against *Candida albicans* (better than amphotericin B). At the same time Fusetani and co-workers [160, 161] isolated five closely related compounds, kabiramides A–E, e.g. (**171**), from Okinawan eggmasses of *Hexabranchus* (*sanguineus?*). Subsequently Faulkner and co-workers [162] isolated kabiramide -B and -C (**171**) from a Palauan sponge, *Halichondria* sp. and a related compound, halichondramide, from a *Halichondria* sp. from Kwajalein, thus suggesting a dietary link to the eggmass compounds. The eggmasses of *Hypselodoris infucata* did not contain the defense allomones of the adult nudibranch (nakafuran-8 and -9) but rather an as yet unidentified terpenoid aldehyde [22].

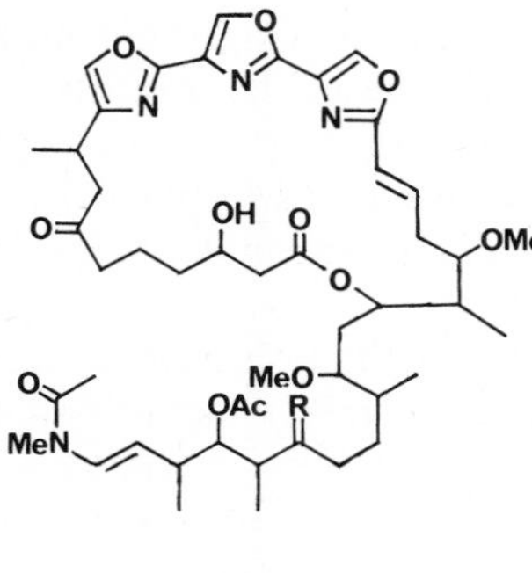

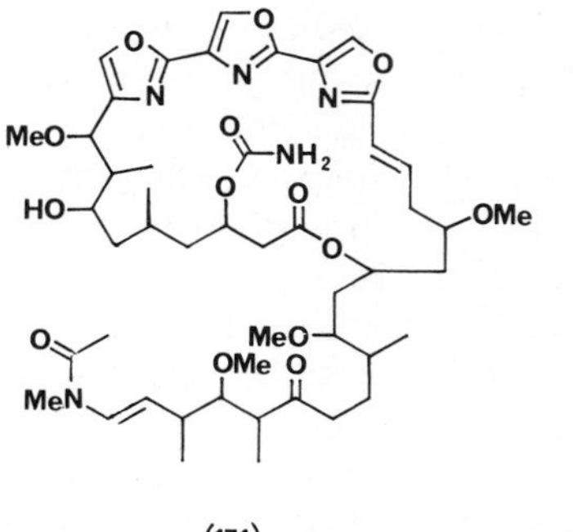

(169)
(170)
(171)

R = O   (169)

R = 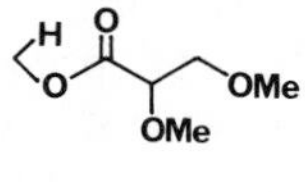   (170)

4 Chemoreception

4.1 Adults

As most nudibranchs have a highly specialized diet, they need some mechanism to locate their food. In a marine environment the most reasonable solution would be chemoreception, equivalent to our sense of smell. Many nudibranchs have been shown to be attracted to their specific food. For instance, *Rostanga rubicunda* feeds only on three sponges, *Microciona coccinea, Haloplocamium neozelandicum* and *Ophlitospongia seriata*. By slowly siphoning water from separate trays containing either sponges, algae and controls into an aquarium containing *R. rubicunda*, Ayling [163] showed that the nudibranch was attracted only to water from the sponges *M. coccinea, O. seriata* and *H. neozelandicum* in that order. Many other examples of the chemical attraction of nudibranchs to their prey are known, but the chemicals involved have rarely been identified [164]–[166]. As mentioned earlier, the nudibranch *Peltodoris atromaculata* and its food source, the sponge *Petrosia ficiformis*, share the same set of high molecular weight polyacetylenes. Experiments by Castiello *et al.* [167] showed that the nudibranch is attracted by chemotaxis to the sponge or to water squeezed from the sponge, but is not attracted to a crude lipid extract containing the polyacetylenes. This suggests that the attractant is either denatured by extraction or not lipid-soluble.

Aeolida papillosa preys on a number of anemones, but prefers *Anthopleura elegantissima* [168] and locates its prey by chemotaxis [169]. The anemone in turn responds to the presence of the nudibranch by withdrawing its tentacles, bulging of the column, by crawling away and sometimes by detaching itself from the substrate. This defensive behaviour combined with an intertidal habitat cause the nudibranch to be consistently found associated with one of its least preferred prey, the subtidal anemone *Metridium senile* [170]. The kairomone signal that evokes the defensive behavior in *Anthopleura* was identified as the betaine anthopleurine (**172**) [171]. This compound is accumulated by *Aeolidia* from *Anthopleura* and apparently leaches from the skin of the nudibranch. *Anthopleura* lives in clone groups and releases anthopleurine only when attacked or damaged. Anthopleurine causes neighboring anemones to undergo the stereotype alarm response, thus giving some measure of protection to nearby members of the group [172, 173]. The compound thereby behaves as an alarm substance, but cannot be classified as an alarm pheromone as no intentional signal has evolved to establish communication between conspecifics.

Carté and Faulkner have recently shown that the polycerid *Tambja eliora* is attracted to its food source, the bryozoan *Sessebugula translucens* [128]. The tambjamines, which had previously been isolated from the defensive secretions of *T. eliora, T. abdere* [127] and the bryozoan, were responsible for this chemotaxis. Water containing 10^{-10}M of tambjamines A and B attracted the nudibranch while 10^{-11}M had no effect and 10^{-8}M caused avoidance. The authors postulate that the avoidance of higher concentrations of tambjamines may assist *T. eliora* in avoiding predation as an encounter between *Tambja* spp. and a predator may result in the localized release of tambjamines, thus "warning" other *Tambja* of the presence of a predator. A simpler explanation would be that at higher concentrations the tambjamines are noxious to *T. eliora*.

When heavily molested, *Navanax inermis* will secrete a yellow mixture of navenones into its slime trail [152]. *Navanax* follow slime trails to locate their opisthobranch prey as well as for purposes of reproduction. When a *Navanax* encounters a trail that contains navenones it turns sharply in avoidance. This was interpreted by Sleeper *et al.* [152] as an intraspecific communication of danger (alarm pheromone). Since the victim itself does not benefit from the metabolic expense incurred in maintaining and secreting the navanones into its trail, we are dealing with a case of altruism. Such behavior has only been demonstrated in schooling organisms (e.g. fish [174]) or clone groups (e.g. anemones [173]), which often behave as a single organism rather than as a group. As *Navanax* are solitary organisms, a more plausible explanation for navenone secretion is simply that of a defense strategy against predators. This possibility was acknowledged by the authors [175].

Chemoreception not only plays a role in locating food but in eliciting feeding behavior (phagostimulation). Recent work by Sakata *et al.* [176]–[178] has shown that 1,2-diacylglyceryl-4'-*O*-(*N*,*N*,*N*-trimethyl)homoserine (**173**) isolated from *Ulva petusa* is the phagostimulant of the sea hare *Aplysia juliana*, which feeds preferentially on *U. petusa*. Similarly, phosphatylcholines (**174**) and digalactosyldiacylglycerols (**175**) from the brown alga *Undaria pinnatifida* were found to be efficient phagostimulants for the abalone *Haliotis discus*. No analogous work has been attempted with nudibranchs but should be of great interest.

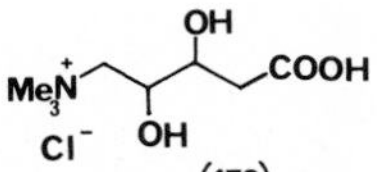

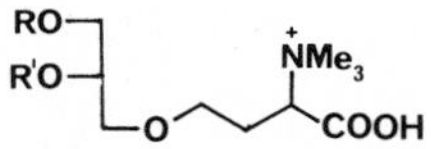

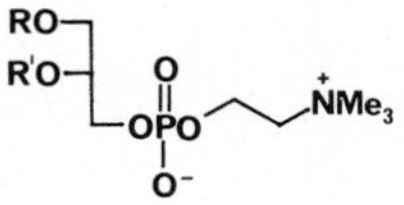

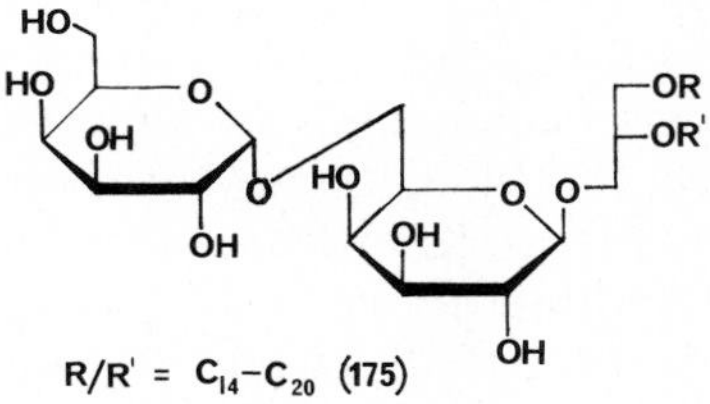

4.2 Larvae

As mentioned earlier, nudibranchs after copulation lay eggs. Some nudibranchs are direct developers, and juveniles hatch directly from the eggs. More typically, they hatch as veliger larvae, which are either planktotrophic (having a long feeding planktonic phase), or lecithotrophic (having a short non-feeding planktonic phase). This larval phase serves to disperse the species, as nudibranchs are semelparous (i.e. the adults undergo a single spawning period and die [179]) and not migratory. New populations are established by the ability of the searching veligers to recognize suitable food sources. The trigger to settle from the water column and metamorphose is usually found in some element of the food source of the adults [180]. Thus *Rostranga pulchra* larvae can only be induced to meta-morphose in the presence of the adult prey sponge, *Ophlitospongia pennata* [181]. The most intensively studied metamorphic induction in nudibranchs is that of the larvae of *Phestilla lugibris* (≡ *P. sibogae*). Hadfield and Karlson [182] first noticed that the *Phestilla* larvae required a chemical inducer from the adult prey, the stony coral *Porites compressa*. Subsequent work has revealed this compound to be a water-soluble, small (< 500 MW) organic molecule found in the growing tips of *P. compressa* [183, 184]. The molecule seems to be active at hormonal levels; large scale extractions of coral are required to obtain enough crude inducer to al-low separation and structural elucidation.

In other molluscs, γ-aminobutyric acid (GABA) has been reported as the metamorphic inducer of abalone larvae [185]. It should be noted that GABA has neurological roles in many kinds of animals; it is therefore unlikely that it is the natural inducer in the red algae that stimulates the settlement of abalone in nature (*Phestilla* can be induced to metamorphose by choline, also a neurotransmitter). The red alga *Delesseria sanguinea* provides the metamorphic stimulus for the scal-lop *Pecten maximus*. The bioactive compound jacaranone (**176**) [186] was pre-viously known as a constituent of the terrestrial plant *Jacaranda caucana* [187].

(**176**)

There are many other chemical questions left to be answered in the field of nudibranch chemical ecology. For instance, do nudibranchs make use of phero-mones: sex attractants, trail markers, alarm compounds or territory markers? These types of compounds are almost entirely unidentified from marine inverte-brates [188, 189] though many have been characterized from insects and other ter-restrial organisms. Knowledge of the inter- and intra-specific chemical communi-cation of nudibranchs could be used to test the theories of terrestrial ecologists on allelochemics [190], largely developed on the basis of plant/insect relation-ships. This type of work requires the close collaboration between biologists and chemists. Future chemical research on the defense allomones of nudibranchs will continue, but the emphasis is already shifting to other more complex fundamental questions, among them biosynthesis of metabolites or chemical recognition of prey, and larval ecology.

Finally, can any of the information gained in chemical studies of nudibranchs be applied to human endeavors? Perhaps, in the future, some of the nudibranch metabolites may be used to repel agricultural (or aquacultural) pests. Perhaps some of the substances are of potential use as pharmaceuticals or as marine anti-fouling agents.

References

1. Faulkner DJ, Ghiselin MT (1983) Mar. Ecol. Prog. Ser. 13:295
2. Thompson TE (1960) J. Mar. Biol. Ass. U.K. 39:115
3. Thompson TE (1960) J. Mar. Biol. Ass. U.K. 39:123
4. Ros J (1976) Oecol. Aquat. 2:41
5. Todd CD (1981) Oceanogr. Mar. Biol. Ann. Rev. 19:141
6. Todd CD (1977) Ph. D. thesis, University of Leeds
7. Anderson ES (1971) Ph. D. thesis, University of California, Santa Cruz
8. Willan RC, Coleman N (1984) Nudibranchs of Australasia, Sydney, Australian Marine Photographic Index
9. Rudman WB (1984) Zool. J. Linn. Soc. 81:115
10. Rudman WB (1981) J. Zool. Lond. 193:421
11. Bertsch H, Johnson S (1981) Hawaiian Nudibranchs, p 44, Honolulu, Oriental Publishing Co.
12. Garstang W (1890) J. Mar. Biol. Ass. U.K. 1:399
13. Cott HB (1940) Adaptive Coloration in animals, p 508, London, Mathuen
14. Harris LG (1973) In: Cheng TC (ed) Current topics of comparative pathobiology, vol 2, Baltimore, Academic Press, p 213
15. Johannes RE (1963) Veliger, 5:104
16. Owen DF (1982) Camouflage and Mimicry, p 115, Chicago, University of Chicago Press 1982
17. Rudman WB (1982) Zool. J. Linn. Soc. 76:183
18. Rudman WB (1983) Zool. J. Linn. Soc. 78:105
19. Rudman WB (1986) Zool. J. Linn. Soc. 86:101
20. Willows AOD (1967) Science 157:570
21. Edmunds M (1986) Proc. Malac. Soc. Lond. 38:121
22. Karuso P unpublished results
23. Eliot CNE (1904) Proc. Zool. Soc. Lond. 2:354
24. Paine RT (1963) Veliger 6:1
25. Haefelfinger HR (1961) Rev. Suisse Zool. 68:331
26. Lewin RA (1970) Pac. Sci. 24:356
27. Thompson TE, Bennett I (1969) Science 166:1532
28. Thompson TE, Bennett I (1970) Zool. J. Linn. Soc. 49:187
29. Herdman WA, Clubb JA (1892) Proc. Lpool Biol. Soc. 4:131
30. Crossland C (1911) Proc. Zool. Soc. Lond. 79:1062
31. Long S (1970) Tabulata 10
32. Yamamura S, Hirata Y (1963) Tetrahedron 19:1485
33. Masuda H, Tomie Y, Yamamura S, Hirata Y (1967) J. Chem. Soc., Chem. Commun. 898
34. Faulkner DJ (1984) Nat. Prod. Rep. 1:251
35. Doty MS, Aguilar-Santos G (1970) Pac. Sci. 24:351
36. Risbec J (1928) Colon. Franc. 2:1
37. Burreson BJ, Clardy J, Finer J, Scheuer PJ (1975) J. Am. Chem. Soc. 97:4763
38. Hagadone MR, Burreson BJ, Scheuer PJ (1979) Helv. Chim. Acta 62:2484
39. Roll D (1984) Ph. D. thesis, University of Hawaii
40. Cimino G, De Rosa S, De Stefano S, Sodano G (1982) Comp. Biochem. Physiol. 73B:471
41. Cafieri F, Fattorusso E, Magno S, Santacroce C, Sica D (1973) Tetrahedron 29:4259
42. Fattorusso E, Magno S, Mayol L, Santacroce C, Sica D (1975) Tetrahedron 31:269
43. Di Blasio B, Fattorusso E, Magno S, Mayol L, Pedone C, Santacroce C, Sica D (1976) Tetrahedron 32:473
44. Gulavita NK, de Silva DE, Hagadone MR, Karuso P, Scheuer PJ, Van Duyne GD, Clardy J (1986) J. Org. Chem. 51:5136
45. Thompson JE, Walker RP, Wratten SJ, Faulkner DJ (1982) Tetrahedron 38:1865
46. Thompson JE, Walker RP, Faulkner DJ (1985) Mar. Biol. 88:11
47. Cimino G, De Stefano S, Guerriero A, Minale L (1975) Tetrahedron Lett. 16:1425
48. Hochlowski JE, Roger PW, Ireland C, Faulkner DJ (1984) J. Org. Chem. 49:241
49. Cimino G, De Stefano S, Guerriero A, Minale L (1975) Tetrahedron Lett. 16:1417

50. Dunlop RW, Kazlauskas R, March G, Murphy PT, Wells RJ (1982) Aust. J. Chem. 35:95
51. Kazlauskas R, Murphy PT, Wells RJ, Daly JJ, Schoenholzer P (1978) Tetrahedron Lett. 19:4951
52. Cimino G, De Stefano S, Minale L, Trivellone E (1972) Tetrahedron 28:4761
53. Vanderah DJ, Schmitz FJ (1975) Lloydia 38:271
54. Walker RP, Thompson JE, Faulkner DJ (1980) J. Org. Chem. 45:4976
55. Hellou J, Andersen RJ, Thompson JE (1982) Tetrahedron 38:1875
56. Hellou J, Andersen RJ (1981) Tetrahedron Lett. 22:4173
57. Cimino G, De Stefano S, Guerriero A, Minale L (1975) Tetrahedron Lett. 16:3723
58. Cimino G, De Stefano S, Minale L, Fattorusso E (1972) Tetrahedron 28:267
59. Gustafson K, Andersen RJ, He C-H, Clardy J (1985) Tetrahedron Lett. 26:2521
60. Capelle N, Braekman JC, Daloze D, Tursch B (1980) Bull. Soc. Chim. Belg. 89:399
61. Shulte GR, Sheuer PJ, McConnell OJ (1980) Helv. Chim. Acta 63:2159
62. Okuda RK (1983) Ph. D. thesis, University of Hawaii
63. Walker RP, Faulkner DJ (1981) J. Org. Chem. 46:1098
64. Schulte GR, Scheuer PJ (1982) Tetrahedron 38:1857
65. Cimino G, De Stefano S, Minale L, Trivellone E (1975) Tetrahedron Lett. 16:3727
66. Cimino G, De Stefano S, De Rosa S, Sodano G, Villani G (1980) Bull. Soc. Chim. Belg. 89:1069
67. Hochlowski JE, Faulkner DJ (1981) Tetrahedron Lett. 22:271
68. Cimino G, De Stefano S, Minale L (1974) Experientia 30:846
69. Bertsch H, Johnson S Ref. 11 p 59
70. Terem B, unpublished results
71. Cafieri F, De Napoli L, Fattorusso E, Santacroce C, Sica D (1977) Gazz. Chim. Ital. 107:71
72. Yasuda F, Tada H (1981) Experientia 37:110
73. Hochlowski JE, Faulkner DJ, Bass LS, Clardy J (1983) J. Org. Chem. 48:1738
74. Kazlauskas R, Murphy PT, Wells RJ, Daly JJ (1980) Aust. J. Chem. 33:1783
75. Kazlauskas R, Murphy PT, Wells RJ (1982) Aust. J. Chem. 35:51
76. Terem B, Scheuer PJ (1986) Tetrahedron 42:4409
77. Hochlowski JE, Faulkner DJ, Clardy J, Matsumoto GK (1983) J. Org. Chem. 48:1141
78. Bergquist PR, Buckelton JS, Cambie RC, Clarke GR, Craw PA, Karuso P, Poiner A, Rickard CEF, Taylor WC, Manuscript in preparation
79. Kazlauskas R, Murphy PT, Wells RJ, Noack K, Oberhänsli WE, Schonholzer P (1979) Aust. J. Chem. 32:867
80. de Silva ED, Scheuer PJ (1982) Heterocycles 17:167
81. Molinski TF, Faulkner DJ (1986) J. Org. Chem. 51:2601
82. Molinski TF, Faulkner DJ, personal communication
83. Karuso P, Skelton BW, Taylor WC, White AH (1984) Aust. J. Chem. 37:1081
84. Karuso P (1984) Ph. D. thesis University of Sydney
85. Karuso P, Taylor WC (1986) Aust. J. Chem. 39:1629
86. Mebs D (1985) J. Chem. Ecol. 11:713
87. Neeman I, Fishelson L, Kashman Y (1975) Mar. Biol. 39:293
88. Kashman Y, Groweiss A, Carmely S, Kinamoni Z, Czarkie D, Rotem M (1982) Pure Appl. Chem. 54:1995
89. Okuda RK, Scheuer PJ (1985) Experientia 41:1355
90. Cafieri F, Fattorusso E, Santacroce C, Minale L (1972) Tetrahedron 28:1579
91. Cimino G, De Rosa S, De Stefano S, Morrone R, Sodano G (1985) Tetrahedron 41:1093
92. Cimino G (1977) In: Faulkner DJ, Fenical WHG (eds) Marine natural products' NATO conference series IV; marine sciences Vol. 1 New York, Plenum Press, p 71
93. Cimino G, De Stefano S, Minale L, Fattorusso E (1971) Tetrahedron 27:4673
94. Cimino G, De Stefano S, Minale L, Fattorusso E (1972) Tetrahedron 28:267
95. Cimino G, De Stefano S, Minale L (1972) Tetrahedron 28:5983
96. Faulkner DJ (1984) Nat. Prod. Rep. 1:551
97. Amade P, Chevolot L, Perzanowski HP, Scheuer PJ (1983) Helv. Chim. Acta 66:1672

98. Cimino G, De Stefano S, Minale L (1972) Tetrahedron 28:1315
99. Cimino G, De Stefano S, Minale L (1972) Experientia 28:1401
100. Kubo I, Pettei M, Pilkiewicz F, Nakanishi K (1976) J. Chem. Soc., Chem. Commun. 1013
101. Kubo I, Ganjion I (1981) Experientia 37:1063
102. Cimino G, De Rosa S, De Stefano S, Sodano G (1981) Tetrahedron Lett. 22:1271
103. Cimino G, De Rosa S, De Stefano S, Sodano G (1986) Pure Appl. Chem. 58:375
104. Cimino G, De Rosa S, De Stefano S, Sodano G, Vallani G (1983) Science 219:1237
105. Cimino G, De Rosa S, De Stefano S, Sodano G (1985) Experientia 41:1335
106. Okuda RK, Scheuer PJ, Hochlowski JE, Walker RP, Faulkner DJ (1983) J. Org. Chem. 48:1866
107. Andersen RJ, Sum FW (1980) Tetrahedron Lett. 21:797
108. Gustafson K, Andersen RJ (1985) Tetrahedron 41:1101
109. Gustafson K, Andersen RJ, Chan MHM, Clardy J, Hochlowski JE (1984) Tetrahedron Lett. 25:11
110. Castiello D, Cimino S, De Rosa S, De Stefano S, Sodano G (1980) Tetrahedron Lett. 21:5047
111. Walker RP, Faulkner DJ (1981) J. Org. Chem. 46:1475
112. Rotem M, Kashman Y (1979) Tetrahedron Lett. 20:3193
113. Cimino G, De Stefano S (1977) Tetrahedron Lett. 18:1325
114. Quinn RJ, Tucker DJ (1985) Tetrahedron Lett. 26:1671
115. Fuhrman FA, Fuhrman GJ, Nachman RJ, Mosher HS (1981) Science 212:55
116. Kim YH, Nachman RJ, Pavelka L, Mosher HS, Fuhrman FA, Fuhrman GJ (1981) J. Nat. Prod. 44:206
117. Quinn RJ, Gregson RP, Cook AF, Bartlett RJ (1980) Tetrahedron Lett. 21:567
118. Millen SV, Gosliner TM (1985) Zool. J. Linn. Soc. 84:195
119. Ayer SW, Andersen RJ (1982) Tetrahedron Lett. 23:1039
120. Gulavita NK, De Silva D, Scheuer PJ, manuscript in preparation
121. Stonard RJ, Petrovich JC, Andersen RJ (1975) Steroids 36:5160
122. Ayer SW, Hellou J, Tischler M, Andersen RJ (1984) Tetrahedron Lett. 25:141
123. Ayer SW, Andersen RJ, He C, Clardy J (1984) J. Org. Chem. 49:2653
124. Gustafson K, Andersen RJ 81982) J. Org. Chem. 47:2167
125. Piers E, Chong MJ, Gustafson K, Andersen RJ (1984) Can. J. Chem. 62:1
126. Farmer WM (1978) Veliger 20:375
127. Carté B, Faulkner DJ (1983) J. Org. Chem. 48:2315
128. Carté B, Faulkner DJ (1986) J. Chem. Ecol. 12:795
129. Rudman WB (1981) Zool. J. Linn. Soc. 72:219
130. Okuda RK, Klein D, Kinnel RB, Li M, Scheuer PJ (1982) Pure Appl. Chem. 54:1907
131. Djura P, Faulkner DJ (1980) J. Org. Chem. 45:735
132. Kazlauskas R, Murphy PT, Quinn RJ, Wells RJ (1977) Tetrahedron Lett. 18:61
133. Hollenbeak KH, Schmitz FJ (1977) Lloydia 40:479
134. Bergquist PR, Wells RJ (1983) Scheuer PJ (ed) Marine natural products; chemical and biological perspectives, Vol. V, New York, Academic Press, p 1
135. Tymiak AA, Rinehart Jr KL, Bakus GJ (1985) Tetrahedron 41:1039
136. Christophersen C (1985) Marine alkaloids. In: Manske RHF (ed) The Alkaloids, New York, Academic Press, p 25
137. Rudman WB (1982) Zool. J. Linn. Soc. 74:147
138. Klein DM (1986) Ph. D. thesis University of Hawaii
139. Targett NM, Bishop SS, McConnell OJ, Yoder JA (1983) J. Chem. Ecol. 9:817
140. Coll JC, Bowden BF, Tapiolas DM, Willis RH, Djura P, Streamer M, Trott L (1985) Tetrahedron 41:1085
141. Cimino G, De Rosa S, De Stefano S, Sodano G (1980) Tetrahedron Lett. 21:3303
142. Andersen RJ, Ayer SW (1983) Experientia 39:255
143. Nybakken J, Ajeska RA (1976) Veliger 19:19
144. Okuda RK, Hagadone MR, Unpublished results
145. Baker BJ (1986) Ph. D. thesis, University of Hawaii
146. Baker BJ, Okuda RK, Yu PTK, Scheuer PJ (1985) J. Am. Chem. Soc. 107:2976

147. Fukushima M, Kato T (1984) Kyoto Conference on Prostaglandins Abstracts p 56 Kyoto Japan
148. Sodano G, Spinella A (1986) Tetrahedron Lett. 27:2505
149. Naylor S (1984) Chem. Britain 118
150. Mynderse JS, Moore RE, Kashiwagi M, Norton TR (1977) Science 196:538
151. Sleeper HL, Paul VJ, Fenical W (1977) J. Am. Chem. Soc. 99:2367
152. Sleeper HL, Paul VJ, Fenical W (1980) J. Chem. Ecol. 6:57
153. Coval SJ, Scheuer PJ (1985) J. Am. Chem. Soc. 50:3024
154. Ireland CM, Faulkner DJ 81978) Bioorg. Chem. 7:125
155. Rudman WB, personal communication
156. Gosliner TM (1981) Biol. J. Linn. Soc. 16:197
157. Gosliner TM, Ghizelin MT (1984) Syst. Zool. 33:255
158. Crozier WJ (1977) Nautilus 30:103
159. Roesener JA, Scheuer PJ (1986) J. Am. Chem. Soc. 108:846
160. Matsunaga S, Fusetani N, Hashimoto K, Koseki K, Noma M (1986) J. Am. Chem. Soc. 108:847
161. Matsunaga S, Fusetani N, Hashimoto K, Koseki K, Noma M (1985) Tennen Yuku Kago-butsu Toronkai Koen Yoshisha 27th, 375
162. Faulkner DJ, Kernan MR, Molinski TF (1986) Japan-U.S. Seminar on Bio-organic Marine Chemistry Abstracts p 20 Okinawa Japan
163. Ayling AM (1968) Tane 14:25
164. Nybakken J, Eastman J (1977) Veliger 19:279
165. Cook EF (1962) Veliger 4:194
166. Braams WG, Geelen HFM (1953) Arch. Neer. Zool. 10:241
167. Castiello D, Cimino G, De Rosa S, De Stefano S, Izzo G, Sodano G (1979) Studies on the chemistry of the relationship between the opisthobranch *Peltodoris atromaculata* and the sponge *Petrosia ficiformis*. In: Levi C, Boury-Estnault N (eds) Biologie des Spongiares, Paris, CNRS, p 413
168. Waters VL (1973) Veliger 15:174
169. Edmunds M, Potts GW, Swinfen RC, Waters VL (1974) J. Mar. Biol. Ass. U.K. 54:939
170. Edmunds M, Potts GW, Swinfen RC, Waters VL (1976) J. Mar. Biol. Ass. U.K. 56:65
171. Howe NR, Sheikh YM (1975) Science 189:386
172. Harris LG, Howe NR (1979) Biol. Bull. 157:138
173. Howe NR, Harris LG (1978) J. Chem. Ecol. 4:551
174. von Frisch K (1941) Naturwissenschaften 29:321
175. Fenical W, Sleeper HL, Paul VJ, Stallard MO, Sun HH (1979) Pure Appl. Chem. 51:1865
176. Sakata K, Ina K (1983) Agric. Biol. Chem. 47:2957
177. Sakata K, Ina K (1985) Bull. Jpn. Soc. Sci. Fish. 51:659
178. Sakata K, Tsuge M, Kamiya Y, Ina K (1985) Agric. Biol. Chem. 49:1905
179. Comfort A (1957) Proc. Malac. Soc. Lond. 32:219
180. Burke RD (1983) Can. J. Zool. 61:1701
181. Chia F-S, Koss R (1978) Mar. Biol. 46:109
182. Hadfield MG, Karlson RH (1969) Am. Zool. 9:317
183. Hadfield MG (1978) Metamorphosis in marine molluscan larvae: an analysis of stimulus an response. In: Chia F-S, Rice ME (eds) Settlement and metamorphosis of marine invertebrate larvae, Elsevier, New York, p 165
184. Hadfield MG, Scheuer D (1985) Bull. Mar. Sci. 37:556
185. Morse DE, Hooker N, Duncan H, Jensen L (1979) Science 204:407
186. Yvin JC, Chevolot L, Chevolot-Magueur AM, Cochard JC (1985) J. Nat. Prod. 48:814
187. Ogura M, Cordell GA, Farnsworth NR (1977) Lloydia 40:157
188. Scheuer PJ (1977) BioScience 27:664
189. Grant PT, Mackie AM (1974) (eds) Chemoreception in marine organisms Academic Press, London New York
190. Whittaker RH, Feeny PP (1971) Science 171:757

Marine Metabolites Which Inhibit Development of Echinoderm Embryos

Nobuhiro Fusetani [1]

Contents

Abstract

The present article describes the procedure and selectivity of the assay system using the fertilized starfish or sea urchin eggs, as well as marine metabolites discovered by using this particular assay system and their mode of actions. The structure elucidation by spectral analyses of some interesting metabolites are also mentioned.

The author emphasizes the usefulness of the assay system in search for bioactive substances from marine organisms, which is shown through the author's works. The author is confident that

1 The University of Tokyo, Faculty of Agriculture, Laboratory of Marine Biochemistry, Bunkyo-ku, Tokyo, Japan.

Bioorganic Marine Chemistry, Volume 1

those who are involved in natural products chemistry research may find this article helpful by a great deal. It is also hopeful that this article will be a driving force for developments in marine natural products chemistry.

Introduction

Discovery of unexpectedly high quantities of prostaglandins in the Caribbean gorgonian *Plexaura homomalla* by Weinheimer and Spraggins in 1969 [1] has reinforced the research on marine natural products, which resulted in isolation of enormous numbers of new metabolites [2, 3]. However, those which are promising for development of new drugs are quite few [4, 5]; only didemnins [6], punaglandins [7] and halichondrins [8] can be listed as candidates. This discouraging result is partly due to assay methods employed. Antimicrobial and cytotoxic assays have frequently been used for screening of new bioactive metabolites. Unfortunately, these methods are non-specific, and numbers of compounds showing broad spectrum of activity, most often accompanied by toxicity toward mammals, are apt to be picked up.

It is therefore desired to adopt an assay method which is specific or selective to a particular activity as well as simple. The fertilized sea urchin egg assay can be reccomended as one such methods. Probably, Ruggieri and Nigrelli [9] were the first researchers who used this assay to evaluate the activity of marine natural products. In recent years, Jacobs's group [10, 11] have revived the fertilized sea urchin egg assay to discover potential antitumor drugs from marine organisms. This assay method can detect such selective agents as DNA synthesis inhibitors, RNA synthesis inhibitors, microtubule assembly inhibitors, and protein synthesis inhibitors, which may lead to development of anticancer drugs.

More useful would be the fertilized starfish egg assay, which was devised by Ikegami et al. [12] in 1979. This is comparable to the fertilized sea urchin egg assay in specificity to active compounds. However, it appears to be more selective.

In this paper I describe the fertilized starfish egg assay method, screening of marine invertebrate extracts by this method, and active metabolites found by the starfish egg assay as well as by the sea urchin egg assay.

1. Starfish Egg Assay

1.1 Assay Method

In 1978 Ikegami et al. [13] re-isolated aphidicolin as an inhibitor of DNA synthesis, and showed that this diterpene prevents mitotic cell division of fertilized starfish eggs. Subsequently, the same group [12] established the rapid assay method using both fertilized starfish and sea urchin eggs, by which several biological activities can be distinguished.

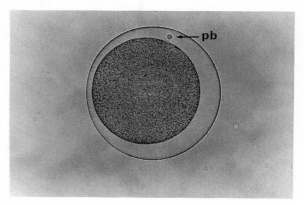

Fig. 1. Mature oocyte of starfish *A. pectinifera*. pb: Polar body

The eggs of the starfish *Asterina pectinifera* are quite desirable for looking into cell divisions, since they are larger than sea urchin eggs, and can be obtained throughout the year in Japan. Their cell membrane possesses high permeability to a variety of substances. Moreover, they have a characteristic process of oocyte maturation, in which 1-methyladenine acts as hormone [14]; within 40 min after addition of 1-methyladenine ($10^{-6} \sim {}^{-8}$ M) to oocytes obtained from a well-matured ovary (each oocyte possesses a germinal vesicle at this stage), oocytes maturation is completed upon breakdown of the germinal vesicle. In sixty minutes the first polar body is released, followed by release of the second polar body in a hundred minutes (Fig. 1). During this process meiosis takes place, so that agents which inhibit protein synthesis affect the oocyte maturation.

Insemination can be done at any time of the maturation process, and the first cell division occurs in precise time. Blastomeres of the embryo divide synchronously until the eighth cleavage. When the fertilized eggs are exposed to low concentrations of aphidicolin (~ 10 μg/ml), a selective inhibitor of DNA polymerase α, they continue to cleave eight or nine times and die; delay of development and various sizes of blastomeres were observed [15] (Fig. 2). However, this diterpene does not affect the oocyte maturation process [15].

Inhibitors of microtubule assembly, such as vinblastine, arrest both the release of the first polar body and the first mitotic division. Two chromosomes (one from egg and the other from sperm) can be seen in each fertilized egg exposed to this alkaloid when stained with a lactoorcein solution [15] (Fig. 3). Cycloheximide, a protein synthesis inhibitor, not only affects the first mitotic division, but also the breakdown of germinal vesicle (meitotic division).

When inhibitors of RNA synthesis like actinomycin D are added to the fertilized eggs, they develop to 64–128 cell stage (in the case of the sea urchin eggs, the development continues to the morula stage). These agents show no effects on the maturation process. Polynucleated unicellar embryos (Fig. 4) are formed upon addition of cytochalasin B, an inhibitor of microfilaments, to the fertilized eggs [16]. Of course, those which cause cytolysis inhibit meitotic and mitotic divisions.

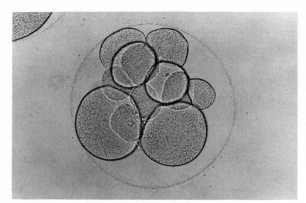

Fig. 2. Effect of aphidicolin on cell division of the fertilized starfish egg

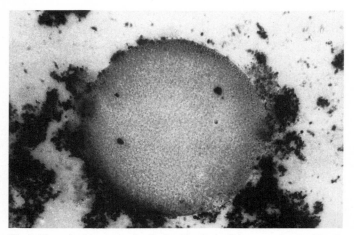

Fig. 3. Two chromosomes in the fertilized starfish egg stained with a lactoorcein solution

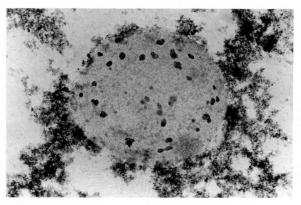

Fig. 4. Polynucleated cell induced by cytochalasin B. Black spots are nuclei stained with a lactoor-cein solution

Table 1. Results of screening by the starfish egg assay

Phylum	Specimens number	Active specimens, number (percent)	Active fraction, number	
			Lipophilic	Aqueous
Porifera	286	92 (32)	64	51
Coelenterata	133	55 (41)	48	23
Plathelminthes	2	0	0	0
Nemertina	4	0	0	0
Mollusca	45	12 (27)	10	2
Annelida	8	1 (13)	1	1
Arthropoda	3	0	0	0
Ectoprocta	41	9 (22)	3	7
Echinodermata	15	5 (33)	5	0
Protochordata	85	15 (18)	5	11
Total	622	189 (30)	136	95

1.2 Screening of Marine Invertebrate Extracts by the Starfish Egg Assay

To prepare the test solution the fresh or frozen specimen is extracted with hot methanol, and the extract is partitioned between water and chloroform. The water-soluble materials are dissolved in seawater at concentrations of 20, 10, 5 and 1 mg equivalent to fresh specimen in 1 ml. The organic-soluble materials are taken up in ethanol and added to seawater. The final concentration of ethanol is less than 1%.

To each well of a multiwell plate which contains sample solution are pipetted about 100 fertilized eggs, and the plate is kept at 20 °C. At appropriate periods, eggs are examined under a microscope.

Table 1 shows a result of our screening. Apparently, sponges are the best targets for isolation of new bioactive metabolites. Further examination indicated that many specimens contain substances affecting protein synthesis or the microfilament system. Coelenterates looked promising, but were found to possess cytolytic compounds in high percentages, which was later confirmed by our isolation work. Mollusks, mainly nudibranchs, gave good results, though enough specimens are rarely available.

These results appear to be similar to those reported for antimicrobial and cytotoxic activity [17–20].

2. Marine Metabolites Active in the Echinoderm Egg Assay

2.1 Polyketides

2.1.1 Polyacetylene Alcohols

In the early stages of our screening, we first encountered a strongly active specimen, a large brown sponge collected in Hachijo-jima island 300 km south of Tokyo. From this sponge of the genus *Petrosia,* an active substance was isolated as a colorless oil in a yield of 0.023 percent on the basis of wet sample. The compound is quite labile at room temperature. Detailed ^{1}H and ^{13}C NMR spectral analyses along with other spectral data led to the unusual polyacetylene alcohol (**1**) [21]. It is a symmetrical compound, which gave ^{13}C NMR signals corresponding to a half the number carbons. The tetraacetate afforded a molecular ion, supporting the structure. To elucidate its stereochemistry we tried a number of reactions, but instability of this metabolite hampered our attempts; configuration of four hydroxyl groups and a double bond in the middle of the molecule remain unknown.

$$HC\equiv CCHCH=CH[CH_2]_6C\equiv CCHCH=CHCHC\equiv C[CH_2]_6CH=CHCHC\equiv CH$$
$$\quad\ \ \overset{|}{OH}\qquad\qquad\qquad\qquad \overset{|}{OH}\quad\ \overset{|}{OH}\qquad\qquad\qquad\qquad \overset{|}{OH}$$

1

$$HC\equiv CCHCH=CHCH_2-R^1-CH_2CH=CHC\equiv CCHC\equiv CCH_2-R^2-CH_2CH=CHCHC\equiv CH$$
$$\quad\ \ \overset{|}{OH}\qquad\qquad\qquad\qquad\qquad \overset{|}{OH}\qquad\qquad\qquad\qquad\qquad \overset{|}{OH}$$

2: $R^1+R^2=C_nH_{2n-6}$; n=25 or 28
3: $R^1+R^2=C_nH_{2n-4}$; n=28, 31, or 34

$$HC\equiv C-CH-CH=CH_2-R^1-CH_2-CH=CH-C\equiv C-CH-C\equiv C-CH_2-R^2-CH_2-CH=CH-C\equiv CH$$
$$\qquad \overset{|}{OH}\qquad\qquad\qquad\qquad\qquad\qquad\quad \overset{|}{OH}$$

4: $R^1+R^2=C_nH_{2n-4}$; n=26, 29

$$HC\equiv C-CH-CH=CH-CH_2-R^1-CH_2-CH=CH-C\equiv C-CH-C\equiv C-CH_2-R^2-CH_2-C\equiv CH$$
$$\qquad \overset{|}{OH}\qquad\qquad\qquad\qquad\qquad\qquad\qquad\quad \overset{|}{OH}$$

5: $R^1+R^2=C_nH_{2n-4}$; n=28, 31, 34

$$HC\equiv CCHCH=CHCHCH_2-R^1-CH_2CH=CHC\equiv CCHC\equiv CCH_2-R^2-CH_2CH=CHCHC\equiv CH$$
$$\quad\ \ \overset{|}{OH}\quad\ \overset{|}{OH}\qquad\qquad\qquad\qquad \overset{|}{OH}\qquad\qquad\qquad\qquad \overset{|}{OH}$$

6: $R^1+R^2=C_{24}H_{44}$

The polyacetylene alcohol inhibits fertilized starfish eggs at a concentration of 1 μg/ml; it causes lysis of cells. Upon acetylation it is no longer active, suggesting that enol alcohols may participate in the activity.

Up to the present, more than 20 polyacetylenes have been isolated from sponges and nudibranchs [2, 3]. Those which are related to our compound are quite few; Cimino and coworkers [22–24] reported several polyacetylene alcohols (**2–6**) from the sponge *Petrosia ficiformis* and its predator, the nudibranch *Peltodoris atromaculata*. These metabolites most likely inhibit cell division of the fertilized echinoderm eggs. However, their structures need to be fully characterized.

2.1.2 Polypropionates

The pulmonates are air-breathing mollusks that populate the intertidal zone. In recent years these organisms have attracted attention of both chemists and biologists, because they produce a majority of polypropionate metabolites [3], many of which display defensive properties such as ichthyotoxic, antimicrobial and cytotoxic activity. Diemenensin-A (**7**) and -B (**8**) were isolated from the Australian pulmonate *Siphonaria diemenensis* as antimicrobial metabolites [25]. They inhibit cell division at a concentration of 1 μg/ml in the sea urchin egg assay. Among many other related metabolites known from pulmonates [3], the diemenensins are the only examples that have been screened for this activity. No information of their mode of action is available.

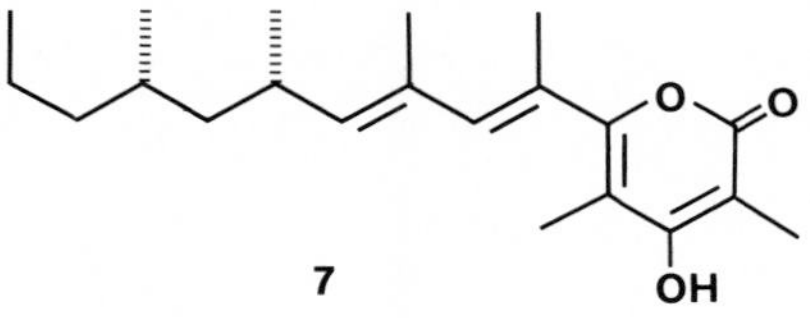

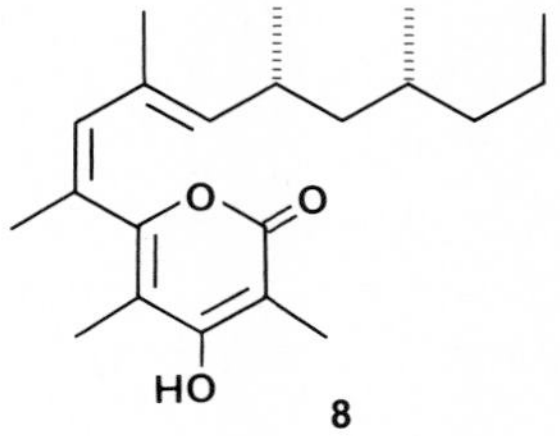

2.1.3 Brevetoxins

Brevetoxin B (**9**) [26], an extraordinary polyether metabolite produced by the dinoflagellate *Gymnodinium breve* that is a well known red tide organism in the Gulf of Mexico, was reported to arrest cell division of the fertilized sea urchin eggs at an ED_{50} of 8.9 μg/ml without causing cellular lysis [27]. Incidentally, several related toxins [3, 28] are known from the same organism and exhibit a wide range of biological activity from ichthyotoxicity to cytotoxicity.

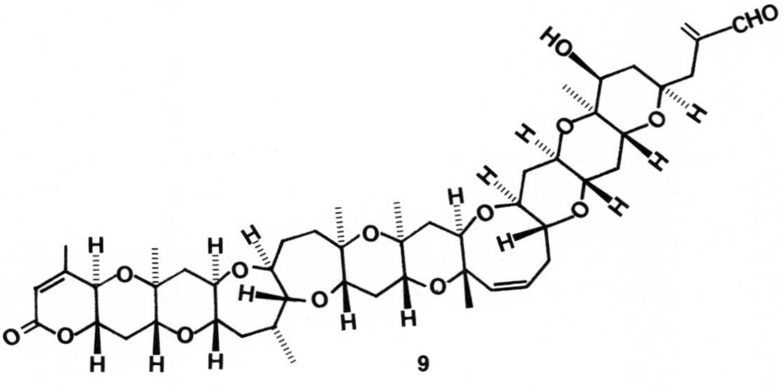

2.2 Terpenoids and Steroids

2.2.1 Monoterpenes

A number of linear and cyclic polyhalogenated monoterpenes are known from marine red algae, especially of the genus *Plocamium* [2, 3]. Only two such compounds, a linear monoterpene (**10**) [29] from *Plocamium cartilagineum* and dichloroöchtodene (11) from *Chondrococcus hornemanni,* were reported to inhibit cell division of sea urchin eggs (78 and 16 percent inhibition at a concentration of 16 µg/ml, respectively) [30].

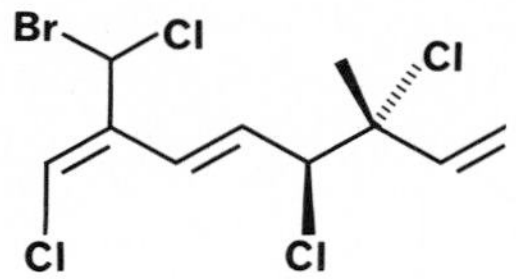

2.2.2 Sesquiterpenes

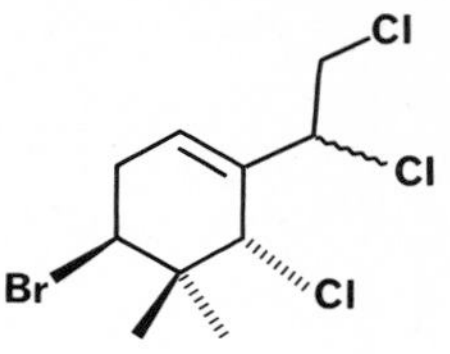

Halogenated chamigrene derivatives are common constituents of the red alga *Laurenica* spp. [2, 3]. Elatol (12), isolated from *L. elata* by Sims and coworkers [31] in 1974, was later found to be active in the sea urchin egg assay (ED$_{100}$ 16 µg/ml) by Jacobs et al. [10]. The corresponding oxidation product, elatone (13), is more potent (ED$_{100}$ 8 µg/ml) and strongly inhibits assembly of the beef brain microtubules. It is noted that elatone also circumscribes the incorporation of thymidine into DNA during the first cleavage of sea urchin embryos.

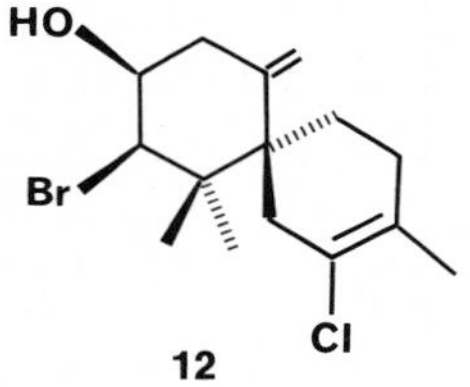

(−)-Curcuphenol (14), (−)-curcuquinone (15) and (−)-curcuhydroquinone (16) were obtained as antibacterial metabolites of the Caribbean gorgonian *Pseudopterogorgia rigida* [32]. Curcuhydroquinone is most active both in the sea urchin egg assay (ED$_{100}$ 8 µg/ml) and the beef brain microtubule assay (100 percent inhibition at 8 µg/ml) [10]. Curcuquinone is less active, while cucuphenol is inactive in both assays.

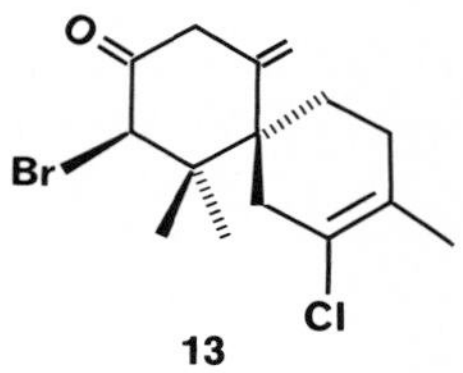

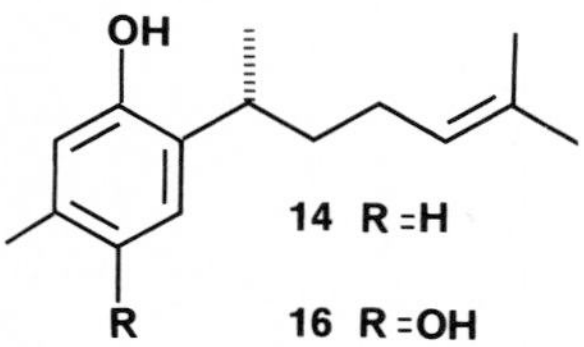

14 R =H

16 R =OH

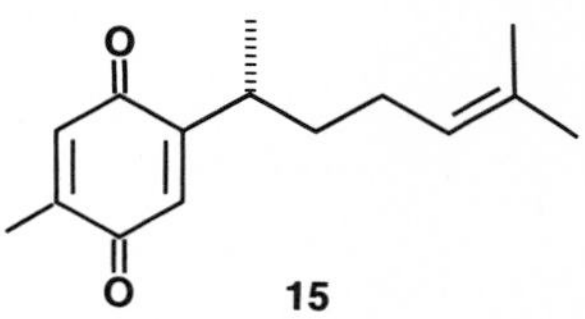

It was also described that two sesquiterpenoid amino quinones (17 and 18) derived from the marine sponge *Dysidea avara* inhibit cell cleavage of sea urchin eggs [33].

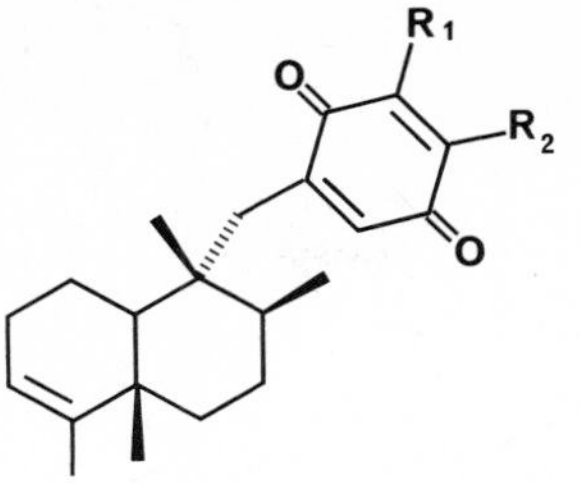

17   R₁ = H, R₂=NHCH₃

18   R₁=NHCH₃, R₂=H

2.2.3 1,4-Diacetoxybutadienes and Related Terpenes

Marine green algae of the family Caulerpaceae elaborate a series of terpenes having a 1,4-diacetoxybutadiene moiety [2, 3], some of which have been reported to arrest development of the fertilized sea urchin eggs. The acetoxy-aldehydes [3] are also active in the sea urchin egg assay. Those which have been reported to be active are listed in Table 2.

Table 2. The 1,4-diacetoxybutadienes and related terpenes active in sea urchin egg assay

Compound	Trivial name	Activity (µg/ml)	Algal species	Refs.
19	Flexilin	≦16	*Caulerpa flexilis*	[34]
			Udotea conglutinata	[35]
20		≦16	*Penicillus capitatus*	[35]
			Udotea cyathiformis	
21		2	*Caulerpa bikiniensis*	[36]
22		≦16	*Udotea flabellum*	[35]
			Penicillus dumetosus	
23		8	*Chlorodesmis fastigiata*	[37]
24		8	*Tydemania expeditionis*	[37]
25		8	*Udotea argentea*	[37]
26		≦16	*Halimeda* spp.	[38]
27		≦16	*P. capitatus*	[35]
			U. cyanthiformis	
28		1	*C. bikiniensis*	[36]
29	Petiodial	≦16	*U. flabellum*	[35]
			Udotea petiolata	[39]
30		≦16	*U. flabellum*	[35]

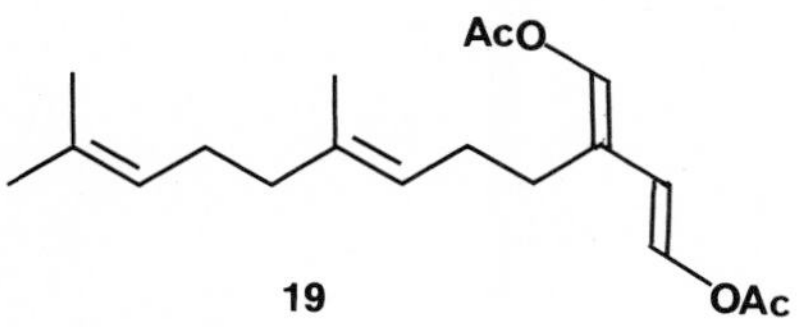

19

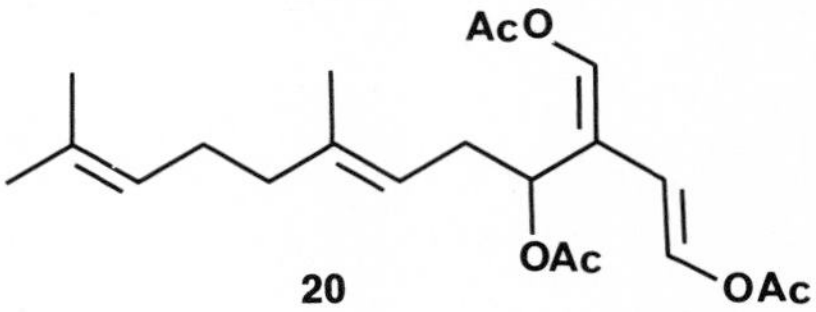

20

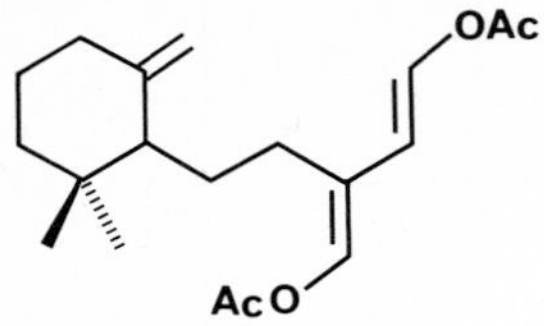

21

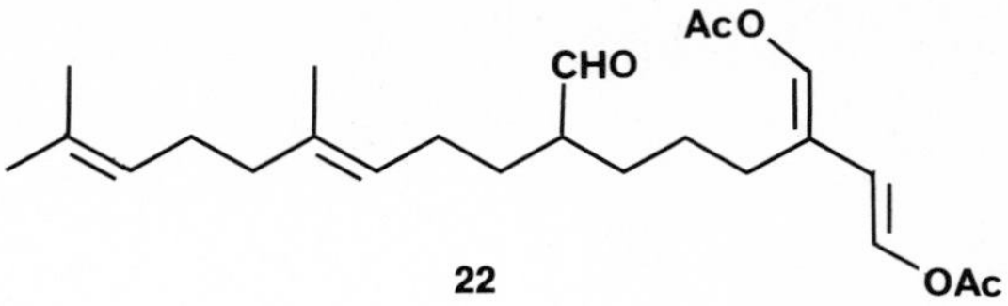

22

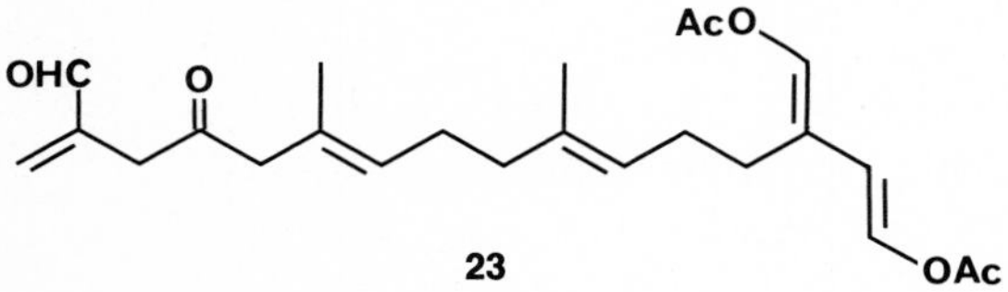

23

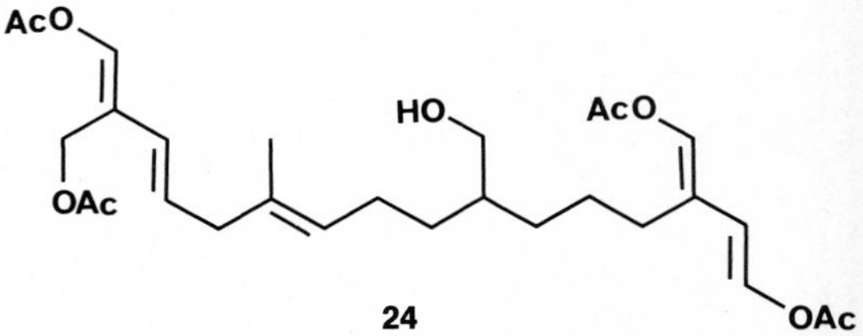

24

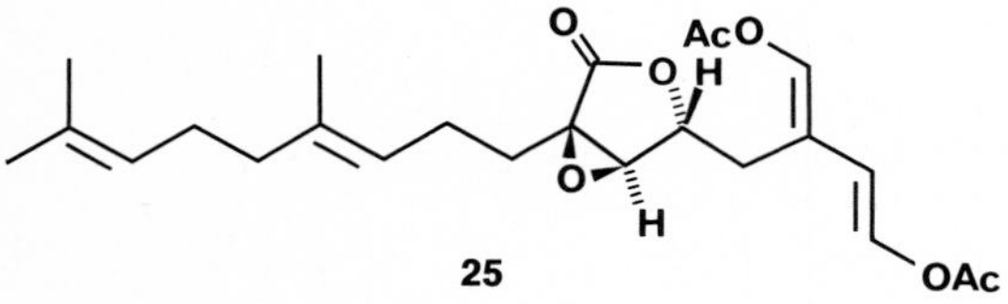

25

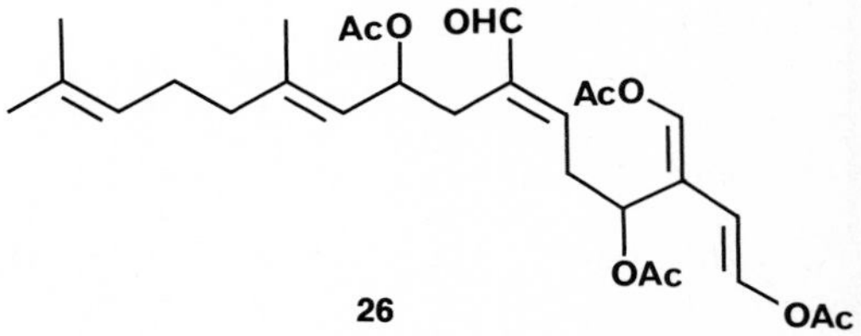

26

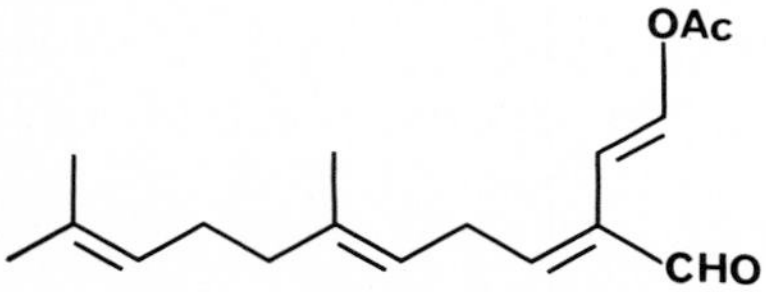

27

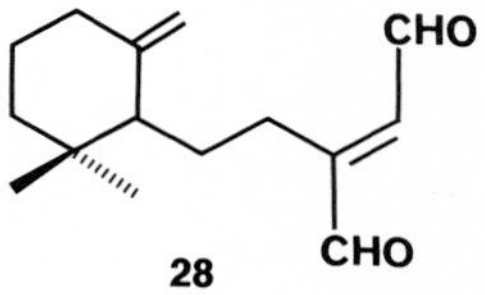

28

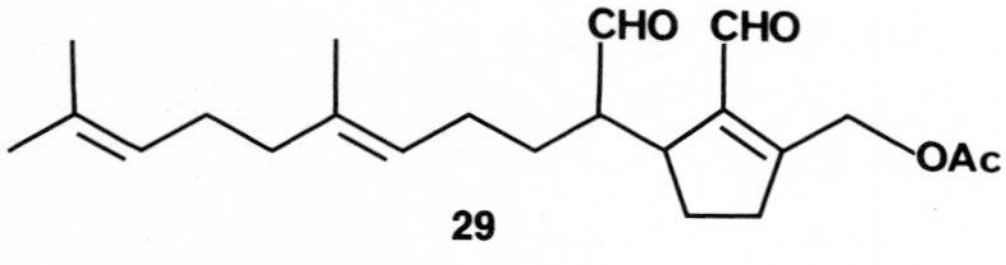

29

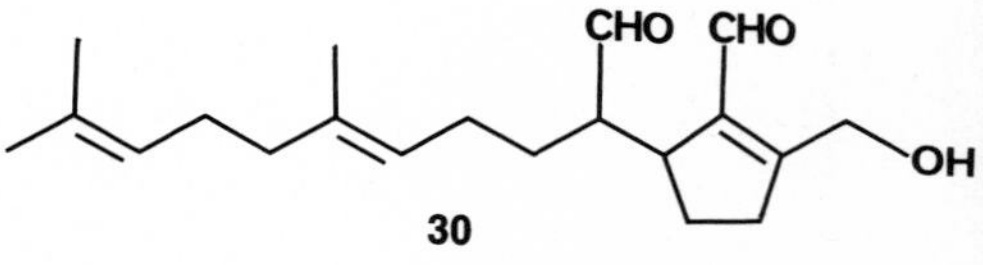

30

Halimedatrial (**31**) [38, 40], which was isolated from the calcareous reef-building algae of the genus *Halimeda*, inhibits the first cleavage of the fertilized sea urchin eggs at a concentration of 1 μg/ml. This terpene shows a variety of biological activities including antimicrobial, ichthyotoxic and antifeedant. *Halimeda* spp. also contain halimedalactone (**32**) [38] and the benzaldehyde (**33**), [38] which are active at a concentration of 16 μg/ml. Though these compounds show relatively strong activity, their mode of action remains to be examined.

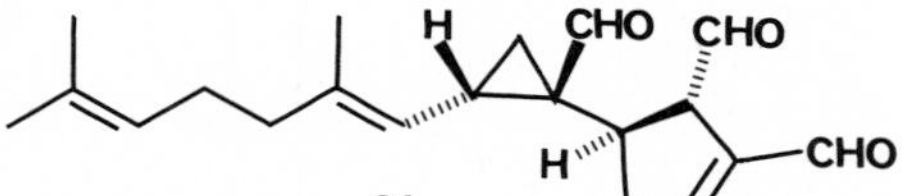

31

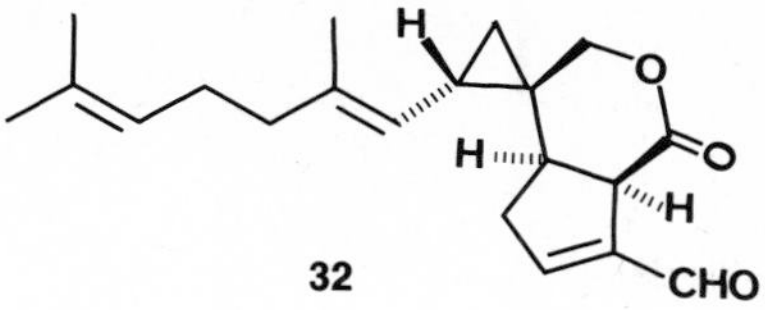

32

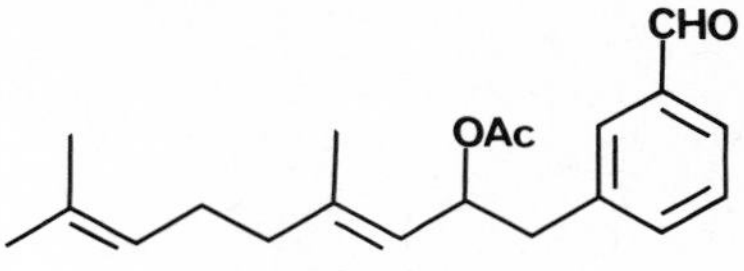

33

2.2.4 Norditerpenes

Although cyclic peroxides of polyketides are common members of sponge metabolites, peroxy-acids of nor-terpenoid skeletons are known from *Prianos* sp. [2, 3].
Muqubilin (**34**) was first isolated as an antimicrobial metabolite of a Red Sea species [41]. Later, it was re-isolated from a Tongan species of *Prianos* [42]. It inhibits
100 percent of cell division of the fertilized sea urchin eggs at a concentration of
16 µg/ml. The stereochemistry of muqubilin has been questioned by Manes *et al.*
[42].

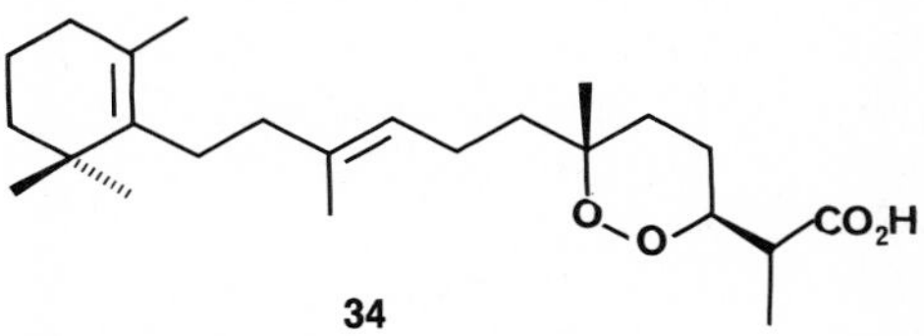

34

2.2.5 Spatane Diterpenes

The spatane group of diterpenes have been isolated from marine brown algae,
Spatoglossum schmittii [43, 44], *S. howleii* [44], *Stoechospermum marginatum* [45],
and *Dilophus marginatum* [46]. Spatol (**35**), obtained from *S. schmittii*, was the
first reported metabolite of this series [43]. The structure was determined by X-ray
diffraction. It inhibits the first cleavage of the fertilized sea urchin eggs at a concentration of 1.2 µg/ml, due to inhibiting tubulin polymerization. Spatol also ex-
hibits a wide spectrum of biological activity such as cytotoxicity toward human
melanoma. These activities are believed to be associated with the 1,3-diepoxide
moiety in the side chain.

   S. schmittii contains the diol **36** and the triol **37** in addition to spatol, while
S. howleii produces the tetraol **38** together with four other spatane diterpenes [44].
It was reported that the diol and tetraol are active at a concentration of 16 µg/ml
in the sea urchin egg assay.

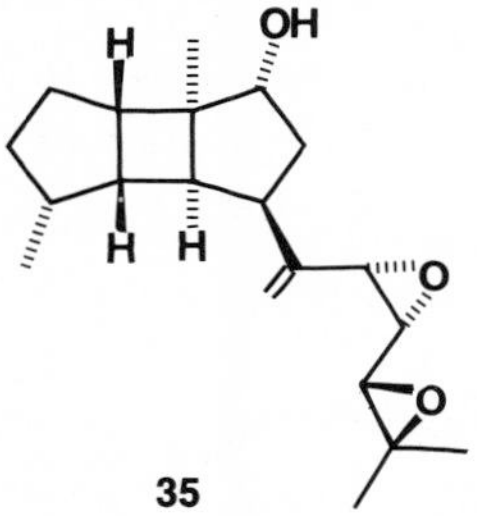

35

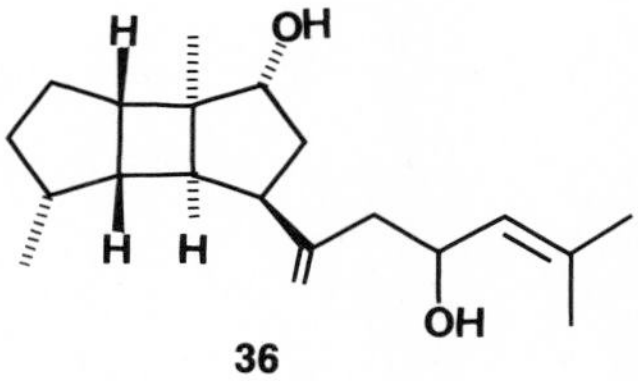

36

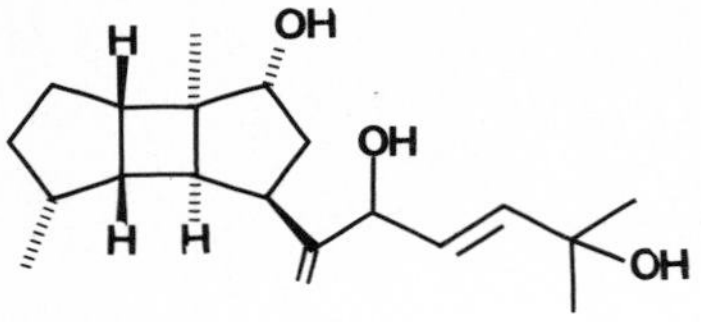

37

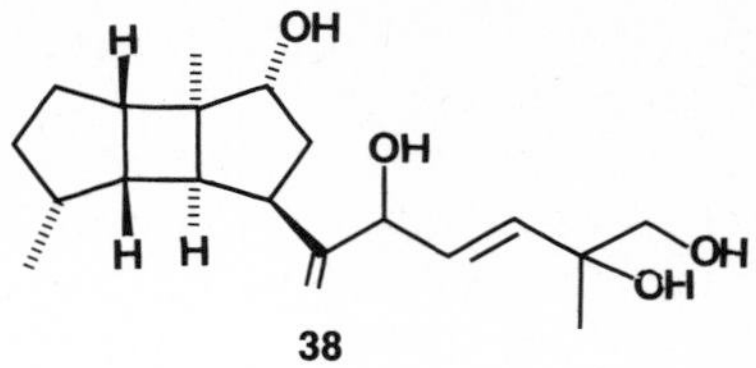

38

2.2.6 Cembranolides

Soft corals and gorgonians are rich sources of cembranoid diterpenes, which possess a diversity of biological activity [2, 3]. Pseudopterolide (**39**), which was obtained from the Caribbean sea whip *Pseudopterogorgia acerosa* [47], is an unusual example of these diterpenes. The structure was determined by X-ray analysis of the cyclic urethane derivative. This terpene inhibits overall cell cleavage but does not inhibit nuclear division in the sea urchin egg assay. It produces multinucleated cells, suggesting inhibition of microfilament formation.

The related metabolites, kallolide A (**40**) and B (**41**) [48], pukalide (**42**) [49] and lophotoxin (**43**) [50] have been reported from the Caribbean gorgonian *Pseudopterogorgia kallos,* the Hawaiian soft coral *Sinularia abrupta,* and four species of gorgonians of the genus *Lophogorgia,* respectively.

Jacobs and coworkers [10] reported that crassin acetate (**44**) [51], a well known gorgonian metabolite, inhibits cell division of the fertilized sea urchin eggs at 16 µg/ml, but not beef brain microtubule assembly. It is therefore likely that many cembranolides are active in this assay.

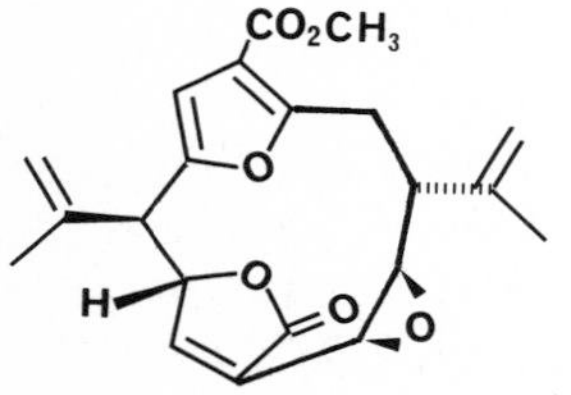

39

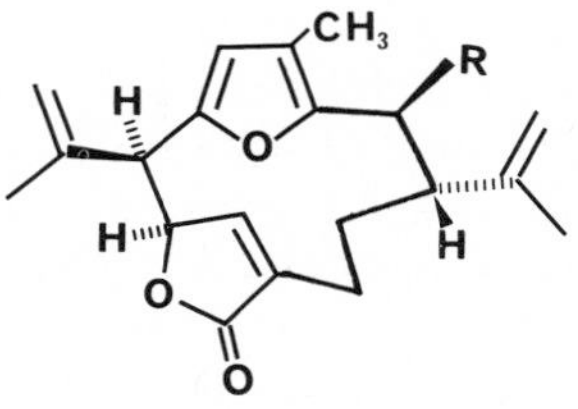

40 R = OH

41 R = H

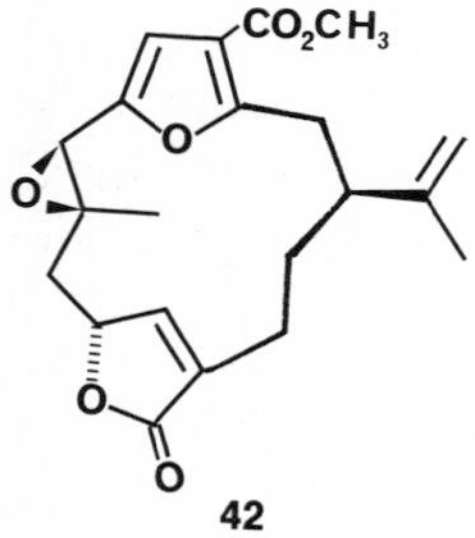

42

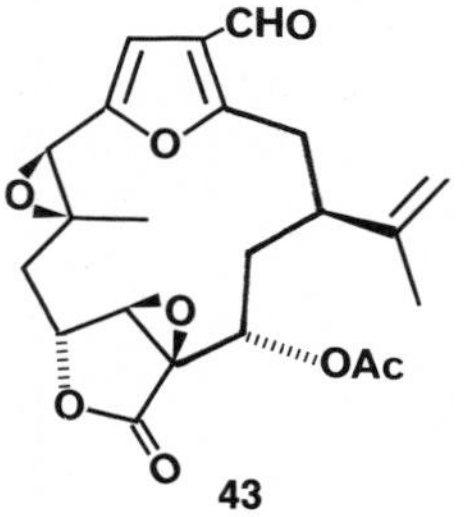

43

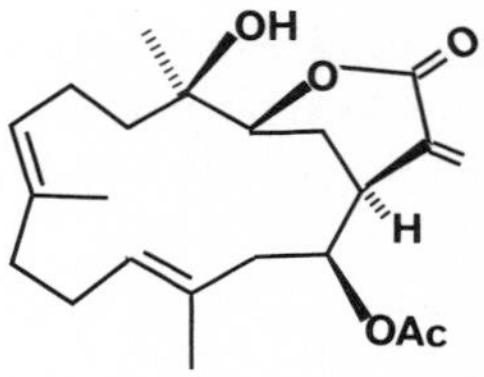

44

2.2.7 Prenylated Quinones and Hydroquinones

A large number of prenylated quinones, hydroquinones and related compounds have been known from brown algae [2, 3]. The Caribbean brown alga *Stypopo-*

dium zonale produces seven such compounds [52], among which a red-colored compound, named stypoldione (**45**) [52, 53] was shown to be a potent inhibitor of synchronous cell division in the fertilized sea urchin egg assay (ED_{50} 1.1 μg/ ml). This activity is attributable to inhibition of microtubule polymerization by binding to a low affinity site on the tubulin dimer [11, 54]. It was indicated that stypoldione blocks cell cycle progression by inhibiting progression through G2 phase [55]. This metabolite is also ichthyotoxic and has antitumor properties. It is known that stypoldione was produced by air oxidation of stypotriol (**46**) [52].

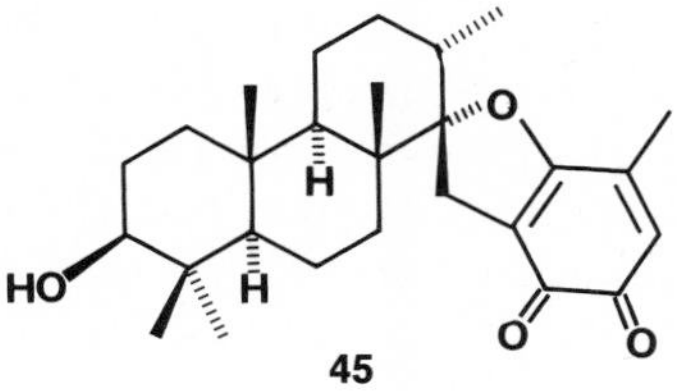

45 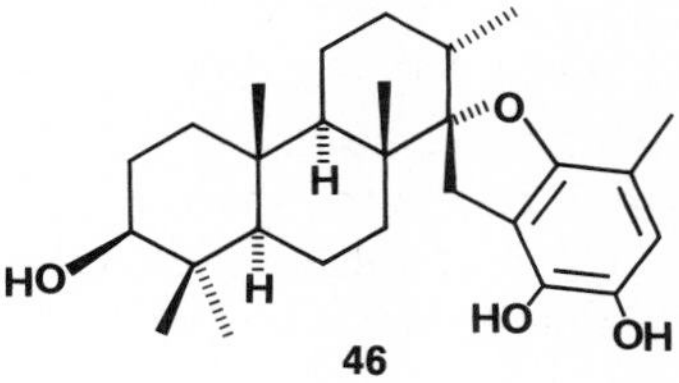   **46**

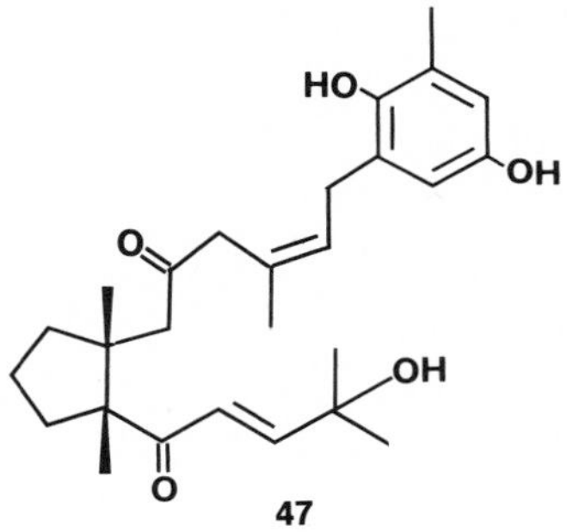

47

Bifurcarenone (**47**) was isolated from the brown alga *Bifurcaria galapagensis,* which is avoided even by marine iguanas [56]. The structure of this unusual compound was elucidated by spectral analysis and chemical degradation. Bifurcarenone not only inhibits cell division of the fertilized sea urchin eggs at an ED_{50} 4.0 μg/ml, but also the growth of microorganisms. Six closely related metabolites have been described from *Cystoseira algeriensis* [3].

The related metabolites, mediterraneols A–D (**48–51**), have recently been obtained from the brown alga *Cystoseira mediterranea* [57, 58]. The structures of these novel diterpenes were determined mainly by spectroscopic analyses including 2D long range heteronuclear correlation NMR. The relative stereochemistry was established by NOE difference spectroscopy. Mediterraneols inhibit both the

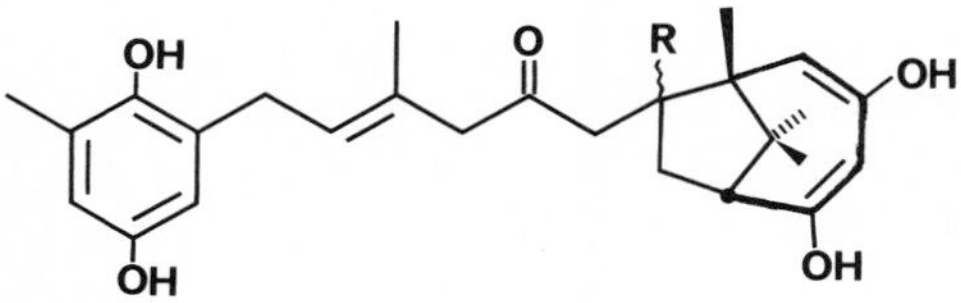

48 R = β–CH₃

49 R = α–CH₃

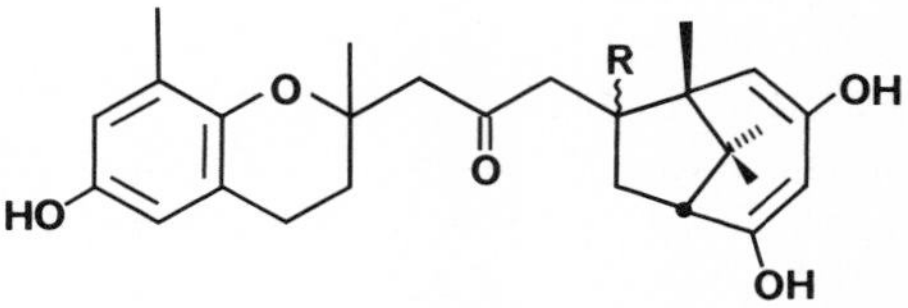

50 R = α–CH₃

51 R = β–CH₃

mobility of sea urchin sperm and miotic cell division of the fertilized sea urchin eggs (ED_{50} 2 µg/ml). Mediterraneols A and B show antitumor activity against P388 leukemia (T/C 128 percent at 32 mg/kg).

2.2.8 Furanosesterterpenes

Furanosesterterpenes are a prominent class of metabolites in marine sponges of the family Thorectidae [2, 3]. Minale and coworkers [59] first reported in 1972 isolation of ircinin-1 (52) and -2 (53) from the Mediterranean sponge *Ircinia oros,* which showed antimicrobial property. Thereafter, more than fifteen furanosesterterpenes have been isolated [2, 3]. Only antimicrobial activity is reported for these metabolites except for suvanin, (54) which was obtained from a Fijian sponge *Ircinia* sp. as an inhibitor of the fertilized sea urchin eggs with an ED_{50} of 16 µg/ml [60]. This unique structure appears to be incorrect [3 b].

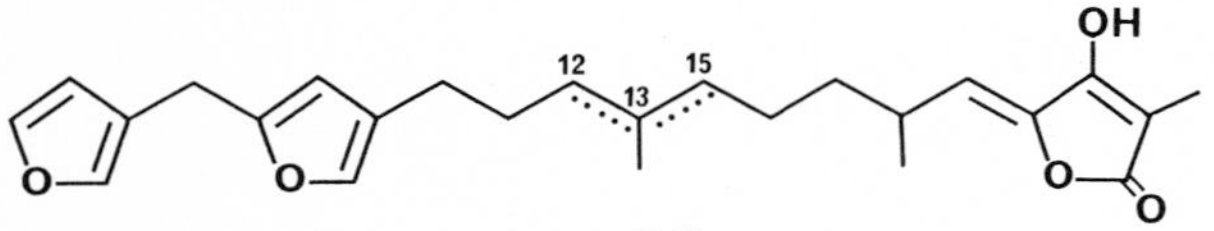

$$52: \Delta^{12,13}$$
$$53: \Delta^{13,15}$$

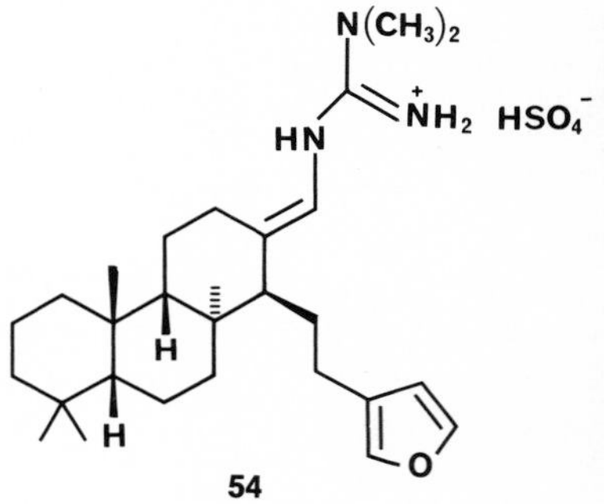

54

We have isolated six new furanosesterterpenes from Japanese sponges by using the starfish egg assay as a guide to isolation. *Cacospongia scalaris,* the second target organism in our research, which was collected in the Gulf of Sagami, gave two furanosesterterpenes 55 and 56, along with ircinins [61]. It was apparent from spectral data that our compounds were closely related to ircinins; the presence of conjugated tetronic acid and butadiene functionalities was evident by UV absorptions at 234 and 265 sh nm, the latter of which was shifted to 312 nm upon addition of alkali. The gross structure including geometry of double bonds was assigned by ^{1}H NMR experiments, including double resonance and difference NOE determinations. These terpenes inhibit cell division of the fertilized starfish eggs at a concentration of 1.0 µg/ml.

Additionally, two linear furanosesterterpenes, spongionellin (57) and dehydrospongionellin (58), were isolated from a sponge of the genus *Spongionella* collected in the Gulf of Suruga [62]. These compounds possess a nonconjugated tetronic acid and difurano methylene constellation in the molecule. The geometry of the 17, 18 double bond was assigned on the basis of C-19 methyl signals (δ_H 1.75, δ_c 24.1) as well as the C-20 methylene signal (δ_c 34.5). The presence of a triene system in dehydrospongionellin was implied by UV absorptions at 290 (ε9820), 279 (11 000), 270 (9100) and 221 nm (5200). These terpenes inhibit cell division of the fertilized starfish eggs at 2.0 µg/ml.

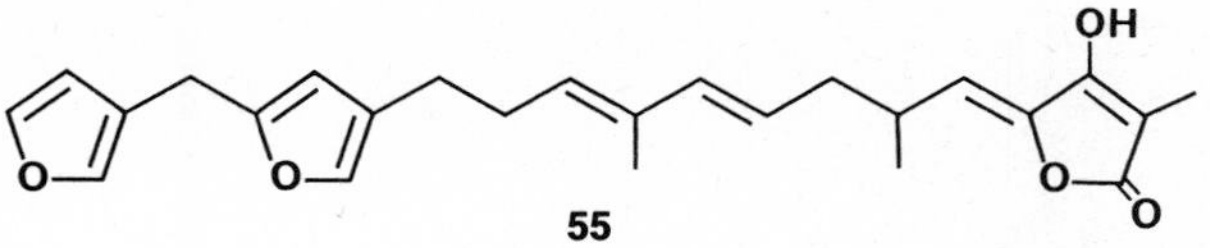

Finally, another species of *Spongionella* (Professor P. R. Bergquist, University of Aukland recently informed us that this sponge is very similar to *Fasciospongia* in the family Thorectidae), which was collected at Okinoshima island 700 km south-west of Tokyo during the cruise on R/V Toyoshio Maru of Hiroshima University, afforded two novel furanasesterterpenes, okinonellin A **(59)** and B **(60)** [63]. They were quite labile and decomposed rapidly during measurements of the NMR spectra. The structure was established mainly by using 2 D NMR techniques and NOE experiments. Okinonellin A has in addition to tetronic acid and terminal methylene functionalities a fused furanocyclohexene moiety linked to a conjugated diene system, which was evidenced both by EIMS fragment ion at *m/z* 135 and NOE experiments. The position of a terminal methylene group was elu-

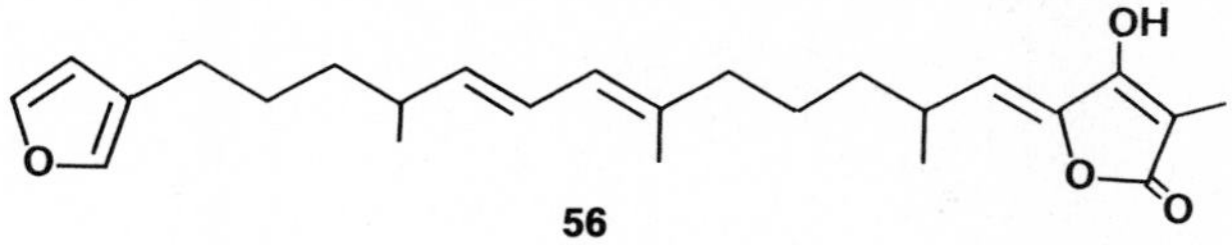

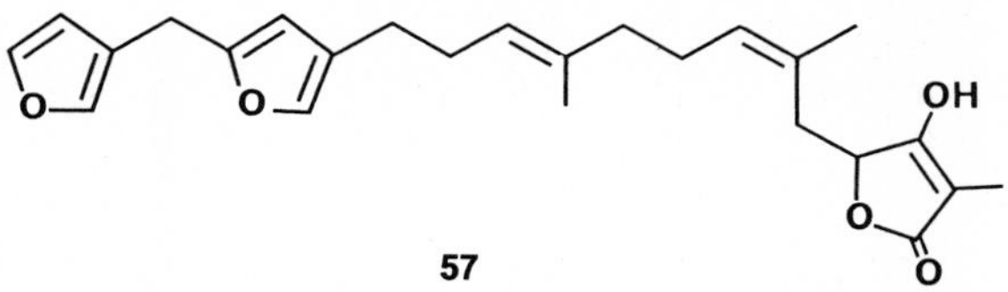

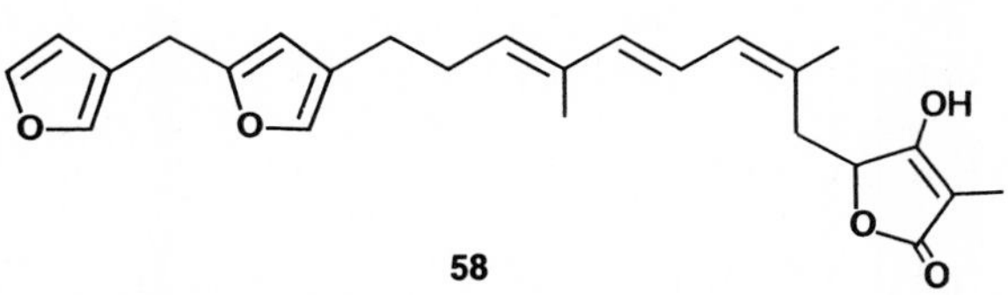

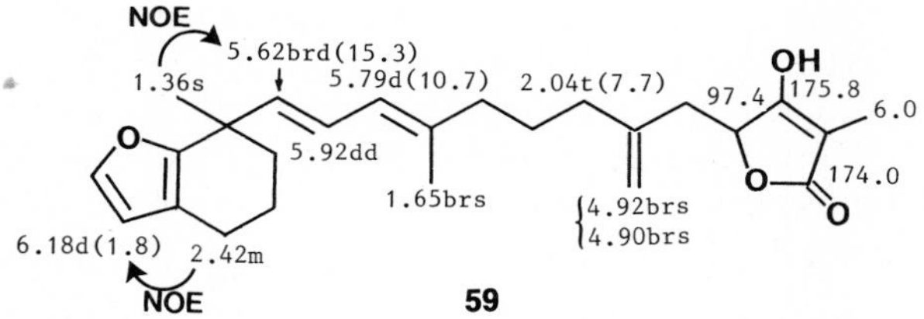

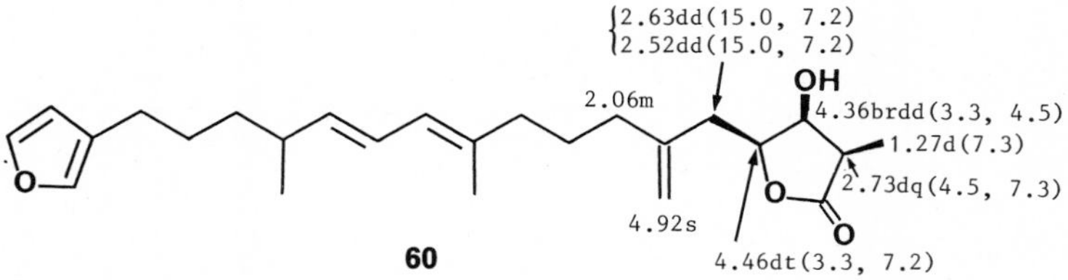

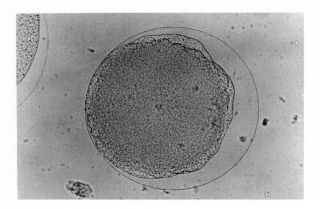

Fig. 5. Starfish embryo affected by the furanosesterterpene (**56**)

cidated by a cross peak observed between the signals at $\delta 2.04$ (H-17) and 4.90 (H-19'). Okinonellin B reveals spectral features similar to those of okinonellin A except for a terminal furan and a dihydrotetronic acid moiety. Relative stereochemistry of the dihyrotetronic acid ring was deduced by NOE experiments. Both terpenes inhibit cell division of the fertilized starfish eggs at a concentration of 5.0 µg/ml.

Unfortunately, full elucidation of the stereochemistry of the furanosesterterpenes remains to be done. Our terpenes showed similar modes of action in starfish egg assay. Incomplete cytokinesis with shallow cleavage furrows were observed (Fig. 5). This may indicate that furanosesterterpenes affect the microfilament system. Jacobs *et al.* [10] found that variabilin (**61**), an antibiotic metabolite of *Ircinia variabilis* [64], inhibits cell division of the fertilized sea urchin eggs at an ED_{100} 16 µg/ml, but does not affect assembly of beef brain microtubule.

61

2.2.9 Glycosides

Ruggieri and Nigrelli [9] first applied the sea urchin egg assay to marine natural products in 1960. They found that the development of the fertilized sea urchin eggs was inhibited by holothurins at 10 ppm, while abnormal embryos were observed at lower concentrations. Later, Anisimov *et al.* [65–68] reported similar results as well as some structure-activity relationships by applying the sea urchin egg assay to sea cucumber saponins. They also found that DNA synthesis in the sea urchin embryos was affected by these glycosides [67].

Asterosaponins [2, 3] are also active in the echinoderm embryo assay. We examined the effect of seventeen starfish saponins on the development of the fertilized starfish and sea urchin eggs [69]. Though starfish eggs are much more resistant to the saponins, the structure-activity relationships obtained in both tets are comparable (Table 3). Of seventeen saponins tested, thornasteroside B (67) was the most active and compound 78 the least, which suggests the importance of the side chain. It was also observed that the sugar sequence influences the activity to some extent. The sulfate group at C-3 appears not to be important for activity, which is in good agreement with the results obtained by Anisimov et al. [66].

All saponins induce polynucleated cells. A similar observation was made for the crude saponin mixture of *Acanthaster planci* by Ruggieri and Nigrelli [79]. Recently, Ikegami and coworkers [80] suggested that agents which bind to cell membranes block cell division without affecting nuclear division and evoke polynucleated cells.

Another example of glycosides which inhibit development of the starfish embryos is moritoside (79), which was isolated from a Japanese gorgonian, *Euplexaura* sp. [81]. It was obtained as a colorless oil in a yield of 1.15×10^{-3} percent based on wet animal, and inhibits the first cell division of fertilized starfish eggs at a concentration of 1.0 µg/ml; it most likely affects the cell membrane and causes lysis.

Spectral data of moritoside indicated the presence of a hydroxylated farnesyl group and a di-substituted hydroquinone moiety in the molecule. The remaining portion containing three acetyl groups and a ring system was deduced by extensive ^{1}H NMR double resonance experiments to be a 3,4,6-triacetoxyaltrose. The absolute configuration of the sugar was secured by comparison with authentic D-altrose after deacetylation of moritoside, followed by enzymatic hydrolysis. The connection of three segments was established by ^{1}H NMR difference NOE experiments together with high resolution EI mass spectroscopy.

Moritoside is the first example of D-altrose in a natural product. In addition, esterification of the sugar with three acetyl groups is rare in nature.

Table 3. Effect of starfish saponins on fertilized echinoderm eggs

Compound	Sea urchin egg		Starfish egg
	ED_{50} (µg/ml)	LD_{99} (µg/ml)	ED_{50} (µg/ml)
62	50	50	200
63	10	10	100
64	15	5	100
65	15	2	200
66	10	5	100
67	5	2	50
68	50	10	100
69	30	10	200
70	10	5	200
71	15	5	200
72	15	10	200
73	15	5	200
74	5	2	100
75	15	5	200
76	10	5	50
77	5	5	50
78	50	50	100

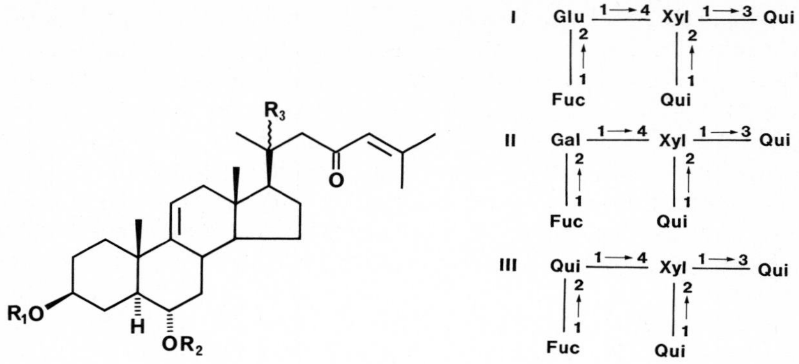

R3
O
R1O
H
OR2
I Glu 1→4 Xyl 1→3 Qui
2 1 Fuc
2 1 Qui
II Gal 1→4 Xyl 1→3 Qui
2 1 Fuc
2 1 Qui
III Qui 1→4 Xyl 1→3 Qui
2 1 Fuc
2 1 Qui

Compound	R_1	R_2	R_3	Trivial name	Refs.
62	SO_3Na	III	OH	acanthaglycoside A	[70]
63	H	III	OH		[70]
64	SO_3Na	II	OH	acanthaglycoside D	[71 a]
76	SO_3Na	VIII	H		[72]

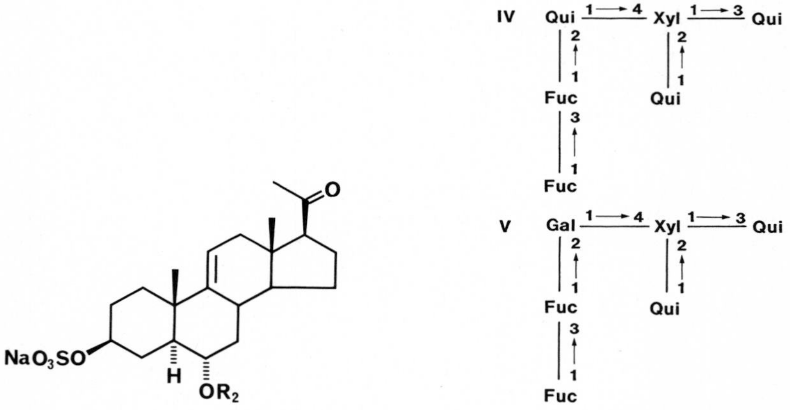

O
NaO3SO
H
OR2
IV Qui 1→4 Xyl 1→3 Qui
2 1 Fuc 3 1 Fuc
2 1 Qui
V Gal 1→4 Xyl 1→3 Qui
2 1 Fuc 3 1 Fuc
2 1 Qui

Compound	R	Refs.
78	II	[74]

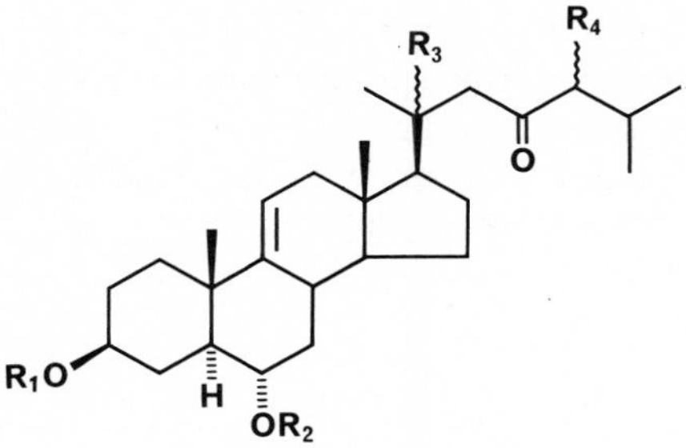

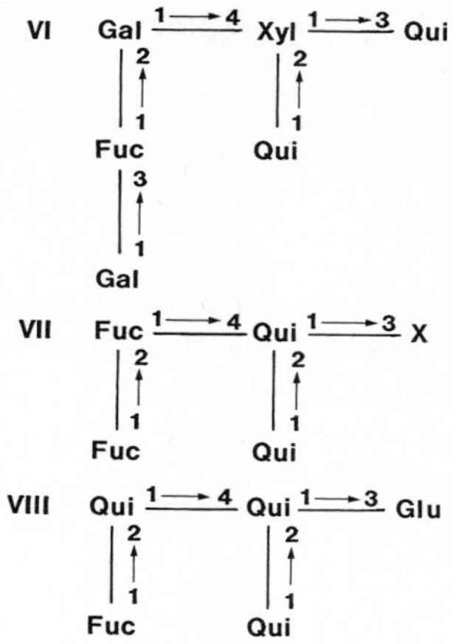

VI   Gal $\xrightarrow{1\longrightarrow 4}$ Xyl $\xrightarrow{1\longrightarrow 3}$ Qui

$\quad\quad\quad\ \overset{2}{|}\quad\quad\quad\quad\overset{2}{|}$

$\quad\quad\quad\ \underset{1}{\uparrow}\quad\quad\quad\quad\underset{1}{\uparrow}$

$\quad\quad\quad$ Fuc $\quad\quad\quad$ Qui

$\quad\quad\quad\ \overset{3}{|}$

$\quad\quad\quad\ \underset{1}{\uparrow}$

$\quad\quad\quad$ Gal

VII   Fuc $\xrightarrow{1\longrightarrow 4}$ Qui $\xrightarrow{1\longrightarrow 3}$ X

$\quad\quad\quad\ \overset{2}{|}\quad\quad\quad\quad\overset{2}{|}$

$\quad\quad\quad\ \underset{1}{\uparrow}\quad\quad\quad\quad\underset{1}{\uparrow}$

$\quad\quad\quad$ Fuc $\quad\quad\quad$ Qui

VIII   Qui $\xrightarrow{1\longrightarrow 4}$ Qui $\xrightarrow{1\longrightarrow 3}$ Glu

$\quad\quad\quad\ \overset{2}{|}\quad\quad\quad\quad\overset{2}{|}$

$\quad\quad\quad\ \underset{1}{\uparrow}\quad\quad\quad\quad\underset{1}{\uparrow}$

$\quad\quad\quad$ Fuc $\quad\quad\quad$ Qui

X = 6−deoxy−β−D−xylo−hexos−4−ulose

Qui = D−quinovose

Fuc = D−fucose

Glu = D−glucose

Gal = D−galactose

Xyl = D−xylose

Compound	R_1	R_2	R_3	R_4	Trivial name	Refs.
65	SO₃Na	I	OH	H	acanthaglycoside C	[73]
66	SO₃Na	II	OH	H	thornasteroside A	[74, 75, 76]
67	SO₃Na	II	OH	CH₃	versicoside C (thornasteroside B)	[75, 76, 71 b]
68	H	VI	OH	H		[77]
69	SO₃Na	VI	OH	H	versicoside A	[77]
70	SO₃Na	VI	OH	CH₃	versicoside B	[71 b]
71	SO₃NH₄	III	OH	H	luidiaglycoside B	[78]
72	SO₃Na	IV	OH	H	luidiaglycoside A	[78]
73	SO₃Na	V	OH	H	marthasteroside A₁	[74]
74	SO₃Na	V	OH	CH₃	acanthaglycoside F	[71 a]
75	SO₃Na	VII	OH	H	asterosaponin 1	[71 b]
77	SO₃Na	VIII	H	H		[72]

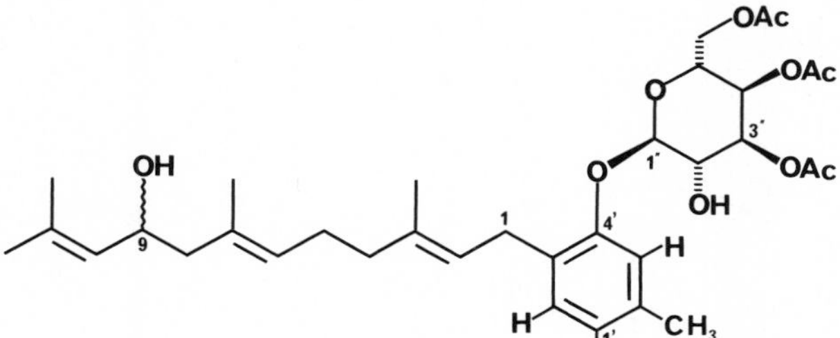

79

2.2.10 Sterol Sulfates

Though sterols themselves are inactive, their sulfates inhibit development of echinoderm embryos. Halistanol sulfate (**80**), a tris-sodium sulfate salt, which was isolated from an Okinawan sponge, *Halichondria* cf. *moorei* [82], arrests development of the fertilized starfish eggs at a concentration of 1.0 µg/ml. Three closely
related sterol sulfates (**81–83**) obtained from the marine sponge *Toxadocia zumi*
are active at a concentration of 5 µg/ml in the sea urchin egg assay [83]. Halistanol
sulfate induces polynucleated cells, and halistanol is no longer active, suggesting
that activity of these sulfates is due to their surfactant nature.

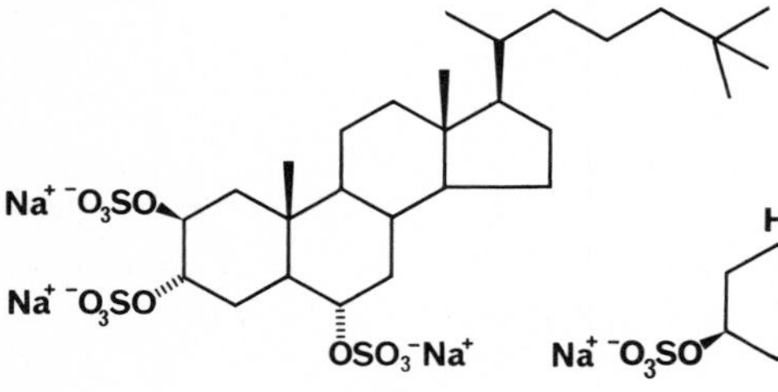

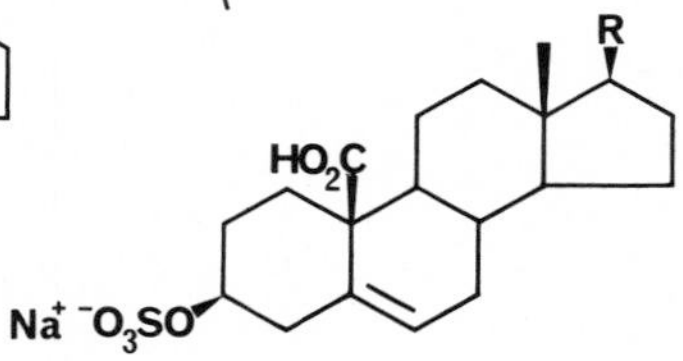

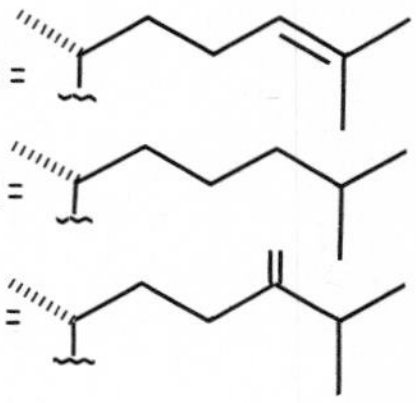

3 Nitrogenous Compounds

3.1 Pyrrole Derivatives

Compounds containing pyrrole are uncommon among marine metabolites except
for chlorophyll derivatives [2, 3]. Only brominated pyrroles are known from
sponges [2, 3].

   Four closely related bipyrroles, named tambjamine A–D (**84–87**), were described from the bryozoan *Sessibugula translucens* and from three species of nembrothid nudibranchs that prey on the bryozoan [84]. These metabolites inhibit cell
division of fertilized sea urchin eggs at a concentration of 1 µg/ml. Similarly, the
corresponding aldehydes which were produced during isolation are also active.
Recently, we have obtained a tetrapyrrole (**88**) related to tambjamines from a blue
bryozoan, *Bugula dentata* [85], which shows no activity in the echinoderm embryo
assay. This may indicate that the enamine portion of the tambjamines is responsible for the activity.

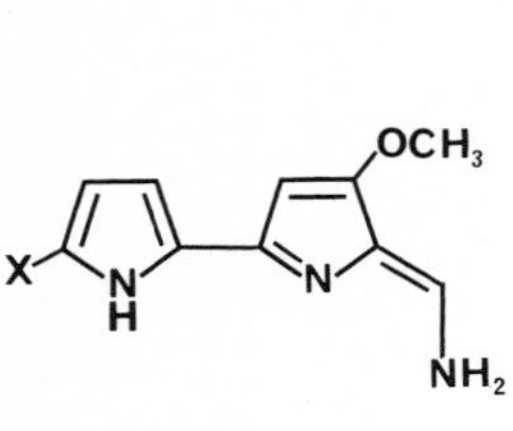

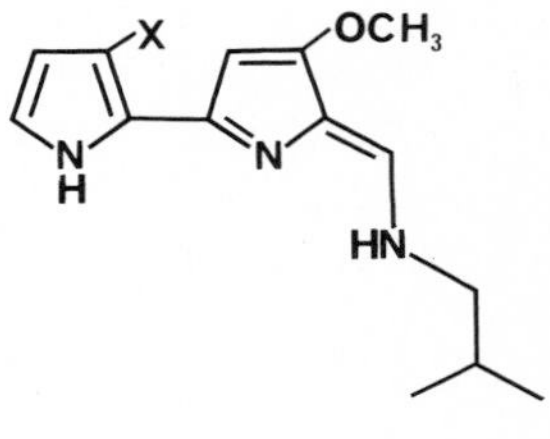

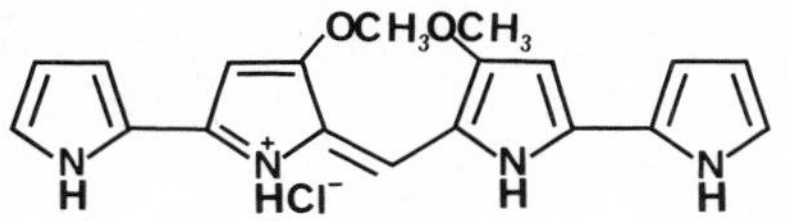

88

Bonellin has long been known as a toxin produced by the echiuroid worm *Bonellia viridis,* which induces larvae to become male worms [86]. The structure of this green pigment (**89**) was determined in 1967 by spectral analysis and X-ray diffraction of the anhydrobonellin methyl ester [87, 88]. Later, four amino acid conjugates of bonellin (**90–93**) were isolated from the body wall of *B. viridis* [89, 90]. The isoleucine conjugate, named neobonellin [**91**], and bonellin cause cellular lysis of the fertilized sea urchin eggs at a concentration of 10^{-6} M [90]. At lower concentrations both compounds arrest the development of the embryos at the two-cell stage. Neobonellin shows one half of the activity of bonellin. It is known that bonellin manifests its antimicrobial activity in a manner similar to that of hematoporphyrin [91], a photodynamic agent, which has attracted attention as an antitumor agent [92]. Therefore, bonellin appears to display its activity through photodynamic action.

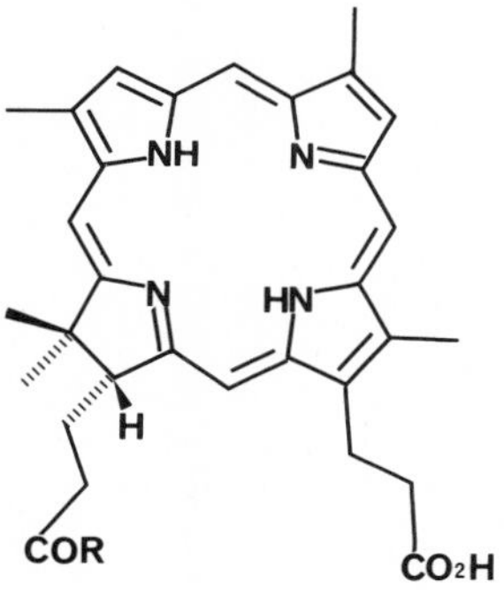

89	R = OH
90	R = N-valyl
91	R = N-isoleucyl
92	R = N-leucyl
93	R = N-alloisoleucyl

3.2 Indole Derivatives

Indole-containing metabolites are a prominent group of marine natural products; they are known from sponges, bryozoans, mollusks, tunicates and algae [2, 3]. Aplysinopsin (**94**) is a well-known antitumor tryptophan derivative isolated from several species of sponges [93–96]. Recently, we re-isolated it from the scleractinian coral *Tubastrea aurea* as an inhibitor in the fertilized starfish egg assay (ED$_{100}$ 2.5 µg/ml) [97]. This coral possesses bright yellow-orange polyps whose color may be due to this alkaloid. Aplysinopsin and its derivatives have been reported from sponges [95, 96] and a Mediterannean coral *Astroides calycularis* [98].

Bromine-containing gramines (**95** and **96**) obtained from the bryozoan *Zoobotryon verticillatum* inhibit cell division of the fertilized sea urchin eggs at an ED$_{50}$ of 16 µg/ml [99].

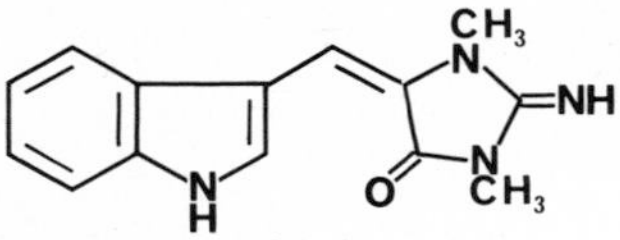

94 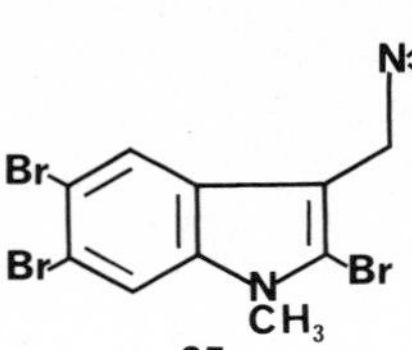   95 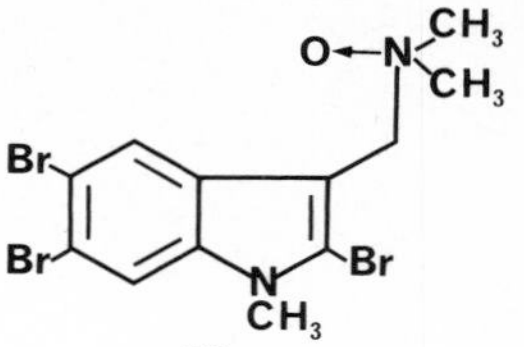   96

Lamellarin A–D are unusual metabolites of the marine prosobranch mollusk *Lamellaria* sp. collected in Palau [100]. Their structures were determined by spectral and X-ray analyses. Of four metabolites, lamellarin C (**97**) and D (**98**) show inhibitory activity at a concentration of 19 µg/ml (15 and 78 percent inhibition, respectively) in the sea urchin egg assay.

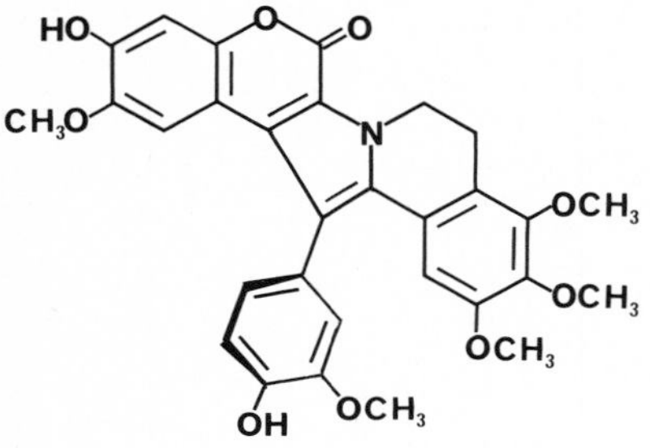

97 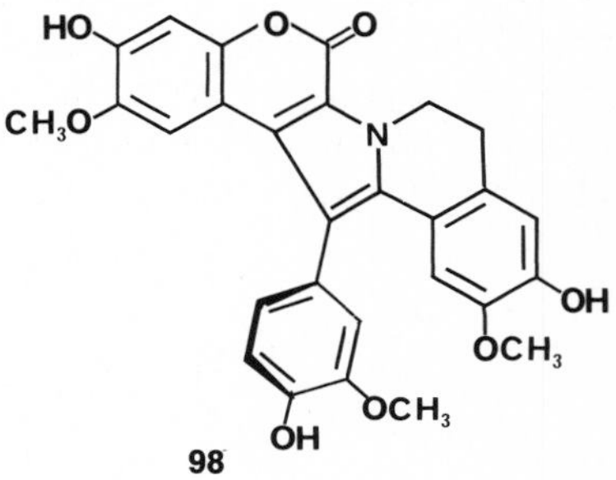   98

3.3 Mycalisines

Marine sponges produce unusual nucleosides, represented by spongothymidine [101], spongouridine [101], spongosine [102] and 1-methyl-isoguanosine [103], which exhibit a wide variety of biological activities [104]. We isolated two novel nucleosides, mycalisine A (**99**) and B (**100**), from a shallow water species of a sponge of the genus *Mycale* that is common along the Pacific coast of Japan [105]. Maycalisine A showed spectral features similar to those of toyocamycin (**102**) which was isolated from *Streptomyces* sp. [106]. Instead of D-ribose in toyocamycin, mycalisine A contains 3-*O*-methyl-5-deoxy-*erythro*-pent-4-eno-furanose, which was confirmed by ^{13}C and ^{1}H NMR spectra including double resonance experiments. In order to secure this structure, we attempted to correlate mycalisine A to toyocamycin by chemical conversion. However, hydroboration gave an inseparable mixture of products, and other reactions were all unsuccessful; even the glycoside linkage was resistant to hydrolysis. Only the acetate was helpful to deduce the structure.

Mycalisine B was apparently closely related to mycalisine A. Upon oxidation with alkaline hydrogen peroxide it gave rise to the corresponding carboxamide derivative **101**. The stereochemistry of the sugar portion was deduced to be the D-form by $^3J_{C-H}$ value of 4 Hz for C-6, NOE experiments and CD spectrum of the *p*-bromobenzoate.

99

100

101

102

Mycalisine A inhibits cell division of the fertilized starfish eggs at a concentration of 0.5 µg/ml, while mycalisine B inhibits at a concentration of 200 µg/ml. The mycalisines affect synchronous cell division of the fertilized starfish eggs, and various sizes of blastomeres were produced as shown in Fig. 6. However, development of embryos proceeds to the morula stage. A similar phenomenon is observed for 5-fluorouracil, a famous inhibitor of DNA synthesis. At present, it is not known how the mycalisines act on the fertilized starfish eggs.

Closely related antibiotics, such as tubercidin [107], toyocamycin [106] and sangivanycin [108] have been isolated from *Streptomyces* spp. Interestingly, the mycalisines show no antimicrobial activity.

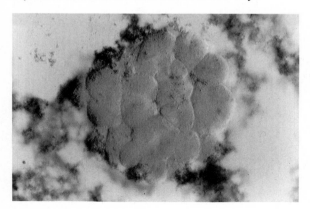

Fig. 6. Starfish embryo affected by mycalisine A. Each blastomere contains a nucleus stained with a lactoorcein solution

3.4 Kabiramides and Ulapualides

Eggmasses of subtropical and tropical nudibranchs have frequently striking colors ranging from orange to red. They resemble rosebuds on rocks or dead corals. Despite their attractiveness, these "organisms" are virtually immune to predation. In our screening for biological active substances, the lipophilic extracts of the eggmasses of an unidentified nudibranch collected at Kabira Bay of Ishigaki island in the Ryukyus showed strong antifungal activity. Later, we recognized that the extracts also inhibit the development of echinoderm embryos. We collected twelve specimens of eggmasses, from which five active substances, named kabiramides A–E after the collection site, were isolated in pure form by means of HPLC on ODS. The major component, kabiramide C, was analyzed by extensive 500 MHz NMR spectroscopy consisting of 2 D homonuclear and heteronuclear correlation techniques, difference NOE and LSPD experiments. These results together with other spectral data led to assign structure **103** to the 28-membered lactone encompassing three contiguous oxazoles [109]. The presence of a trisoxazole moiety (**108**) was established by low field signals in ^{1}H and ^{13}C NMR, large $^1J_{C-H}$ value (211 Hz), $^{2,3}J_{C-H}$ values and LSPD experiments. The connection of four segments obtained from COSY analyses were performed by NOE and

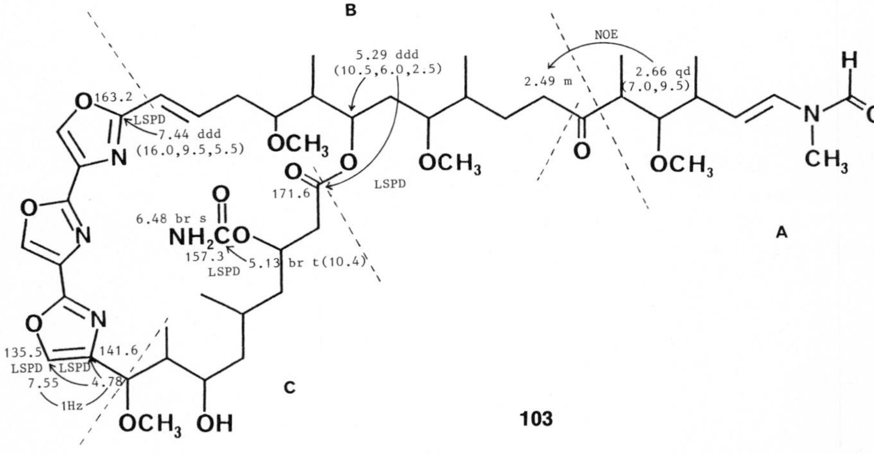

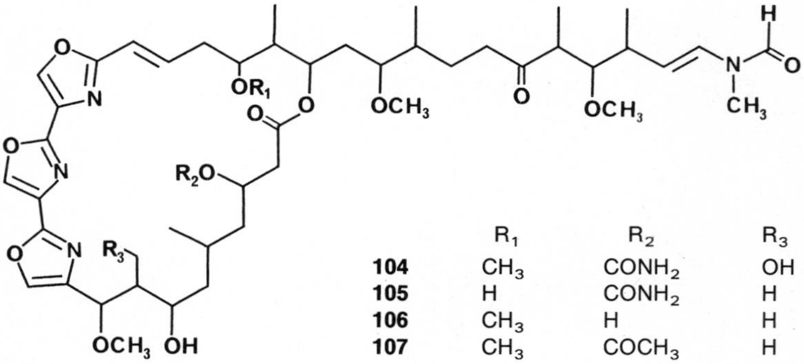

	R_1	R_2	R_3
104	CH_3	$CONH_2$	OH
105	H	$CONH_2$	H
106	CH_3	H	H
107	CH_3	$COCH_3$	H

LSPD experiments. Similarly, structures of kabiramide A (**104**), B (**105**), D (**106**) and E (**107**) were elucidated [110].

At the same time we completed the structure determination of kabiramide C, the Scheuer group [111] also assigned the gross structures of closely related metabolites, ulapualide-A (**109**) and -B (**110**), which were isolated from the eggmasses of the Hawaiian nudibranch *Hexabranchus sanguineus*. They used not only 2 D NMR techniques but also chemical degradation. Confirmation of a trisoxazole moiety was made by ozonolysis of ulapualide-B, followed by $NaBH_4$ reduction and acetylation, which afforded the bisoxazole **111**. This was then transformed to the bisamide **112**, thus securing the trisoxazole structure.

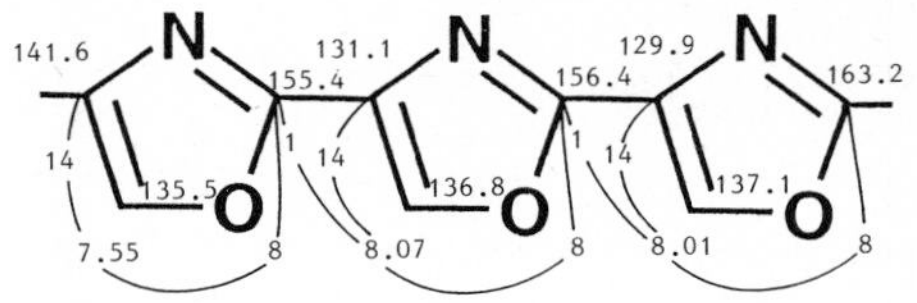

108

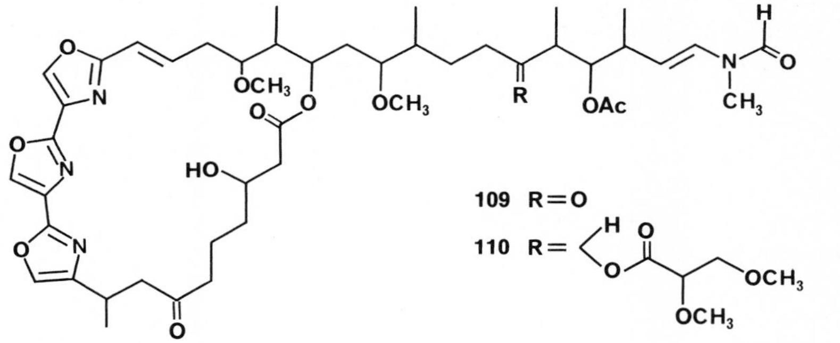

109 R=O

110 R=

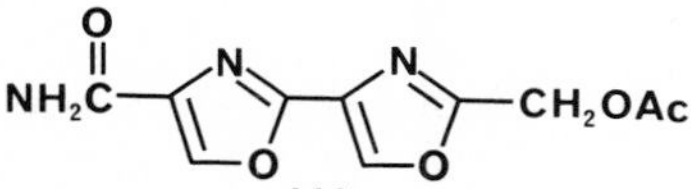

111

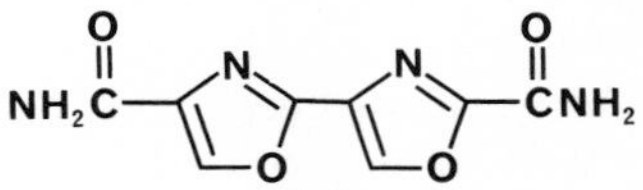

112

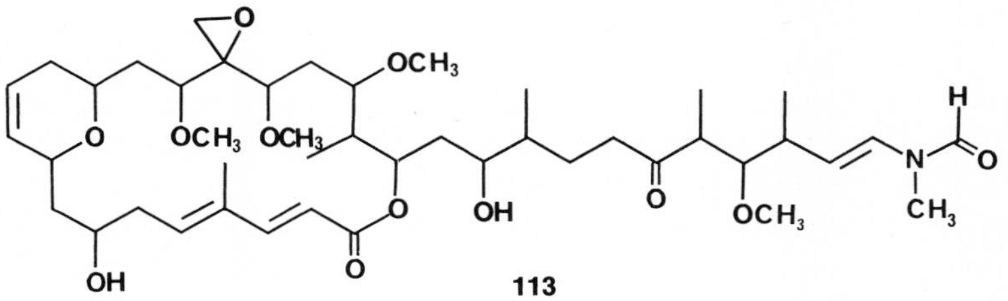

113

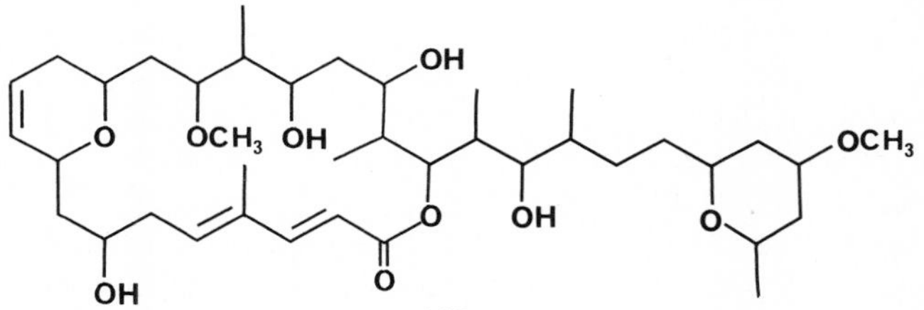

114

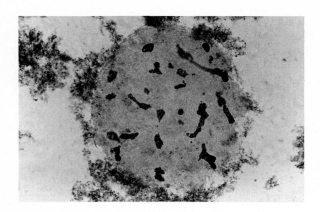

Fig. 7. Polynucleated starfish embryo induced by kabiramide C. Unusual nuclei (black rods) are seen in a cell stained with a lactoorcein solution

Stereochemistry of thirteen chiral centers in the kabiramides remains to be determined. Surprisingly, these nudibranch macrolides are structurely similar to scytophycin B (**113**) [112] and swinholide A (**114**) [113], which were reported from a freshwater blue-green alga, *Scytonema pseudohofmanni,* and from the Red Sea sponge *Theonella swinhoei,* respectively. Origins of these macrolides including the nudibranch metabolites are interesting.

Kabiramides exhibit strong activity in the starfish egg assay (kabiramides A–E, LD_{100} 1.0, 0.2, 0.2, 0.2 and 0.2 μg/ml, respectively). They arrest cell division completely at higher concentrations, and at lower concentration cell division proceeds up to the eight-cell stage, which includes polynucleated cells possessing unusually shaped nuclei (Fig. 7).

3.5 Calyculins

Another extraordinary marine metabolite, which inhibits the development of echinoderm embryos, is calyculin A (**115**) [114]. This has been isolated from the Japanese sponge *Discodermia calyx,* which revealed the strongest activity in our starfish egg assay. Calyculin A is the major active component (yield, 0.15 percent on the basis of wet sponge) of the sponge, which also produces three related metabolites that can be separated by a series of HPLC runs. Extensive spectroscopic analyses including 500 MHz 2 D NMR and ^{31}P NMR implied that calyculin A contains an all (*E*)-tetraene, a spiroketal, an oxazole, a phosphate and a nitrile in the molecule. To connect these fragments, NOE and other available techniques were employed. Since the gross structure which we assigned was not fully unambiguous, especially connectivity of a spiroketal portion and the position of a phosphate ester, we decided to grow crystals for X-ray diffraction. After nine months we finally obtained enough crystals for our purpose from a mixture of n-hexane, acetone and ether. The structure obtained by X-ray diffraction (Fig. 8) was virtually identical with that elucidated by spectral analyses.

Calyculin A embraces an octamethyl, polyhydroxylated C_{28} fatty acid which is linked to two γ-amino acids. The fatty acid moiety embodies a conjugated all-(*E*)-tetraene, a β-hydroxytetrahyrofuran which is esterified by phosphoric acid

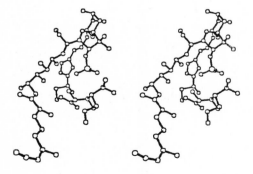

115

and whose hetero oxygen forms a spiroketal with a γ-hydroxytetrahydropyran. The phosphate ester is resistant to chemical or enzymatic hydrolysis, which may be due to hydrogen bonds as seen in the crystal structure.

Calyculin A inhibits the first cell division of the fertilized starfish eggs at a concentration of 0.001 µg/ml. When the unfertilized eggs are exposed to the metabolite, the polar bodys appear without delay, suggesting that calyculin A does not affect protein synthesis. In the case of the fertilized eggs, they displayed an unexpected feature: enlarged nucleus was released from the cell as shown in Fig. 9. Moreover, calyculin causes fragmentation of DNA in the fertilized sea urchin eggs (Fig. 10). At present, we cannot interpret this unusual phenomenon.

Calyculin A is also strongly cytotoxic against L1210 with an ID_{50} of 1.75 ng/ml. DNA, RNA and protein synthesis in tumor cells were inhibited by our compound. It is likely that calyculin A affects biological system(s) other than DNA, RNA and protein syntheses.

Fig. 8. Computer-generated perspective drawing of calyculin A

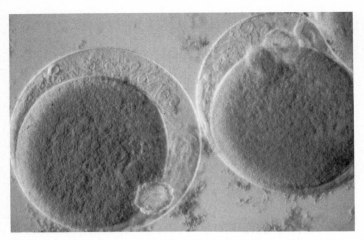

Fig. 9. Enlarged nucleus released from the fertilized sea urchin egg exposed to calyculin A

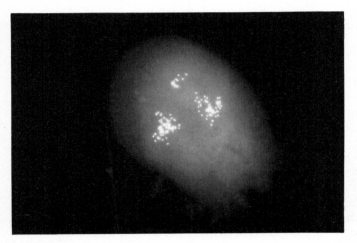

Fig. 10. Fragmentation of DNA in the fertilized sea urchin egg caused by calyculin A. Visualized by immunofluorescence technique

3.6 Discodermins

Discodermin A–D (**116–119**) are cyclic tetradecapeptides isolated from the Japanese sponge *Discodermia kiiensis* as antimicrobial metabolites [115, 116]. They in-

HCO-D-Ala-L-Phe-L-Pro-X-D-Trp-L-Arg-D-Cys(O₃H)-L-Thr-L-MeGln-D-Leu-L-Asn-L-Thr-Sar ⌐

116	X = D-*t*-Leu-L-*t*-Leu
117	X = D-Val-L-*t*-Leu
118	X = D-*t*-Leu-L-Val
119	X = D-Val-L-Val

hibit development of starfish embryos at concentrations ranging from 2 to 50 µg/ml; discodermin A is most potent, while discodermin D least potent. This is also true for their antimicrobial activity. These peptides cause cellular lysis.

Conclusion

A selective assay system using fertilized starfish or sea urchin eggs appears to be a useful probe for detecting bioactive marine metabolites that affect biochemical events taking place in eukaryotes; *e.g.*, protein synthesis, nucleic acid synthesis, microtubule formation, and others. This has been shown by our isolation of a variety of novel metabolites from marine invertebrates including kabiramides and caluculin A which possess unprecedented structural features. Though many classes of marine metabolites were active in the assay, careful examination of the affected eggs allowed one to speculate about their mode of action. Of course, further studies, in which more sophisticated systems are involved, are required to pinpoint the site and mode of action. It is desirable to use the echinoderm egg assay in combination with antibacterial and antifungal assays and cytotoxicity tests. In this way, one can discover substances which may have activity against prokaryotes or tumor cells.

References

1. Weinheimer AJ, Spraggins RL (1969) Tetrahedron Lett 5185
2. Scheuer PJ (1978, 1978, 1980, 1981, 1983) "Marine Natural Products – Chemical and Biological Perspectives, Vols I–V, Academic Press, New York
3. a. Faulkner DJ (1984) Nat Prod Rep 1:251, 551. b. (1986) Nat Prod Rep 3:1
4. Pettit GR, Cragg GM, Herald CL (1984) "Biosynthetic Products for Cancer Chemotherapy" Vol 4, Elsevier, Amsterdam
5. Baker JT (1984) in "Natural Products and Drug Development" (Krogsgaard-Larsen P, Christensen SB, Koford H, ed), p 145, Munksgaard, Copenhagen
6. Rinehart Jr KL, Gloer JB, Cook Jr JC, Mizsak SA, Scahill TA (1981) J Am Chem Soc 103:1857
7. Baker BJ, Okuda RK, Yu PTK, Scheuer PF (1985) J Am Chem Soc 107:2976
8. Uemura D, Takahashi K, Yamamoto T, Katayama C, Tanaka J, Okumura Y, Hirata Y (1985) J Am Chem Soc 107:4796
9. Ruggieri GD, Nigrelli RF (1960) Zoologica 45:1
10. Jacobs RS, White S, Wilson L (1981) Fed Proc 40:26
11. Jacobs RS, Culver P, Langdon R, O'Brien T, White S (1985) Tetrahedron 41:981
12. Ikegami S, Kawada K, Kimura Y, Suzuki A (1979) Agric Biol Chem 43:161
13. Ikegami S, Taguchi T, Ohashi M, Oguro M, Nagano H, Mano Y (1978) Nature 275:458
14. Kanatani H (1973) Int Rev Cytol 35:253
15. Nagano H, Hirai S, Okano K, Ikegami S (1981) Develop Biol 85:409
16. Nemoto S, Yoneda M, Uemura I (1980) Develop Growth Differ 22:315
17. Burkholder PR (1973) in "Biology and Geology of Coral Reefs" (Jones OA, Endean R, ed), Vol II: Biology 1, p 117, Academic Press, New York
18. Shaw PD, McClure WO, Van Blaricom G, Sims J, Fenical W, Rude J (1976) in "Food-Drugs from the Sea Proceedings 1974" (Webber HH, Ruggieri GD, ed), p 429, Marine Technol Soc, Washington, DC

19. Weinheimer AJ, Karns TKB (1976) in "Food-Drugs from the Sea Proceedings 1974" (Webber HH, Ruggieri GD, ed), p 491, Marine Technol Soc, Washington, DC
20. Rinehart Jr KL, Shaw PD, Shield LS, Gloer JB, Harbour GC, Koker MES, Samain D, Schwartz RE, Tymiak AA, Weller DL, Carter GT, Munro MHG, Hughes Jr RG, Renis HE, Swynenberg EB, Stringfellow DA, Vavra JJ, Coats JH, Zurenko GE, Kuentzel SL, Li LH, Bakus GJ, Brusca RC, Craft LL, Young DN, Connor JL (1981) Pure Appl Chem 53:795
21. Fusetani N, Kato Y, Matsunaga S, Hashimoto K (1983) Tetrahedron Lett 24:2771
22. Castiello D, Cimino G, de Rosa S, de Stefano S, Sodano G (1980) Tetrahedron Lett 21:5047
23. Cimino G, Crispino A, de Rosa S, de Stefano S, Sodano G (1981) Experientia 37:924
24. Cimino G, de Giulio A, de Stefano S, Sodano G (1985) J Nat Prod 48:22
25. Hochlowski JE, Faulkner DJ (1983) Tetrahedron Lett 24:1917
26. Nakanishi K (1985) Toxicon 23:473
27. Baden DG, Mende TJ, Lichter W, Ellham L (1981) Toxicon 19:455
28. Shimizu Y, Chou H-N, Bando H, van Duyne G, Clardy JC (1986) J Am Chem Soc 108:514
29. Crews P, Naylor S, Hanke FJ, Hogue ER, Kho E, Braslau R (1984) J Org Chem 49:1371
30. Crews P, Myers BL, Naylor S, Clason EL, Jacobs RS, Staal GB (1984) Phytochemistry 23:1449
31. Sims JJ, Lin GHY, Wing RM (1974) Tetrahedron Lett 3487
32. McEnroe FJ, Fenical W (1978) Tetrahedron 34:1661
33. Cimino G, de Rosa S, de Stefano S, Cariello L, Zanetti L (1982) Experientia 38:896
34. Blackman AJ, Wells RJ (1978) Tetrahedron Lett 3063
35. Paul VJ, Fenical W (1984) Tetrahedron 40:2913
36. Paul VJ, Fenical W (1982) Tetrahedron Lett 23:5017
37. Paul VJ, Fenical W (1985) Phytochemistry 24:2239
38. Paul VJ, Fenical W (1984) Tetrahedron 40:3053
39. Fattorusso E, Magno S, Mayol L, Novellino E (1983) Experientia 39:1275
40. Paul VJ, Fenical W (1983) Science 221:747
41. Kashman Y, Rotem M (1979) Tetrahedron Lett 1707
42. Manes LV, Bakus GJ, Crews P (1984) Tetrahedron Lett 25:931
43. Gerwick WH, Fenical W, van Engen D, Clardy J (1980) J Am Chem Soc 102:7991
44. Gerwick WH, Fenical W (1983) J Org Chem 48:3325
45. Gerwick WH, Fenical W, Sultanbawa MUS (1981) J Org Chem 46:2233
46. Ravi BN, Wells RJ (1982) Aust J Chem 35:129
47. Bandurraga MM, Fenical W, Donovan SF, Clardy J (1983) J Am Chem Soc 104:6463
48. Look SA, Burch MT, Fenical W, Zheng Q-T, Clardy J (1985) J Org Chem 50:5741
49. Missakian MG, Burreson BJ, Scheuer PJ (1975) Tetrahedron 31:2513
50. Fenical W, Okuda RK, Bandurraga MM, Culver P, Jacobs RS (1981) Science 212:1512
51. Ciereszko LS, Sifford DH, Weinheimer AJ (1960) Ann NY Acad Sci 90:917
52. Gerwick WH, Fenical W (1981) J Org Chem 46:22
53. Gerwick WH, Fenical W, Fritsch N, Clardy J (1979) Tetrahedron Lett 145
54. O'Brian T, Jacobs RS, Wilson L (1983) Mol Pharmacol 24:493
55. White SJ, Jacobs RS (1983) Mol Pharmacol 24:500
56. Sun HH, Ferrara NM, McConnell OJ, Fenical W (1980) Tetrahedron Lett 21:3123
57. Francisco C, Banaigs B, Valls R, Codomier L (1985) Tetrahedron Lett 26:2629
58. Francisco C, Banaigs B, Teste J, Cave A (1986) J Org Chem 51:1115
59. Cimino G, De Stefano S, Minale L, Fattorusso E (1972) Tetrahedron 28:333
60. Manes LV, Naylor S, Crews P, Bakus GJ (1985) J Org Chem 50:284
61. Fusetani N, Kato Y, Matsunaga S, Hashimoto K (1984) Tetrahedron Lett 25:4941
62. Kato Y, Fusetani N, Matsunaga S, Hashimoto K (1985) Chem Lett 1521
63. Kato Y, Fusetani N, Matsunaga S, Hashimoto K (1986) Experientia, 42:1299
64. Faulkner DJ (1973) Tetrahedron Lett 3821
65. Anisimov MM, Kuznetsova TA, Shirokov VP, Prokofyeva NG, Elyakov GB (1972) Toxicon 10:187
66. Anisimov MM, Fronert EB, Kuznetsova TA, Elyakov GB (1973) Toxicon 11:109

67. Anisimov MM, Shcheglov VV, Stonik VA, Fronert EB, Elyakov GB (1974) Toxicon 12:327
68. Anisimov MM, Prokofieva NG, Korotkikh LY, Kapustina II, Stonik VA (1980) Toxicon 18:221
69. Fusetani N, Kato Y, Hashimoto K, Komori T, Itakura Y, Kawasaki T (1984) J Nat Prod 47:997
70. Komori T, Nanri H, Itakura Y, Sakamoto K, Taguchi S, Higuchi R, Kawasaki T, Higuchi T (1983) Liebigs Ann Chem 37
71a. Itakura Y (1983) "Studies on the Saponins from the Starfish *Acanthaster planci* L., *Asteropecten latespinosus* Meissner and *Asterias amurensis* [cf] *versicolor* Sladen", Doctoral Thesis, Kyushu University
71b. Itakura Y, Komori T (1986) Liebigs Ann Chem 359
72. Krebs HC, Komori T, Kawasaki T (1984) Liebigs Ann Chem 296
73. Itakura Y, Komori T, Kawasaki T (1983) Liebigs Ann Chem 56
74. Komori T, Matsuo J, Itakura Y, Sakamoto K, Ito Y, Taguchi S, Kawasaki T (1983) Liebigs Ann Chem 24
75. Kitagawa I, Kobayashi M (1978) Chem Pharm Bull 26:1864
76. Kitagawa I, Kobayashi M, Sugawara T (1978) Chem Pharm Bull 26:1852
77. Itakura Y, Komori T, Kawasaki T (1983) Liebigs Ann Chem 2079
78. Komori T, Krebs HC, Itakura Y, Higuchi R, Sakamoto K, Taguchi S, Kawasaki T (1983) Liebigs Ann Chem 2092
79. Ruggieri GD, Nigrelli RF (1966) Am Zool 6:380
80. Tsuchimori N, Ikegami S, Miyashiro S, Tsuji T, Kida T, Shibai H (1986) Comp Biochem Physiol 84C:381
81. Fusetani N, Yasukawa K, Matsunaga S, Hashimoto K (1985) Tetrahedron Lett 26:6449
82. Fusetani N, Matsunaga S, Konosu S (1981) Tetrahedron Lett 22:1985
83. Nakatsu T, Walker RP, Thompson JE, Faulkner DJ (1983) Experientia 39:759
84. Carté B, Faulkner DJ (1983) J Org Chem 48:2314
85. Matsunaga S, Fusetani N, Hashimoto K (1986) Experientia 42:84
86. Nigrelli RF, Stempien Jr MF, Ruggieri GD, Liguori VR (1967) Fed Proc 26:1197
87. Pelter A, Ballantine JA, Ferrito V, Jaccarini V, Psaila AF, Schembri PJ (1976) J Chem Soc Chem Commun. 999
88. Ballantine JA, Psaila AF, Pelter A, Murray-Rust P, Ferrito V, Schembri P, Jaccarini V (1980) J Chem Soc Perkin I, 1080
89. Pelter A, Abela-Medici A, Ballantine JA, Ferrito V, Ford F, Jaccarini V, Psaila AF (1978) Tetrahedron Lett 2017
90. Cariello L, de Nicola Giudici M, Zanetti L, Prota G (1978) Experientia 34:1427
91. Gauthier MJ, de Nicola Giudici M (1983) Curr Microbiol 8:195
92. Berns MW (ed) (1984) "Hematoporphyrin Derivative Photoradiation Therapy of Cancer", Alan R Liss, New York
93. Kazlauskas R, Murphy PT, Quinn RJ, Wells RJ (1977) Tetrahedron Lett 61
94. Hollenbeak KH, Schmitz FJ (1977) Lloydia 40:479
95. Djura P, Stierle DB, Sullivan B, Faulkner DJ, Arnold E, Clardy J (1980) J Org Chem 45:1435
96. Tymiak AA, Rinehart Jr KL, Bakus GJ (1985) Tetrahedron 41:1039
97. Fusetani N, Asano M, Matsunaga S, Hashimoto K (1986) Comp Biochem Physiol 85B:845
98. Fattorusso E, Lanzotti V, Magno S, Novellino E (1985) J Nat Prod 48:924
99. Sato A, Fenical W (1983) Tetrahedron Lett 24:481
100. Andersen RJ, Faulkner DJ, He C-H, van Duyne GD, Clardy J (1985) J Am Chem Soc 107:5492
101. Bergmann W, Burke DC (1955) J Org Chem 20:1501
102. Bergmann W, Burke DC (1956) J Org Chem 21:226
103. Quinn RJ, Gregson RP, Cook AF, Bartlett RT (1980) Tetrahedron Lett 21:567
104. Baslow MH (1969) "Marine Pharmacology", Williams & Wilkins, Baltimore
105. Kato Y, Fusetani N, Matsunaga S, Hashimoto K (1985) Tetrahedron Lett 26:3483
106. Ohkuma K (1961) J Antibiot Ser A 14:343
107. Anzai K, Nakamura G, Suzuki S (1957) J Antibiot Ser A 10:201

108. Rao KV, Renn DW (1963) Antimicrob Agents Chemother 77
109. Matsunaga S, Fusetani N, Hashimoto K, Koseki K, Noma M (1986) J Am Chem Soc 108:847
110. Matsunaga S, Fusetani N, Hashimoto K, Koseki K, Noma M to be submitted to J Org Chem
111. Roesener JA, Scheuer PJ (1986) J Am Chem Soc 108:846
112. Moore RE, Patterson GML, Mynderse JS, Barchi J Jr, Norton TR, Furusawa E, Furusawa S (1986) Pure Appl Chem 58:263
113. Carmely S, Kashman Y (1985) Tetrahedron Lett 26:511
114. Kato Y, Fusetani N, Matsunaga S, Hashimoto K, Fujita S, Furuya T (1986) J Am Chem Soc 108:2780
115. Matsunaga S, Fusetani N, Konosu S (1984) Tetrahedron Lett 25:5165
116. Matsunaga S, Fusetani N, Konosu S (1985) Tetrahedron Lett 26:855

The Search for Antiviral and Anticancer Compounds from Marine Organisms

Murray H. G. Munro [1], Richard T. Luibrand [2], and John W. Blunt [1]

Contents

1 Chemistry Department, University of Canterbury, Christchurch, New Zealand.
2 Chemistry Department, California State University, Hayward, CA 94542, USA.

Bioorganic Marine Chemistry, Volume 1

Abstract

A survey is presented of the occurrence of compounds from marine sources, other than micro-organisms, which have been reported to have antiviral, antitumor or cytotoxic properties. The biological data is given for each occurrence, together with a discussion of the methods employed to obtain these data. The taxonomic distribution of all compounds showing these activities is tabulated.

Strategies for the detection of further compounds with potential as leads for antiviral or anti-cancer agents are presented.

1. Introduction

The grouping together of antiviral and anticancer agents from marine sources in one review is not a matter of expediency, but follows from the increasing realization of the importance of the sea as the most likely source of future "leads" in both areas. Both antiviral and anticancer drugs are cell-growth inhibitory substances, which have to act selectively without damage to the host organism. Cancer is the second leading cause of death in the Western World and it is estimated that 60% of all illness in developed countries is a consequence of viral infections. Despite the availability of drugs in each area, more effective formulations are required. Some of these new drugs will undoubtedly come from natural products screening, in combination with synthetic modification and analog production.

In terms of known species and habitats, the oceans are our greatest resource of metabolic products [1] (excluding insects). Thus, putting aside such problems as procurement, marine metabolites represent the statistically most likely source of these vital growth inhibitory substances. However, just to work in marine natural products is not sufficient to ensure the discovery of new drugs. What is required is the implementation of strategies that maximize the probability of finding active compounds. An example of such a strategy would be the examination of **selected** marine phyla and the routine screening of all extracts from these for cytotoxicity, a commonly used guide for the discovery of antineoplastic compounds.

The earlier surveys of marine organisms for cytotoxicity [2–6] were mostly random and not oriented towards given classes, or based on ecological considerations or folklore [7–10]. These surveys indicated that sessile marine invertebrates have the higher probability (9–10%) of providing compounds with cytotoxic properties. A less random search might focus on certain classes of invertebrates and would increase the probability of locating **active** species.

It should be noted, however, that this increase in efficiency arising from selective screening is offset by the probability of missing active species in the other, statistically less favored classes. For instance, excellent antineoplastic activities were reported for surveys of Pacific Basin marine algae [11, 12], and to a lesser extent from Florida [13].

Ecological considerations, based on field observations of biological interactions, can also provide pointers towards the ultimate selection of active species. Observations such as a modification of behavior, overgrowth or inhibition of growth by a neighboring organism suggest a chemical defensive or offensive capability. The chemicals involved, are often cytotoxic [7, 14]. The tambjamines are good examples of such cytotoxic chemicals (Sect. 4.6.3). Emphasizing the collection of the apparently defenseless organisms such as tunicates, soft corals, sponges and sea hares would also represent such a selective screening strategy.

The concept of isolating antiviral substances from marine organisms is not new. Prior to the late 1970's there had been many reports of such activity [1, 15–18], but there had been no systematic surveys for antiviral activity in extracts from marine organisms. In 1978 Rinehart carried out such a search in the Caribbean and concluded that the probability of finding antiviral activity was highest in Cyanophyta, Phaeophyta and Chordata (Ascidiacea), and was significant in Porifera, Cnidaria and Echinodermata [19]. Activity has been observed in a variety of marine algal classes [20–22], and other marine organisms [23]. A high incidence of antiviral activity has also been noted in two recent large surveys. In the Caribbean, an incidence of 9% in sponges has been reported by SeaPharm Inc. [24], while in New Zealand waters the incidence is 8% for a collection of sponges, bryozoans and ascidians [25]. These surveys suggest that the phyla which most frequently possess antiviral activity are also those which precedent has shown are most likely to possess cytotoxic compounds. If this conclusion is now coupled with the recognition that the in vitro method of screening for antiviral activity (see Sect. 2.2) also detects cytotoxicity, then the simultaneous search for antiviral and antineoplastic compounds is a rational approach.

The contrast between a review on marine antineoplastic agents in the mid-1980's with one presented little more than a decade ago is startling. In their review Li, Goldin and Hartwell described the antineoplastic properties of **extracts** from 29 species [26]. The only confirmed structures mentioned in the review were the spongonucleosides isolated by Bergmann et al. from the sponge *Cryptotethya crypta* [27–29]. Progress in the isolation and characterisation of the marine antineoplastic agents has accelerated rapidly in recent years and some of the species, such as *Bugula neritina* and *Ecteinascidea turbinata* cited in the 1974 review, have now provided important antineoplastic leads for this decade. In contrast to that review, only antineoplastic or cytotoxic **compounds** will be presented here, with little emphasis on **extracts.**

This review is presented in 3 parts, dealing with antiviral, antineoplastic and cytotoxic compounds. The latter category includes those compounds for which in vitro cytotoxicity is reported; some of these compounds show no antineoplastic activity in vivo, or have yet to be tested. All compounds reviewed have also been organized along taxonomic lines and this information is presented in tabular form. Compounds from marine microorganisms have not been included.

2. Antiviral Compounds

2.1 Antiviral Chemotherapy

Although vaccines have been very successful in controlling many viral diseases, some diseases are unlikely to be controlled by vaccination. The idea that antiviral compounds with pharmaceutical value could be found has been accepted only slowly, partly because of the toxicity of many of the earlier antiviral agents. The discovery of the role of the interferons as cellular antiviral systems and the elucidation of differences between normal cellular metabolism and viral replication have led to a renewed interest in antiviral chemotherapy.

Viruses are the smallest known biological structures that carry all the genetic information needed for replication. They differ from procaryotes and eucaryotes in that they carry only one type of nucleic acid. Viruses cannot propagate in a lifeless medium and require access to a host cell for replication. Most viruses invade specific types of cells and are therefore usually species-selective. Once within the cell, the virus normally takes over the cellular synthetic machinery to replicate all the necessary viral nucleic acids and proteins. After assembly, the new virus particles must leave the cell. All these processes involve both cellular and viral enzymes. Antiviral agents must inhibit at least one of the steps in this process **without** effect on the host.

In addition to this inhibition, the antiviral agent must have a broad spectrum of activity, inhibit the virus completely, have favorable pharmacodynamic properties and not be immunosuppressive. This last feature is important, as there should be no suppression of the normal immune processes. In the ideal situation the antiviral drug checks the infection while the immune system prepares to destroy the last virus particle [30–33]. This point is critical for those immunocom-

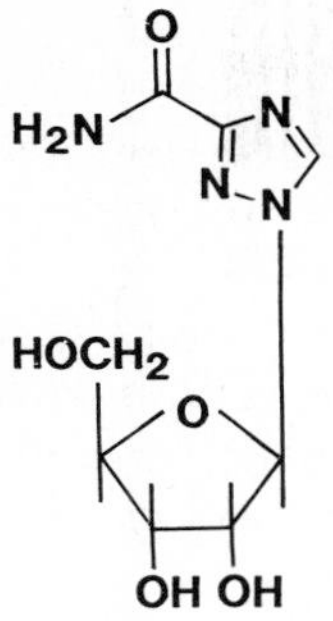

1 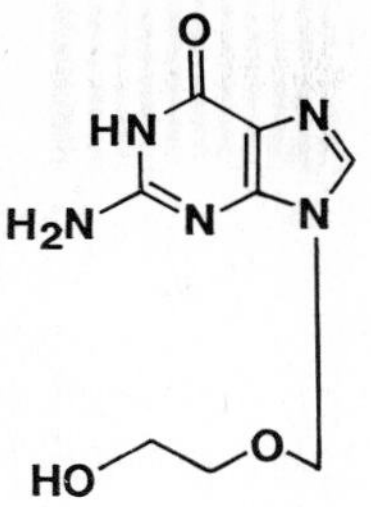   2

promised by illness (cancer, AIDS) or drug therapy (transplants and cancer). A frequent cause of death in these instances is from a viral infection. Adjuvant antiviral chemotherapy is vital in these circumstances [34].

Not surprisingly, only a few antiviral drugs have been developed. In the main they are nucleosides or nucleoside analogs. Among the most successful to date are ribivarin (**1**) and acyclovir (**2**) which act as inhibitors of viral specific enzymes [33, 35].

2.2 Antiviral Testing

Most in vitro antiviral screening is carried out as a modification of the plaque reduction technique first described by Herrmann [36]. In this assay, a monolayer of a continuous mammalian cell line is grown in wells in an appropriate medium. After infection with a known number of plaque forming units of the virus the cells are overlaid with the medium in a viscous base, such as methylcellulose. The liquid test samples are applied to paper discs, which when dry are placed in the well. Depending on the cell line and the virus used, the results can be read within 24–48 hours. If the test sample is antiviral there will be a reduction in the number of plaques formed with respect to the control [37]. Quite a number of different cell lines can be used in this assay, with the most commonly used being monkey kidney cell lines (eg CV-1, BSC, Vero). Testing can be carried out against a wide variety of viruses although *Herpes simplex* Type I (*HSV*-I), *Vesicular stomatitis* (*VSV*) and *Polio* Type I viruses seem to be the most prevalent. The reading of the assay can be by visualisation of the plaques by staining, or by microscopic examination. This system can be used readily in the field. An added advantage is that, as well as giving a measure of the antiviral properties of an extract or compound, it gives an assessment of the cytotoxicity. Such cytotoxic extracts become candidates for testing against tumor cell lines such as P388 or L1210.

The animal models for human viral diseases are not yet well-developed and many compounds that are active in vitro have failed in animal or human tests. For example, there is not a suitable animal model for *Rhinovirus* infections. Better models will ensure the more rapid confirmation of promising in vitro active compounds as potential antiviral drugs.

2.3 Antiviral Compounds

A number of the earlier marine compounds that were described as antiviral had only modest activity, but compounds with significant activity are now appearing.

   As the number of compounds involved is still relatively small they will be reviewed individually. Although antiviral activity has been noted for compounds such as chondriol [38], the dollabellanes [39] and the asterosaponins [40], no further reports have appeared, consequently these will not be reviewed.

2.3.1 Nucleosides

The development of the first successful antiviral drugs and the recent history of marine natural products are inextricably linked. It was Bergmann's discovery of naturally occurring arabinosides as marine natural products that triggered intense interest in these "unnatural" nucleosides and their potential biological properties. The subsequent development of the antiviral drugs ara-A and ara-C has been documented exhaustively [41–45].

   Since Bergmann's original discovery of spongothymidine (**3**), spongosine (**4**) and spongouridine (**5**) from the sponge *Cryptotethya crypta* [27–29], nucleosides have been isolated from other classes of marine organisms. Another arabinoside, ara-A (9-β-D-arabinofuranosyladenine) (**6**), has been isolated from the Mediterranean gorgonian *Eunicella cavolini,* along with the 3′-O-acetyl derivative and spongouridine [46]. Ara-A was by then a known compound, having been synthesised some 24 years earlier [47]. It was recognised as an antiviral agent in 1964 [48, 49] and has been used therapeutically against *Herpes encephalitis* since the late 1970's [50].

   A modified riboside related to the antiviral agent tubercidin has been isolated from the red alga *Hypnea valendiae* [51]. The compound, 5′-deoxy-5-iodotubercidin (**7**) was the first natural 5′-deoxyribosyl nucleoside, and the first example of a specifically iodinated metabolite.

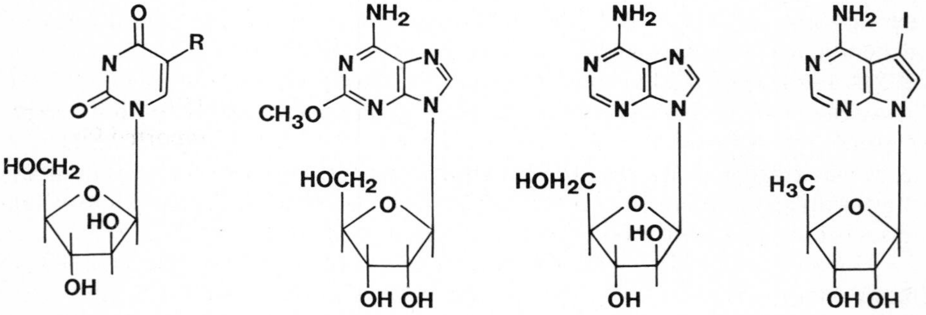

3 R = CH₃

5 R = H

4

6

7

2.3.2 Biopolymers

Antiviral biopolymers have been described from two main sources: glycoproteins from abalone and other molluscs [52], and polysaccharides from various species of marine macroscopic algae [53]. This topic has been recently reviewed [54, 55].

The inhibitors from the algal extracts are polysaccharides with a high degree of sulfation; at least part of the macromolecule is believed to be branched. The degree of homogeneity of the polysaccharides is unknown and it is believed that nucleic acids or proteins are possibly associated with the macromolecule at specific sites.

The mode of action of the algal polysaccharides appears to be interference with virus adsorption or penetration processes, possibly by formation of a non-infectious polysaccharide-virus complex. The antiviral properties of sulfated polysaccharides from other sources have been well-documented and are supported by Ehresmann's work. There may well be a common mechanism of action for all sulfated polysaccharides, with the differences in effect or selectivity reflecting the influence of structural features in each type of polysaccharide [53, 56, 57].

It is possible that these sulfated polysaccharides with antiviral properties share many features in common with the sulfated polysaccharides reported to have antitumor properties (see Sect. 3.2.2).

2.3.3 Aplidiasphingosine, Polyandrocarpidines, Acarnidines

The organisms from which these compounds were isolated are all from the Gulf of California. Various biological activities, including modest antiviral activity, was detected in these species by Rinehart in various expeditions to this area and in each case the major bioactive component(s) has been isolated and characterized.

Aplidiasphingosine (**8**) was isolated from an *Aplidium* sp. of tunicate and can be considered a formal sphingosine derivative from serine and the corresponding diterpene acid. The antiviral properties of the extract were noted, but were not quantified for aplidiasphingosine itself. The pure compound was also effective against a range of Gram-positive and Gram-negative bacteria and fungi, and was cytotoxic towards KB and L1210 tumor cells in vitro (ED_{50} 8.3 and 1.9 µg/ml) respectively [58].

The relative stereochemistry of aplidiasphingosine was ascertained by Mori using elegant stereocontrolled syntheses and is as depicted (**8**) [59–61]; the enantiomeric identity of (**8**) is uncertain as no chiroptical data was reported [58].

The polyandrocarpidines (**9, 10**), guanidine derivatives from a colonial tunicate of the genus *Polyandrocarpa*, were strongly antibacterial, inhibited the growth of the L1210 cell line and were marginally antiviral ([62], see also [62a]). Purification of the polyandrocarpidines was complex, as they were isolated as a 9:1 mixture of homologs and each homolog was a mixture of geometric isomers. The initial structural assignment of the polyandrocarpidines [62] was later corrected by Faulkner to the isomeric γ-methylene γ-lactams (**9, 10**) [63] and corroborated by the synthesis of the hexahydro derivative [64].

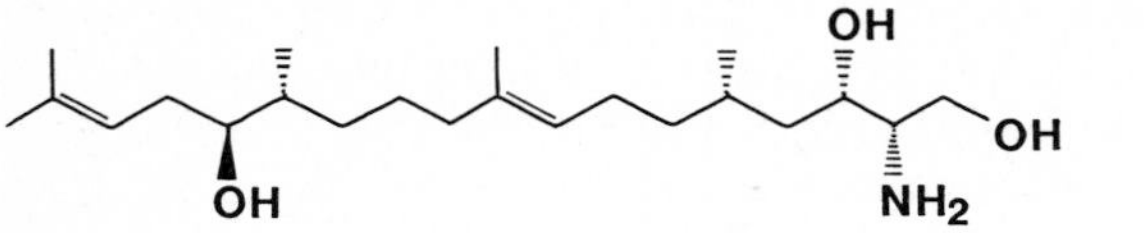

8

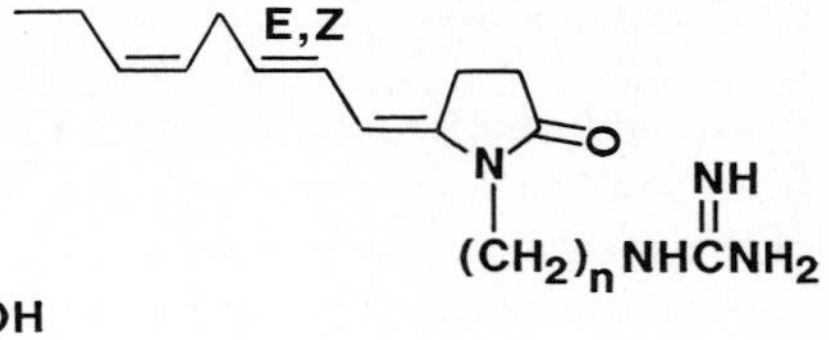

9   n = 5

10 n = 4

The acarnidines (**11–13**) are a trio of guanidine derivatives of the homosper-midine skeleton which were isolated from the sponge *Acarnus erithacus* [65]. The natural acarnidine (**11**) has since been synthesised by two groups [66–68]. The synthetic acarnidine, and all of the 22 analogs synthesised were strongly antibacterial and cytotoxic against the BSC cell line, but were inactive against a range of DNA and RNA viruses [69]. The antiviral activity reported in *A. erithacus* is apparently not resident with the acarnidines. This organism clearly warrants further investigation.

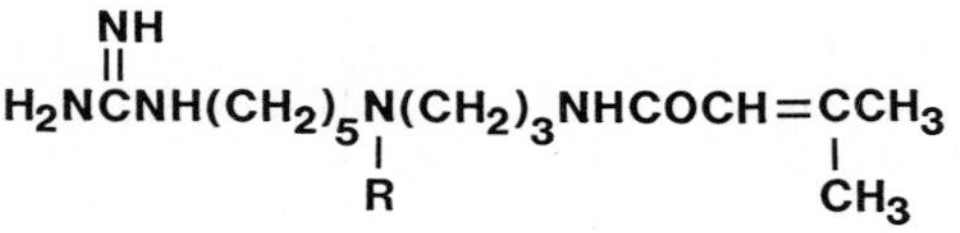

$$\underset{\text{R}}{\overset{\text{NH}}{\text{H}_2\text{NCNH(CH}_2)_5\text{N(CH}_2)_3\text{NHCOCH}=\underset{\text{CH}_3}{\text{CCH}_3}}}$$

11   R = −CO(CH$_2$)$_{10}$CH$_3$

12   R = −CO(CH$_2$)$_3$CH=CH(CH$_2$)$_5$CH$_3$ (Z)

13   R = −COC$_{13}$H$_{21}$

2.3.4 Variabilin

The sesterterpene tetronic acid variabilin (**14**) was first isolated by Faulkner in high yield from the sponge *Ircinia variabilis* [70]. Since then variabilin has been isolated from other sponges of the genus *Ircinia* [71–73] and the initial reports of its antibiotic properties have been expanded to include cytotoxicity [74]. Variabilin is also an effective antiviral agent. In a survey of New Zealand marine invertebrates [25] variabilin was found to be the biologically active component in a total of 9 species from 3 genera of the family Thorectidae. Although the extracts were usually characterized in the screen as being strongly antiviral with little or no accompanying cytotoxicity, this was not true of pure variabilin. Cytotoxic effects overwhelmed any antiviral effect in the in vitro and in vivo assays on variabilin. In the cases investigated, variabilin was the only bioactive molecule in the extract.

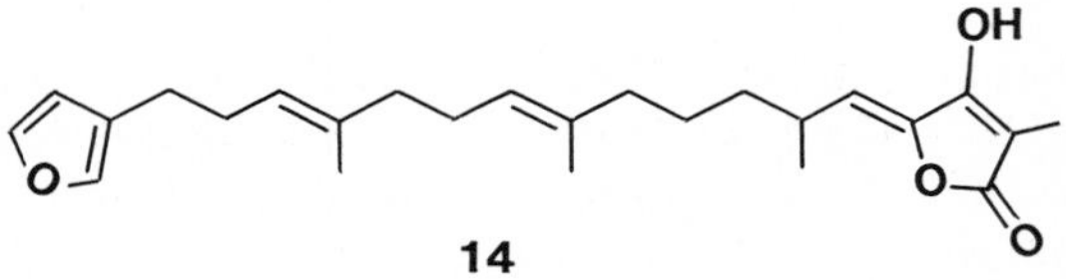

14

The possibility that the strong viral inhibition/low cytotoxicity noted for variabilin is the result of synergism between variabilin and another component in the crude extract has not been overlooked and is being explored [69].

2.3.5 Cyclohexadienones

Higa and co-workers isolated a series of four monoterpene cyclohexadienones (**15–18**) from the red alga *Desmia hornemanni* [75]. Each of these compounds was tested for antitumor activity against L1210 cells and for antiviral activity against *HSV*-1 and *VSV*. Significant cytotoxicity was observed for several of the compounds, as had been observed before for halogenated monoterpenes ([76–78] see Sect. 4), but only the synthetic acetyl derivative (**19**) was antiviral.

The finding of antiviral activity in a simple, non-polar molecule with seemingly specific structure/activity requirements is of particular interest. Halogenated monoterpenes are the major metabolites of many of the genera of the Rho-

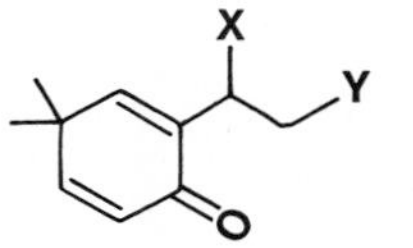

15 X = Cl , Y = OH

16 X = OH, Y = OH

17 X = Cl , Y = Br

18 X = Br , Y = Br

19 X = Cl , Y = OAc

dophyta and the literature abounds with references to the isolation of this class of compound. More extensive screening of red algae for antiviral activity is warranted.

2.3.6 Didemnins

The didemnins are the most promising compounds found from marine organisms as a result of screening for antiviral and antitumor activities. They were characterized following the 1978 "Alpha Helix Caribbean Expedition" (AHCE) organised by Rinehart [19]. The didemnins were isolated from a tunicate of the *Trididemnum* genus of the family Didemnidae, whose extracts were both cytotoxic and very strongly antiviral. The antiviral activity was detected by a shipboard screen. This assay, devised by Hughes, detected the inhibition of a *Herpes* virus in a monkey kidney continuous cell line (see Sect. 2.2) [37]. The *Trididemnum* species has now been collected from a wide variety of locations. Regardless of collection site, the extract has shown biological activity against DNA and RNA viruses and L1210 murine (mouse) leukemia cells.

By following a bioactivity-directed purification scheme the three most abundant didemnins, A (**20**), B (**21**) and C (**22**) were isolated [19, 79, 80]. Didemnin B was a minor component, and only traces of didemnin C were available. The

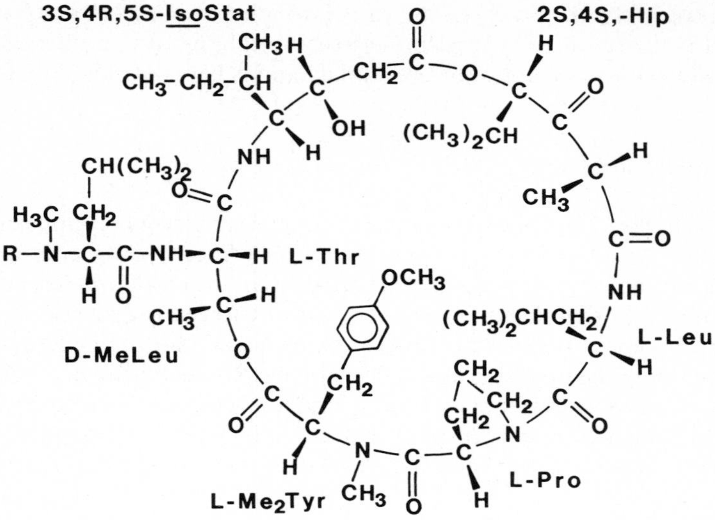

20   R=H

21   R=HO−C−CO−N−C−CO−
(with CH₃, H, CH₂, CH₂, CH₂, Lac, L-Pro labels)

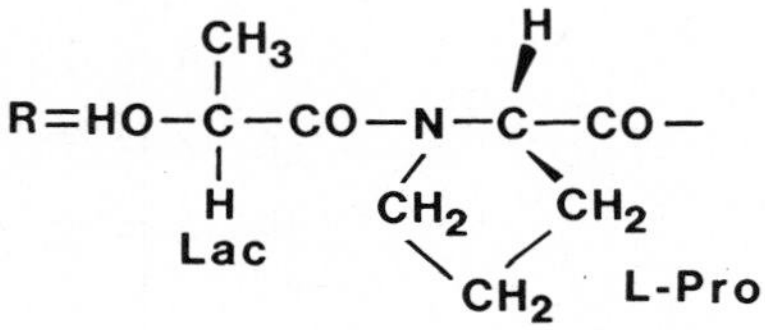

22   R=HO−C−CO
(with CH₃, H, Lac labels)

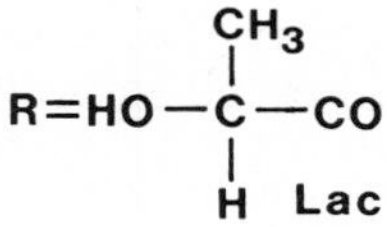

didemnins are cyclic depsipeptides (peptides condensed in part through bonds other than the peptide linkage). The constituent amino acids and their configurations were identified by a combination of mass spectrometry, gc, gc/ms and nmr on the hydrolysis product from didemnin A. The unusual structural units, hydroxyisovalerylpropionate (Hip) and the amino acid 3S,4R,5S-isostatine, were also fully characterized [82]. Sequencing of the six amino acids and the keto acid was achieved primarily by application of the various methods developed for studying peptide structures by mass spectrometry, and as such, represents a classic case study [81]. With the structure of didemnin A established, the structures of the didemnins B (**21**) and C (**22**) quickly followed.

In addition to didemnins A, B, and C, a series of nor-didemnins A–C have also been isolated. In this series, the amino acid isostatine is replaced by a nor-isostatine unit [79]. A total of 12 didemnins have now been detected [82].

Both didemnin A and B have received intensive scrutiny and their general biochemical and cellular effects have been determined. Early experiments concluded that their overall effect was inhibition of protein synthesis [83].

Although structurally similar, the didemnins vary markedly in their effects and the concentrations required for inhibition. The didemnins clearly influence many aspects of cellular metabolism, but the molecular basis of these effects has not been elucidated. Meanwhile, the didemnins are being assessed as potential antiviral, antineoplastic and immunosuppressive agents.

All extracts of the *Trididemnum* species inhibited a variety of DNA and RNA viruses in cell culture at 1 mg/ml, but they were also toxic to the host cells at this concentration [80]. Quantitative data for the pure didemnins A and B indicated good dose-response relationships against *Coxsackie* A21, equine *Rhinovirus, Parainfluenza* and *HSV*-1 and -2 [83]. This in vitro study was later extended to a broad range of RNA viruses representing families which include highly virulent human pathogens. The didemnins A and B were found to have significant activity against Rift Valley fever, Venezuelan equine encephalomyelitis virus and yellow fever virus. In all the cases studied, didemnin B was significantly more active than didemnin A. These studies also confirmed the cytotoxic properties of the didemnins [84].

Subsequent in vivo testing on mice inoculated with lethal vaginal doses of *HSV*-2 showed that intravaginally applied didemnins A and B could protect over 70% of the mice [83]. In another test, the severity of herpes lesions in cutaneous infections of mice with *HSV*-1 virus was significantly decreased by daily topical application. There were limitations, however, in that it was necessary to start the applications 2 days prior to infection. No decrease at all in the lesions was observed if treatment was delayed until 1 hour after infection. An added complication was the skin irritation observed even with the application of low concentrations of the didemnins. No activity was observed if they were injected intraperitoneally [85]. Antiviral activity of the didemnins has also been tested in vivo with two other viruses. The didemnins offered protection to mice challenged with a lethal dose of Rift Valley Fever, but some drug-related deaths were noted [84]. In the case of infection with Semliki Forest virus, neither didemnin showed any activity [85].

It is quite possible that the therapeutic potential of the didemnins as antiviral drugs is limited. The observation that there was little significant difference in the ED_{50}'s against a range of RNA viruses in cell culture and protein synthesis (see above) suggests that the antiviral activity of the didemnins is being mediated through their inhibitory effect on host protein synthesis. Consequently, they are not selectively antiviral and affect the normal host-cell and the virus-infected cell to the same degree [86].

The antitumor and immunosuppressive properties of the didemnins are reviewed in Antitumor Compounds (see Sect. 3.2.2).

2.3.7 Eudistomins

The most active antiviral species detected during Rinehart's 1978 Alpha Helix Caribbean expedition was the colonial tunicate *Eudistoma olivaceum* [19]. The tunicate was collected from a variety of sites from Belize to Florida. The crude extracts from all the samples inhibited plaque formation by *HSV*-1 virus in mon-

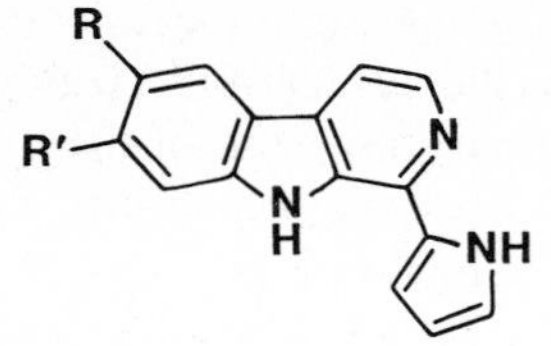

		R	R'
23 (A)	:	OH	Br
24 (B)	:	OH	Br (+C_2H_6O)
35 (M)	:	OH	H

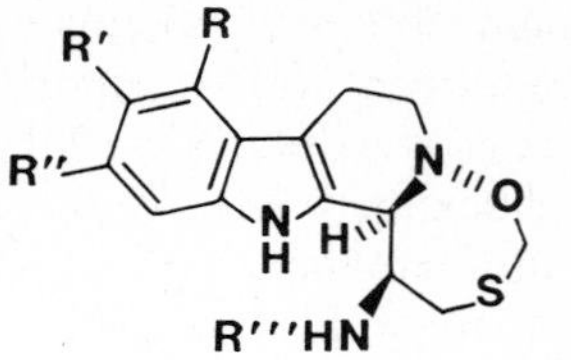

		R	R'	R''	R'''
25 (C)	:	H	OH	Br	H
27 (E)	:	Br	OH	H	H
28 (F)	:	H	OH	Br	C_2H_3O
33 (K)	:	H	H	Br	H
34 (L)	:	H	Br	H	H

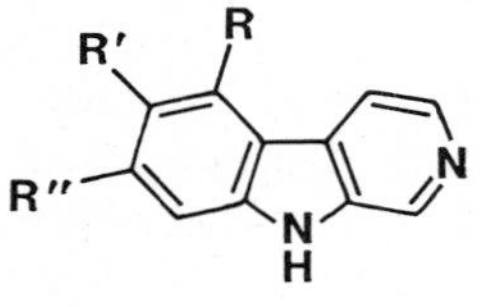

		R	R'	R''
26 (D)	:	Br	OH	H
32 (J)	:	H	OH	Br
36 (N)	:	H	Br	H
37 (O)	:	H	H	Br

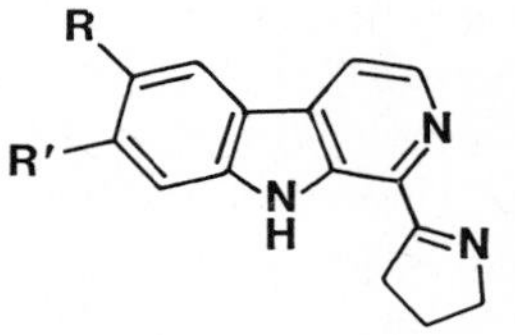

		R	R'
29 (G)	:	H	Br
30 (H)	:	Br	H
31 (I)	:	H	H
38 (P)	:	OH	Br
39 (Q)	:	OH	H

key kidney cells (CV-1) with little cytotoxicity. Partitioning of the extract followed by extensive reverse phase and silica gel chromatography gave the eudistomins A–Q (23–39) as yellow oils [87–89]. The eudistomins are all β-carboline derivatives and fall into four distinct structural categories – Type I unsubstituted (26, 32, 36, 37), Type II, 1-pyrrol-2-yl derivatives (23, 24, 35), Type III, 1-pyrrolin-2-yl derivatives (29–31, 38, 39) and Type IV, 1,2,3,4-tetrahydro-β-carbolines with an oxathiazepine ring fused at C-1 and N-2 (25, 27, 28, 33, 34). The structures of the simpler eudistomins were based on nmr spectroscopy, mass spectroscopy and the elimination of alternative structural possibilities by synthesis. Synthesis of selected examples of the simpler eudistomins (Types I to III) confirmed the assigned structures [89]. The stereochemistry of the Type IV eudistomins was assigned by difference noe spectroscopy [90].

The isolated eudistomins were assayed against *HSV*-1. Each showed antiviral activity, confirming that this was the structural type responsible for the antiviral activity observed in the shipboard assays. However, the level of activity varied markedly. By far the most active were those containing the oxathiazepine ring, Type IV, with activities in the test system as low as 0.005 µg per disc.

As a class, the oxathiazepine containing eudistomins, (Type IV), were very much more active than Types III and I, and these were more active than Type II. The nature and substitution pattern on the benzene ring also influences the antiviral activity. For example, in the oxathiazepine (Type IV) series: E (**27**) (5-Br, 6-OH) > C (**25**) (6-OH, 7-Br) > L (**34**) (6-Br) = K (**33**) (7-Br). In the Type III series: P (**38**) (6-OH, 7-Br) = H (**30**) (6-Br) > G (**29**) (7-Br) ~ Q (**39**) (6-OH) ~ I (**31**) (no substitution).

In addition to the potent activity towards the DNA virus *HSV*-1, the antiviral eudistomins C (**25**) and E (**27**) were also found to be active against *HSV*-2, the *Vaccinia* virus and RNA viruses such as *Coxsackie* A-21 and equine *Rhinovirus*. Interestingly, the activity of C (**25**) is reduced 100-fold if the phenol and amine groups are acetylated. Since the very strong antiviral activity in this series appears to be associated with the oxathiazepine ring and is only modestly influenced by substituents on the benzene ring, the acetylation of the amine function at C-10 presumably has the greater effect on the antiviral activity.

Biosynthetically, all the eudistomins are considered to be derived from tryptophan. The pyrrole and pyrrolinyl ring in Types II and IV are presumed to be of glutamate origin, while the oxathiazepine ring system of Type IV is most reasonably derived from cysteine.

No synthesis has yet been reported for the oxathiazepine eudistomins, but attention has already focused on synthons such as N(2)-hydroxy-1,2,3,4-tetrahydro-β-carbolines [91]. The synthesis of this new ring system presents a distinct challenge, but the challenge will have to be accepted if the full range and type of antiviral activity of the oxathiazepine eudistomins is to be exploited.

3. Antitumor Compounds

3.1 Terminology

In the literature describing naturally occurring growth inhibitory compounds the terms in vitro and in vivo activity are often used loosely, or even incorrectly. Many compounds have been cited as anticancer or antitumor agents, when they are only cytotoxic to tumor cells in vitro. Some broad-spectrum poisons, with no selectivity toward tumor cells and no hope of being useful anticancer agents, are included in this category of compounds. In order to eliminate confusion in the literature the National Cancer Institute (NCI) has defined these terms precisely [92]. **Cytotoxicity** should only refer to toxicity to (tumor) cells in culture. Terms such as antitumor, anticancer, or antineoplastic should **not** be used in referring to in vitro results. For in vivo activity in experimental systems the terms **antitumor** or **antineoplastic** should be used, and the term **anticancer** is reserved for reporting data from clinical trials in humans.

3.2 Screening Methods

The NCI has been the largest and most effective organization to develop a screening program for cytotoxic compounds, and has developed protocols for many procedures in the selection process [93]. The objective of a screening program is to retain extracts having biological activities, which **may** include antitumor activity, and to discard those extracts which appear to have no possibility of containing active compounds [94]. The NCI recommends that each fractionation step in the search for potential antineoplastic agents from natural products be guided by a bioassay. Some of the assays most commonly used in the program will be briefly mentioned here.

The simplest, fastest, most sensitive and least expensive prescreens are in vitro models. An example of an in vitro test system which has been frequently used as a bioassay to guide natural products isolations is the KB cell culture derived from a human epidermoid carcinoma of the mouth. Evaluation is based on the effective dose which inhibits cell growth to 50 percent of the control growth ($ED_{50} = ID_{50}$). Usually the ED_{50} values are expressed in the concentration unit of µg/ml. Natural products extracts with $ED_{50} \leq 30$ µg/ml in the first test, with an average ≤ 20 µg/ml in the first two tests, are considered active. For pure compounds an ED_{50} value ≤ 10 µg/ml is satisfactory to warrant further testing [95].

The present approach of the NCI is to use in vivo tumor models for the screening step. The P388 leukemia system in mice (PS) is used first as a prescreen to discard most negative compounds, but not **any** active compounds. The P388 assay was selected because it is sensitive to over 95 percent of the known clinically effective anticancer agents, yet does not give an unmanageable number of positives for subsequent screening. Once a compound or extract is selected for further testing, it is subjected to a screening panel which consists of four models, the L1210 lymphoid leukemia (LE), (which is less sensitive than the P388 leukemia), B16 melanoma, M5076 sarcoma, and the MX-1 human mammary tumor. For the P388, L1210, B16 and M5076 assays the results are expressed in percent as a ratio of the mean survival time of the test group, compared to the mean survival time of the control group (T/C). For the P388 test an increase in life span of 20% (T/C 120%) or greater is the minimum criterion for a compound or extract to be considered to have a statistically significant level of activity. For each of the other tests the statistically significant level is fixed at T/C $\geq 125\%$. Compounds with a T/C of 150% or greater in any of these assays have a biologically important activity, and become candidates for development to clinical trials [92, 94, 95].

The MX-1 human mammary tumor assay mentioned above is carried out as a xenograft in athymic mice. This test system was introduced in the NCI tumor panel in 1977, and has been responsible for the selection of several compounds which are now in development towards clinical trials, even though they were inactive in five murine tumors in the same screening panel. Activity criteria are expressed in percent as the ratio of mean tumor weight of excised tumors from the test group, compared with the mean tumor weight of the control group (T/C). A T/C value of $\leq 20\%$ is considered significant, while T/C values of $\leq 10\%$ are biologically important [92].

It may not always be practicable to make use of the facilities offered by the NCI. Fortunately, there are alternative approaches to screening crude samples or monitoring separations. An example of an alternate cell system is the sea urchin embryo. The freshly fertilized egg is a cell which normally undergoes cleavage rapidly. The effects of a test compound on this process can be monitored easily [96]. In this assay the extract is added within 5 minutes of fertilization. In controls the first cleavage of the egg occurs after 2 hours. If 80 to 100% inhibition of cleavage occurs at a concentration of 16 µg/ml or less, a substance is considered to be active [74]. The advantages of this method are that it is rapidly carried out and does not require special cell cultures, methods, or equipment.

Two groups have investigated the significance and mechanism of "activity" in the fertilized sea urchin egg assay [74, 97]. Although more work is needed to define the mechanism of action of the substances which inhibit the mitotic cell division, the sea urchin embryos, like cancer cells, do appear to possess selective drug sensitivity. The sea urchin bioassay appears to be a reasonable prescreen to determine which substances merit further evaluation for antineoplastic ability. Compounds which are found to be cytotoxic by the sea urchin egg assay should be further tested using in vivo methods.

Two other assay systems have been used to screen for biological activity. These are the crown-gall potato disc and the brine shrimp bioassays. Crown-gall is a neoplastic disease of plants induced by *Agrobacterium tumefaciens* and once initiated, the tumor tissue grows autonomously, independent of the normal control mechanisms of the plant. Such a plant tumor system has a number of advantages over animal models or tissue culture methods. Studies have indicated that this assay is fairly accurate in predicting in vivo P388 activity. There are some false positives, but few false negatives. The assay will not replace the in vivo P388 assay, but does allow the rapid screening, in house, of a large number of extracts or fractions without the requirement of expensive equipment or highly skilled personnel. The brine shrimp assay is a simple lethality test and is not specific for antitumor activity. However, it is a good indicator of cytotoxicity and is ideal for the rapid evaluation of fractionation procedures [98–101].

3.2.1 Compounds with Marginal Antineoplastic Properties

The marine natural products reviewed in this section have shown only marginal antineoplastic properties.

3.2.1.1 Crassin Acetate

Crassin acetate (**40**), first isolated by Ciereszko in 1960 [102, 103], is a major metabolite of several Caribbean gorgonians of the genus *Pseudoplexaura* [104] and is a minor component of *Eunicea calyculata* [105]. Crassin acetate is the most intensively investigated member of a series of cembranolides which are toxic to various organisms. The toxicity of crassin acetate has been demonstrated for protozoans [106], larvae of nudibranchs [107], and parrot fish [108]. It inhibited cell

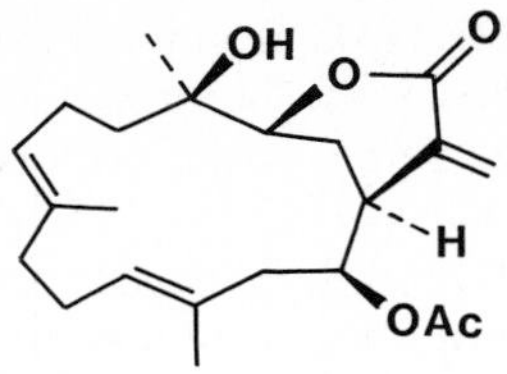

40

division of fertilized sea urchin eggs at a concentration of 16 µg/ml [74, 108]. Cytotoxicity was observed in vitro in both the KB and L1210 cell lines (ED_{50} 2 and 0.2 µg/ml respectively, [19, 104]. When tested for antineoplastic properties using the P388 cell line in vivo, crassin acetate had marginal activity (T/C 130 at 50 mg/kg) [104].

The structure of crassin acetate contains the α-methylene lactone moiety which has been found in numerous cytotoxic terpenoids of terrestrial origin [109]. Kupchan and coworkers have presented evidence that suggests tumor inhibition by α-methylene lactones results from reaction with the sulfhydryl groups of enzymes [110, 111].

Because of the relatively low level of antitumor activity of crassin acetate in the in vivo tests in mice, further evaluation in mammalian systems was not warranted. As is the case with many other toxic compounds, crassin acetate and other cembranolides are not selective enough to be useful as anticancer agents [92].

3.2.1.2 Prenylhydroquinones

The first examples of hemiterpenoid derivatives from the marine environment were isolated by Howard from the Californian colonial tunicate *Aplidium californicum*. These were prenylhydroquinone (**41**), 6-hydroxy-2,2-dimethylchromene (**42**) and prenylquinone (**43**) [112].

Earlier reports of the testing of prenylhydroquinones such as geranylhydroquinone had shown that in test animals the pre-administration of such compounds prevented the induction of some forms of leukemia, Rous sarcoma, and mammary carcinoma [113]. Furthermore, these compounds had low levels of toxicity, were well-tolerated by the test animals and allowed the prolonged administration of large doses [114]. On the basis of the potential cancer protective properties of these compounds, prenylhydroquinone was tested in vivo against the P388 cell line giving a life extension of 38% at a dose rate of 3.12 mg/kg. In addition to providing protection against tumors, prenylhydroquinone also gave protection against mutagens. In a modified Ames assay, both prenylquinone and

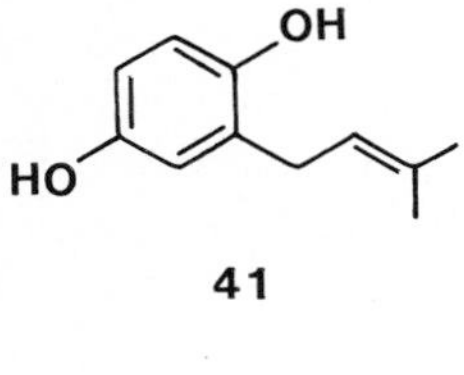

41 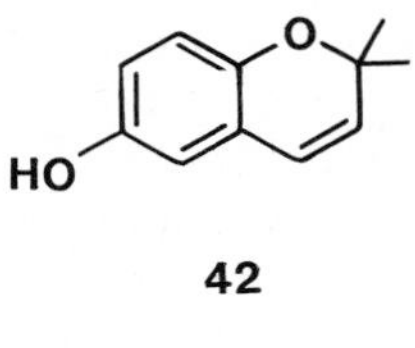   42 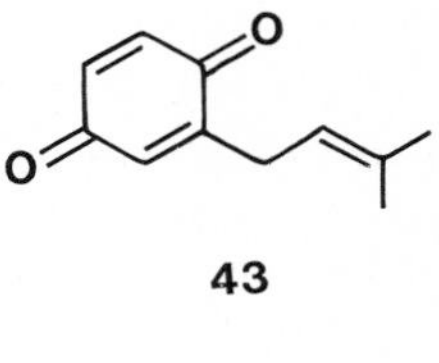   43

the chromene derivative (42) drastically reduced the effects of the carcinogens benzo(a)pyrene and aflatoxin B1 as potent mutagens against *Salmonella typhi-murium*. The mutagenic effects of ultraviolet light were also dramatically reduced in the presence of prenylhydroquinone.

Terpene hydroquinones have been isolated from a variety of marine organisms such as algae, ascidians, octocorals, and sponges [115–117]. Examples of terpenoid hydroquinones and quinones which are cytotoxic will be discussed later (Sect. 4.1).

3.2.1.3 Debromoaplysiatoxin

Kato and Scheuer were the first to characterize debromoaplysiatoxin (44) [118–120], but the absolute stereochemistry was established somewhat later [121]. Although initially it was found in the digestive gland of the sea hare *Stylocheilus longicauda*, the ultimate source is the blue-green alga *Lyngbia majuscula*, upon which the sea hare feeds [122, 123]. A mixture of two other cyanophytes *Schizothrix calcicola* and *Oscillatoria nigroviridis* provided an additional source of debromoaplysiatoxin [124]. Interest in the blue-green algae was stimulated by the discovery of antileukemic activity in lipophilic extracts. In vivo testing of mice

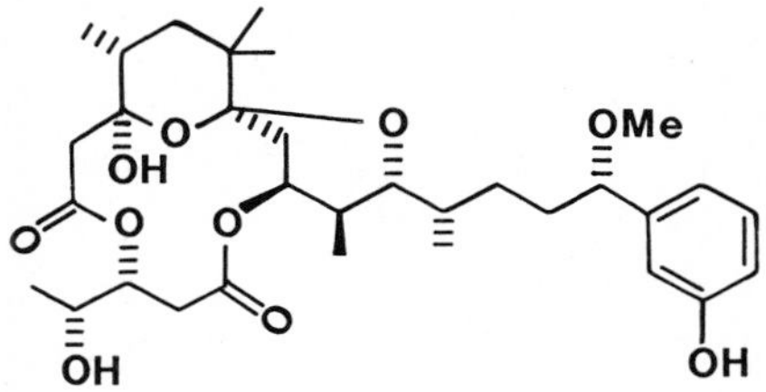

44

with P388 gave a life extension of 67% at 1.5 µg/mouse, but this was also the LD$_{50}$ (lethal dose for 50% of the mice) for the purified compound in mice [122]. Since the highest life extension was obtained at or near the chronic toxicity level, debromoaplysiatoxin cannot be considered an effective antitumor agent. The potent toxin (minimum lethal dose 0.2 mg/kg) is inactive in the B16 melanoma and Lewis lung test systems.

Debromoaplysiatoxin is, however, a tumor promoter [123, 125]. Debromoaplysiatoxin (44) and its brominated analog, aplysiatoxin, are the causative agents of a severe contact dermatitis that sometimes affects swimmers during the summer months in Hawaii [121].

3.2.1.4 Palytoxin

Palytoxin (45), the most lethal non-proteinaceous marine toxin, was first isolated from the zooanthid *Palythoa toxica* by Moore and Scheuer [126]. Other *Palythoa* spp. such as *P. tuberculosa* and *P. mammilosa* are also sources of palytoxin [127]. The concentration of palytoxin is seasonal, and evidence has been obtained that

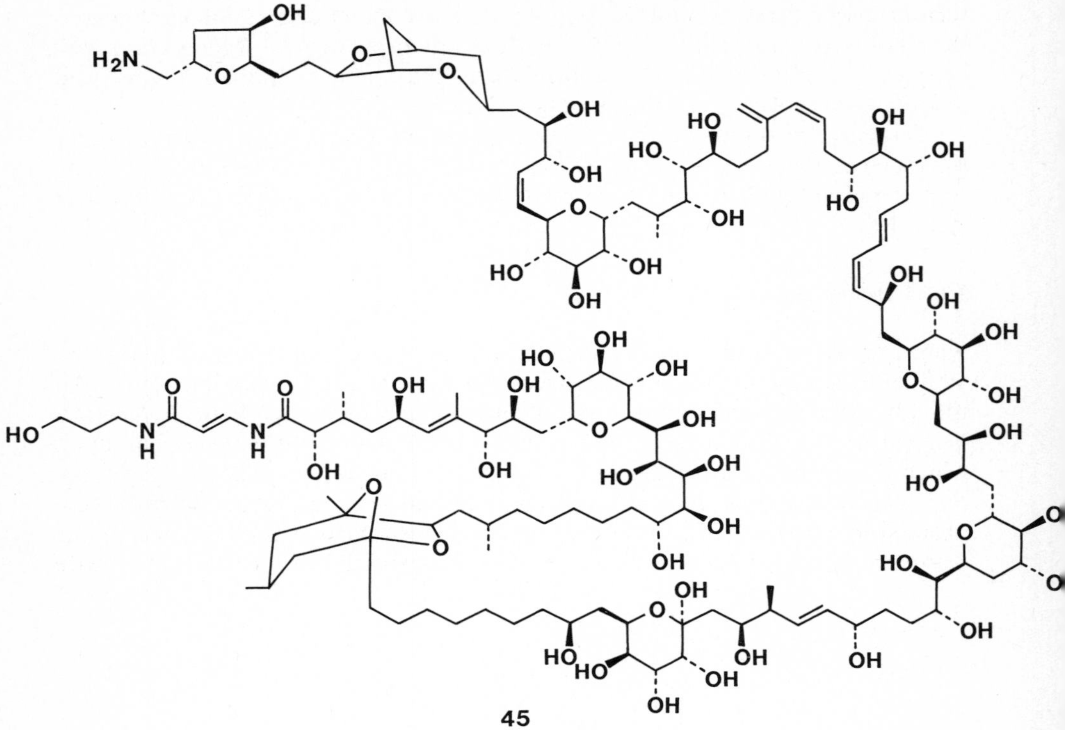

45

it is produced by a symbiotic bacterium [125]. The groups of Moore and Hirata independently determined the skeletal structure using chemical and spectroscopic methods [128, 129]. The stereochemistry at the sixty-four chiral centers was determined by spectroscopic studies [130], and by synthesis of the major degradation fragments by Kishi and coworkers [131–134].

Palytoxin is 25 times more toxic to mice than tetrodotoxin, with an LD_{50} of 0.15 μg/kg by intravenous injection. This toxicity probably arises from its effect on the cardiovascular system [127]. At levels of $^1/_{10}$ the minimum lethal dose, palytoxin completely cured Ehrlich ascites tumors in mice, but the in vivo activity against P388 leukemia was only marginal. The best figure obtained was only a 32% life-extension, and does not warrant future work with this system [135].

3.2.2 Compounds with Strong In vivo Activity

The compounds reviewed in this section have all shown strong in vivo activity, which in most cases is still the subject of investigation.

3.2.2.1 Marine Biopolymers

Several marine biopolymers with antitumor properties have been characterized partially. Recently, Shimizu and Kamiya have surveyed the literature of these

Table 1. Selected data of antitumor activity[a] of macromolecular fractions from marine invertebrates

Species[b]	MW $\times 10^3$		Dose (mg/mouse $\times$ days)	Tumor inhibition ratio (%)	Complete regression
1. *Ecteinascidia turbinata*	10		10×3	92.5	2/3[c]
2. *Holocynthia hilgendorfi*	10		10×3	74.4	0/5
3. *Styela plicata*	10		10×3	60.7	0/6
4. *Anthocidaris* sp.	300	50	10×3	77.0	1/4
5. *Strongylocentrotus drobachiensis*	300		10×3	72.9	1/6[c]
6. *Mercenaria mercenaria*	300		10×3	61.4	0/5
7. *Mercenaria mercenaria*	300		10×3	85.5	1/4
8. *Mercenaria mercenaria*	300	100	10×3	75.5	0/4[c]
9. *Placopecten magellanicus*	300	10	5×2	81.4	1/5[c]
10. *Aequipecten irradians*	300	10	10×1	76.1	2/4[c]
11. *Crassostrea virginica*	300	100	10×3	82.3	0/4[c]
12. *Mya truncata*	300	100	10×3	77.4	0/3[c]
13. *Nordotis* sp.	50		10×3	79.0	0/6

[a] Antitumor activity was determined using Sarcoma solid form in ICR mice.
[b] Common name 1. Sea squirt, 2. Tunicate, 3. Tunicate, 4. Purple sea urchin, 5. Green sea urchin, 6. Quahog, 7. Little neck clam, 8. Cherry stone clam, 9. Sea scallop, 10. Bay scallop, 11. Oyster, 12. Steamer clam, 13. Abalone.
[c] These fractions were toxic.

substances, including sulfated polysaccharides, the sphyrnastatins, strongylo-statin 1, halitoxin, and mercenene, and these will not be covered here [55]. Since that review, Pettit has discovered strongylostatin 2 [136, 137], geodiastatins 1 and 2 [138], lytechinastatin [139], and several palystatins [140]. None of these more recently reported substances has shown antineoplastic activity sufficiently promising to warrant further work [141]. Schmitz obtained a partial structure for hali-toxin [142, 143], which is inactive in vivo against the Lewis Lung carcinoma and B16 melanoma tumor systems [144]. Work on the antitumor polysaccharide extracts from marine algae continues [145–147].

Shimizu has observed favorable antitumor activity against Sarcoma 180 for a series of thirteen purified fractions from marine invertebrates, and these results are summarized in Table 1 [148]. The most striking antineoplastic activity was obtained from a fraction (MW 10,000) isolated from the tunicate *Ecteinascidia turbinata*. Two out of the three mice tested were cured and an average life extension of 92.5% was obtained. Progress on the isolation of other biologically active components from *E. turbinata* has also been made by other groups (see below).

3.2.2.2 *Ecteinascidea turbinata*

The remarkable antineoplastic properties of the ethanol-water extracts of the Caribbean tunicate *Ecteinascidea turbinata* were reported first in 1970 [3]. This crude extract was active against KB cells and gave impressive results in vivo against P388. The T/C values ranged up to 230% and a number of cures were observed (survival for >30 days when mean survival is 10 days) [149, 150]. Regardless of geographic location, *E. turbinata* specimens all possessed similar antineoplastic properties, indicating that the active principle was intrinsic to the species [150].

As well as having antineoplastic properties, the extract also acted as a powerful modulator of immune responses. The immunosuppressive components are of higher molecular weight [55], and may well be similar to the higher molecular weight antineoplastic fractions recently defined by Shimizu [148] (see above). It has been the immunosuppressive properties of the ethanol/water extract that have been most extensively studied over the years [150, 151]. The extract was also reported to be a potent inhibitor of DNA synthesis in various human cell lines [152, 153].

The potential antineoplastic properties of *E. turbinata* were also detected by Rinehart's expedition to the Caribbean in 1978 [19]. The Illinois group now reports the isolation of 5 antineoplastic compounds from the extract. The low concentration and instability of these compounds presented distinct problems. The isolation process was assisted by the development of a bioautographic tlc technique using CV-1 cells in culture to detect the active constituents. The five components had T/C values in the in vivo P388 model of up to 250% [82]. It is likely that these are the lower molecular weight antineoplastic components of the *E. turbinata* extracts [55].

Until more is known about these agents any comment on their mode of action must be speculative. It is interesting to note, however, that the crude extract could stimulate killer functions in macrophages from peritoneal exudate cells [150] (see lemnalol).

3.2.2.3 Lemnalol

Lemnalol (**46**), from the soft coral *Lemnalia tenuis* [154], was the first reported ylangene sesquiterpenoid from a marine organism, and the first oxygenated ylangene from any source.

The antineoplastic activity of lemnalol was discovered as a result of a routine screening of natural products from Japanese marine organisms. An in vitro assay system that detects the activation of murine peritoneal exudate cells (PEC) against a DBA/MC fibrosarcoma cell line was used. Kikuchi observed that although lemnalol alone was not cytotoxic at a concentration of 10–40 µg/ml, in the presence of PEC it activated macrophages, resulting in a selective cytotoxicity toward the tumor cells. In this respect, lemnalol is similar to other agents, such

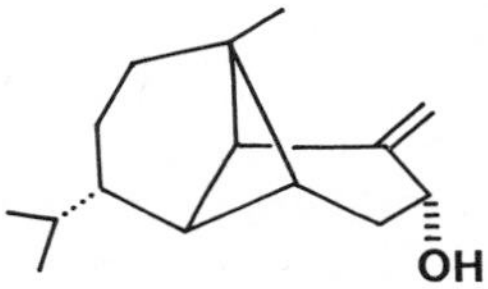

46

as the synthetic nucleosides poly(I)–poly(C), which are also known to activate macrophages in vitro [155]. In contrast, the epimeric alcohol was inactive in the assay, and the α,β-unsaturated ketone was strongly cytotoxic, even in the absence of PEC [156]. This was the first reported example of a marine-derived compound which could activate PEC to exhibit cytotoxicity in a tumor cell line in vitro. By examining the effect of lemnalol on the size of in vivo murine tumors induced by the same cell line, Kikuchi determined that lemnalol was antineoplastic at a concentration of 400 µg/site.

The utilization of the peritoneal exudate cells (PEC) in the in vitro testing for cytotoxicity is an interesting alternative to the use of the usual tests, such as the P388, because it screens for compounds acting through the immune system (biological response modifiers) instead of direct acting agents (cytotoxic compounds). Had only the usual tests been applied in a prescreen to determine which compounds should be tested for in vivo activity, lemnalol (which is not cytotoxic to tumor cells itself) would have been by-passed. Biological response modifiers will probably be most useful in those cancer cases in which the immune system is not suppressed. In advanced cancer the immune system usually functions only minimally or not at all, and immunostimulants generally do not have a large therapeutic effect. Chemotherapy with cytotoxic drugs, perhaps combined with surgery or radiation, is generally more effective in such cases [94]. It is likely that both biological response modifiers and cytotoxic compounds will be valuable because of the ir complementary roles.

3.2.2.4 Stypoldione

Stypoldione (**47**) is a red crystalline diterpenoid *ortho*-quinone obtained from the Caribbean brown seaweed *Stypopodium zonale* [157]. It was found as part of a study of mechanisms of chemical defense against foraging in the marine environment [158]. Stypoldione is ichthyotoxic, and is a potent inhibitor of division of sea urchin eggs (ED_{50} 1.1 µg/ml; ED_{100} 8 µg/ml) [74]. It inhibits microtubule polymerization in vitro by a unique mechanism which is different from the action of the "mitotic spindle" poisons, such as the vinca alkaloid anticancer agents and colchicine [159].

In spite of the fact that stypoldione is toxic to fish, it appears to be relatively non-toxic to mice. Preliminary in vivo studies of the effects on Ehrlich ascites tumor cells were positive at a dose rate of 100 mg/kg, with no sign of acute toxicity in mice within a 24-hour period. In the P388 system a 42% increase in survival time was observed. This level of antitumor effectiveness was considered sufficient

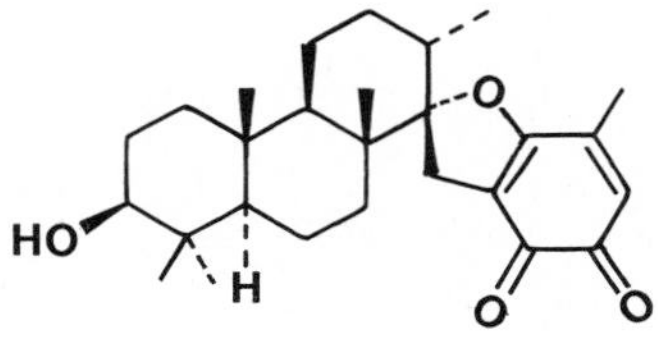

47

to warrant further investigation. While these are only preliminary studies, they do indicate that stypoldione may be effective in inhibiting the growth of dividing cells. Although the dosage level is high, when compared to the microtubule inhibitor vincristine, the therapeutic index (therapeutic/toxic dose ratio) is good [160].

3.2.2.5 Clavulones, Punaglandins and Chlorovulones

Promising antitumor activity has been shown by a number of marine prostaglandins isolated from octocorals. Kikuchi, Yamada, and co-workers were the first to report the isolation of the geometric isomers clavulone I, II, and III (48–50) from the Japanese soft coral *Clavularia viridis* [161, 162]. In the following issue of Tetrahedron Letters, Kitagawa and co-workers reported the same compounds, named claviridenone-d, -c, and -b, respectively, and a fourth geometric isomer claviridenone-a (51) [163, 164]. The duplication of names for these compounds has become confusing, and we recommend that the names claviridenone-b, -c, and -d be retired. Similarly, we recommend that the previously reported names 20-acetoxyclavulones III and II [165] be used instead of 2-acetoxy-claviridenones-b and -c [166]. Claviridenone-a and clavulones I–III were strongly cytotoxic, with growth inhibition activity against L1210 leukemia cultured cells at ED_{50} 0.2–0.4 μg/ml [166].

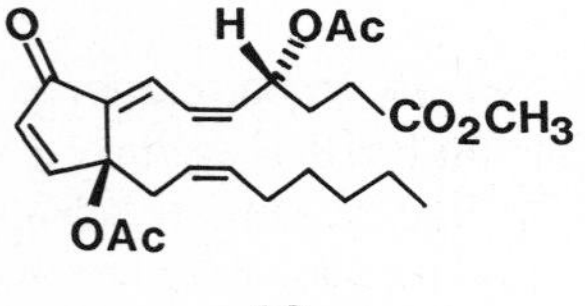

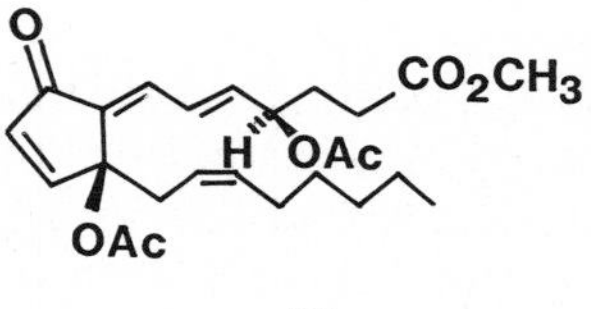

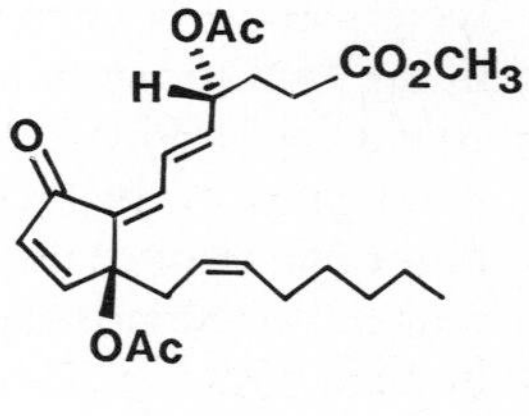

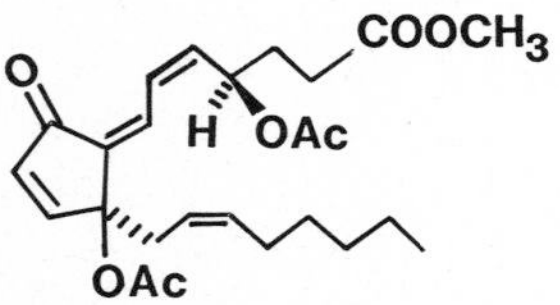

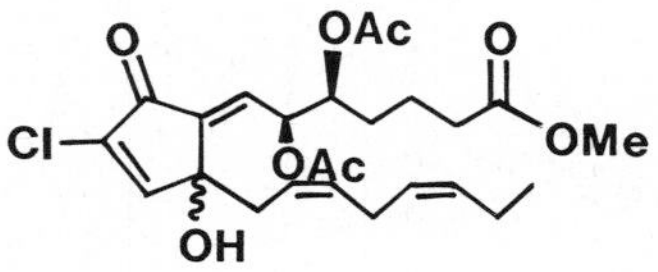

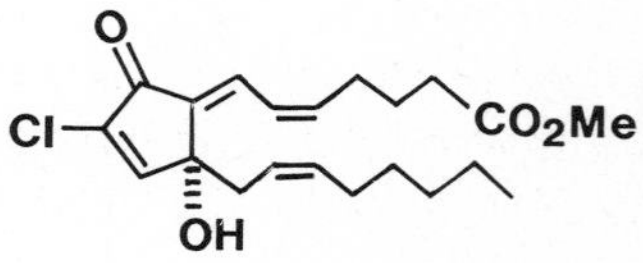

Scheuer and co-workers obtained a series of even more potent eicosanoids, the punaglandins, from the Hawaiian octocoral *Telesto riisei*. This organism lacks symbiotic algae which suggests that the octocoral is the ultimate source of these prostaglandins [167]. The punaglandins are characterized by the C-10 chlorine and C-12 hydroxyl functions. One of them, punaglandin-3 (52), inhibited L1210 leukemia cell proliferation with an ED_{50} value of 0.02 μg/ml, which represented a 15-fold greater activity than was shown by the corresponding clavulone. This remarkable cytotoxicity is approximately equal to that of vincristine and doxorubicin, some of the most effective anticancer agents now in use. In vivo studies were carried out using mice with Ehrlich ascites tumors. Treatment for five consecutive days with 20 mg/kg/day of clavulone II (49) or 5 mg/kg/day of punaglandin-3 (52) resulted in a 60% increase in life span (ILS) or 80% ILS respectively. Both produced long term survivors [168]. The antitumor activity of these marine eicosanoids appears to be remarkably potentiated by the C-10 chloro and C-12 hydroxy substituents.

Both protein synthesis and DNA synthesis have been shown to be affected profoundly by the clavulones and the punaglandins [169].

An additional series of C-10 chloro-eicosanoids was obtained by investigating clavulone congeners from *C. viridis* [170, 171]. Chlorovulone I (53) showed strong antiproliferative activity in human promyelocytic leukemia (HL-60) cells in vitro (ED_{50} 0.01 µg/ml), which was about 13-times stronger than that of clavulone I [172]. A cytotoxic effect was observed in this cell line at concentrations greater than 0.1 µg/ml. At the time of writing this review, in vivo testing results had not been reported. Compounds such as the punaglandins and chlorovulones show strong potential as antitumor agents.

The total syntheses of clavulones [173–176], and chlorovulone II [171] have been reported. The latter synthesis established that the absolute configuration of chlorovulones is opposite to that of clavulones, even though the structurally related compounds co-exist in the same marine organism. The absolute configuration of the punaglandins has not been determined.

3.2.2.6 Bryostatins

In 1968 Pettit and co-workers initiated a systematic search for antineoplastic activity in marine organisms [2]. This search had a wide geographical base, and was launched as a consequence of the discovery by this group of antineoplastic compounds in terrestrial arthropods [177]. They found promising activity in alcohol extracts from several phyla, but especially from the Bryozoa, also called Ectoprocta or Polyzoa. Significant in vivo activity was obtained for extracts from the cosmopolitan bryozoan *Bugula neritina* (68–100% life extension using P388 leukemia).

As a group, the Bryozoa have not yet received a great deal of attention, but it appears that this situation is now rapidly changing. A review of the secondary metabolites from marine bryozoans has recently appeared [178].

Based on the excellent in vivo results for extracts from *B. neritina*, Pettit undertook a detailed study of this, and another bryozoan species with antineoplastic activity, *Amathia convoluta*. In the 16 years since the discovery of the in vivo activity, a total of 17 related compounds, bryostatins 1–17, have been reported. In addition, two further bryostatins, A and B, have been detected in the sponge *Lissodendoryx isodictyalis* invaded by *B. neritina* [179]. At the time of this review the structures and antineoplastic activities of bryostatins 1–13 have been determined. A feature of this work has been the sheer bulk of material necessary to obtain sufficient quantities of the various bryostatins for structural studies. Typically, the bryostatins are present at about the $10^{-6}\%$ range in the whole animal. Collections of 500 kg have been common; indeed, 1000 kg were required to

isolate 0.7 mg of bryostatin 13. The complexity of the situation has been com-
pounded somewhat by the inconsistent yields of the bryostatins and marked geographical variations in the composition of extracts from *B. neritina*. In the process
of dealing with such large quantities, the Arizona group has extended and developed various partitioning techniques for the concentration of the active components. All the separations were bioassay-directed throughout. The isolation of the
bryostatins represents a significant achievement that could only be accomplished
by a determined, long-term commitment.

54

Extraction of 500 kg of *B. neritina* from Californian waters led to sufficient
quantities of bryostatins 1, 2, and 3 for structural studies. The compounds were
isolated by solvent partitioning followed by bioassay-guided fractionation using
gel permeation and silica gel chromatography. The structures and stereochemis-
tries of the whole series are based on an X-ray crystallographic determination of
bryostatin 1 (**55**) [180]. Developments in solution-phase secondary ion mass spectroscopy for the recognition of molecular ions [181–182] have aided the characterization of the remaining identified bryostatins [183–190]. All of the bryostatins
have in common the unique bryopyran ring system (**54**). The shortest path around
the ring involves the pyran oxygens and defines a cavity surrounded by a 20-mem-
bered ring. With the exception of bryostatins 3, 10, 11, and 13, the identified bryostatins are all C-7 and C-20 derivatives of the same skeleton (**55–67**). Bryostatins 10 (**64**), 11 (**65**) and 13 (**67**) are 20-deoxy derivatives, while in bryostatin 3
(**57**) an acetylidene sidechain has been converted into a lactol.

The origin of the bryostatins is of inherent interest and has led to speculation
on various possibilities. A dietary origin is supported by the finding of the octa-
dienoate esters (bryostatins 1–3, 12) only in Californian collections, and also by
the incidence of bryostatin 8 only in *A. convoluta*. Alternatively, the finding of
most of the other characterized bryostatins from widely diverse geographical locations suggests de novo synthesis by *A. convoluta*. However, the recent discovery
of dinosterol and 23S,24R-cholestane, steroids characteristic of dinoflagellates, in
B. neritina is taken as strong evidence for a symbiotic relationship between zoox-
anthellae-type dinoflagellates and the bryozoan. The symbiont may also be the
source of the potent defensive, and/or offensive, bryostatins [188]. This question
of origin will have to await experimental confirmation by biosynthetic experiments, or chemical examination of the microorganisms found in association with

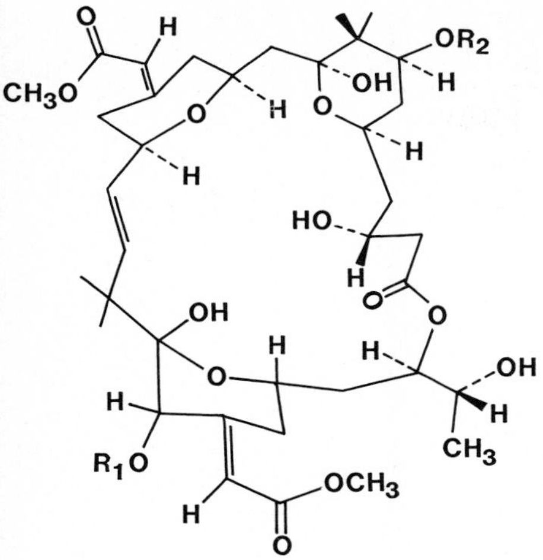

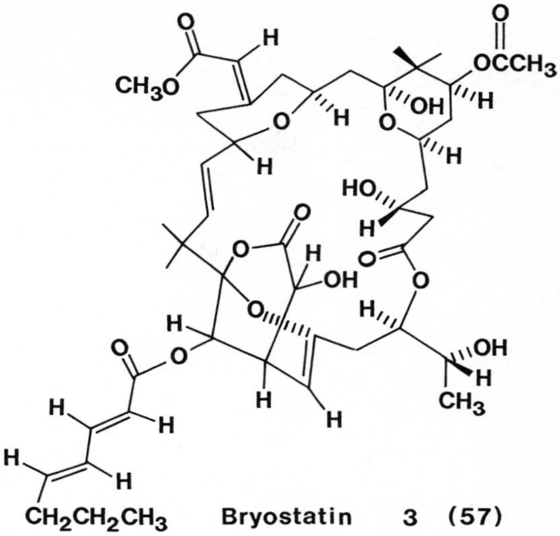

Bryostatin   3   (57)

			R_1	R_2
Bryostatin	1	(55)	2,4−octadienoate	Ac
	2	(56)	2,4−octadienoate	H
	4	(58)	butyrate	isovalerate
	5	(59)	Ac	isovalerate
	6	(60)	Ac	butyrate
	7	(61)	Ac	Ac
	8	(62)	butyrate	butyrate
	9	(63)	butyrate	Ac
	10	(64)	deoxy	isovalerate
	11	(65)	deoxy	Ac
	12	(66)	2,4−octadienoate	butyrate
	13	(67)	deoxy	butyrate

B. neritina and *A. convoluta*. Preliminary work to answer this question is under-way [191].

All of the bryostatins reported to date display impressive activity in both the in vivo and in vitro P388 screens (see Table 2) [180, 183–190] and bryostatin 1 has been the subject of successful preclinical testing by the NCI.

Examination of the bryostatins for structure/activity relationships leads to in-teresting conclusions. For example, reversal of the 7/20 ester groups as in bryo-statins 6 and 9 does not appear to significantly affect the P388 activity (Table 2). The activities of the 20-deoxy bryostatins, 10, 11, and 13, are comparable to those reported for the analogous 7,20-substituted bryostatins cf. bryostatin 10 with 4,

Table 2. Origin and P388 activity of the bryostatins

Byro- statin	Organism[a, b]	P388 Lymphocytic leukemia	
		In vivo % Life extension/ dose (μg/kg)	In vitro ED_{50} (μg/ml)
1	1A	52–96/10–70	0.89
2	1A	60/30	–
3	1A	63/30	–
4	1BCD, 2C	62/46	10^{-3}–10^{-4}
5	1BCD	88/185	1.3×10^{-3}–2.6×10^{-4}
6	1BCD, 2C	39–82/46–185	3.0×10^{-3}
7	1BCD	77/92	2.6×10^{-5}
8	1A, 2C	74/110	1.3×10^{-3}
9	1AC	64/92.5	1.2×10^{-3}
10	1BC	–	7.6×10^{-4}
11	1BC	64/92.5	1.8×10^{-5}
12	1A	47–68/30/50	1.4×10^{-2}
13	1A	–	5.4×10^{-3}

[a] 1, *Bugula neritina;* 2, *Amathia convoluta.*
[b] A, California; B, Gulf of California (Mexico); C, Gulf of Mexico (Florida); D, Gulf of Sagami (Japan).
Refs. [180, 183–190].

bryostatin 11 with 7, bryostatin 13 with 6 and 8. This essentially eliminates the need for a 20-ester function in the bryopyran system as an absolute prerequisite for antineoplastic activity, and will simplify the problems of synthesis. The recorded P388 data (Table 2) suggests that the presence of smaller groups such as acetate, or even hydrogen, at the C-7 oxygen enhances the activity.

Reports on the mode of action of the bryostatins are only just starting to appear. Although the bryostatins do not appear to have antimicrobial properties, they inhibit RNA synthesis, exhibit strong binding to protein kinases, stimulate protein phosphorylation and activate intact human polymorphonuclear leukocytes [190, 192, 193].

Although the comment has been made that the bryostatins do not appear to have a broad spectrum of activity [194], this potent series will continue to attract attention. There is considerable potential in the bryopyran skeleton for analog production, but since natural bryostatins are a rare resource, further developments in that area will have to await successful synthetic routes. Synthetic approaches to the bryostatins appear to be well under way [191, 195].

3.2.2.7 Dolastatins

The toxic properties of an Indian Ocean sea hare of the genus *Dolabella* were first described nearly 2000 years ago by Pliny the Elder [196], but the medical potential of *Dolabella auricularia* was not recognised until 1976, when the antineoplastic properties of constituents of this mollusc were discovered [197]. Pettit and his co-workers later described the isolation and preliminary characterization of a series of nine peptides present in this mollusc in very low concentration [198]. Bioassay-guided fractionation of an ethanol extract of the mollusc, provided the dolastatins 1–9. From 100 kg of wet sea hare, only about 1 mg of each of these potent cell growth inhibitory substances was obtained. Results of in vivo testing of dolastatin 1 against the P388 lymphocytic leukemia showed an 88% life extension at a dose of 11 µg/kg; against the murine B16 melanoma it provided a curative response (33%) with a dose of 11 µg/kg (T/C 240 to T/C 139 at 1.37 µg/kg). The very high life extension values and the very low dose levels required for the responses make dolastatin 1 one of the most active known antineoplastic agents from any source [141, 198].

Because of the very low yield and the amorphous nature of the solids, which precluded X-ray analysis, a structure for only one of the peptides, dolastatin 3, has thus far been proposed [199]. The extensive use of spectroscopic techniques and a series of microchemical experiments led to the proposal that *cyclo* [Pro-Leu-Val-(gln)Thz-(gly)Thz] (68) was the structure of dolastatin 3 [199]. In addition to the regular amino acids, dolastatin 3 contains two new thiazole amino acids (gly)Thz (69) and (gln)Thz (70). The configurations of the amino acids were not assigned.

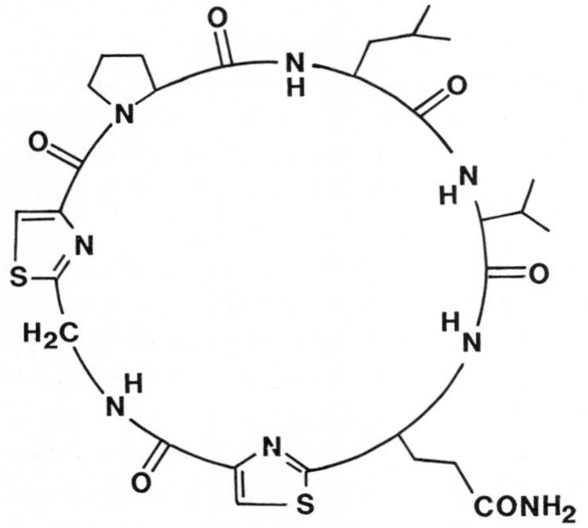

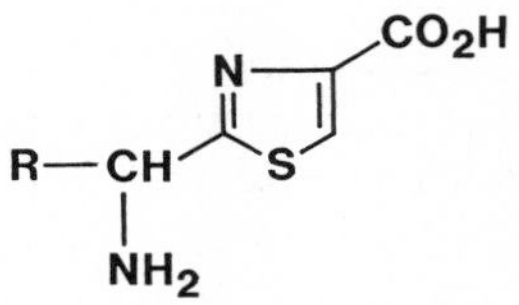

Since the natural product could be obtained only with great difficulty, final structural confirmation required comparison with a synthetic product of known configuration. Unambiguous synthesis of 68 by three independent groups proved that the proposed structure for dolastatin 3 (68), with its all-L configuration of amino acids, was untenable. By using reaction sequences that ensured high optical purity in the product, a remarkable 12 of the possible 16 stereoisomers of each of the proposed structure and reversed bonding options were synthesised [200-202]. The synthetic products differed from the natural dolastatin 3 by such criteria as melting point, optical rotation and spectral data and were inactive against L1210 leukemia cells at concentrations as high as 250 µg/ml.

While it is disappointing that none of the cyclic peptides synthesised were biologically active, efficient synthetic routes have been established. The synthesis of remaining diastereomers, which possibly include the biologically active isomer(s), will undoubtedly be reported in the near future unless the original published structure is incorrect.

Thiazole and thiazolidine containing biosynthetic products are relatively rare and are associated usually with strong biological activity (e.g. bleomycin and thiostrepton). Several cytotoxic or toxic compounds with thiazole and thiazolidine rings have been isolated from marine organisms in recent years (see Sect. 4.6). The potent Red Sea sponge toxins latrunculins A–D, which contain a thiazolidinone moiety, are also in this category [203].

The assignment of further structures in the dolastatin series will be of considerable assistance in ascertaining the importance and possible role of the thiazole unit in the development of antineoplastic agents.

3.2.2.8 Didemnins

The discovery, characterisation and discussion of the antiviral properties of the didemnins has been covered in Sect. 2.3. Although the didemnins were initially of great interest because of their antiviral properties, it was soon apparent that they were active moderators of many biochemical functions.

The discovery of the immunosuppressive properties of didemnin B (21) came as an extension of the finding of the strongly antiproliferative properties of the didemnins towards a variety of cell lines and viruses. Montgomery and Zukoski found that didemnin B had potent immunosuppressive activity in several in vitro assays of lymphocyte immune function [204]. In one aspect of the in vitro work, they noted that the potency was at a level equivalent to endogenous hormone activity [205].

When didemnin B was assayed in vivo in a murine model using the graft-versus-host reaction, it was again found to be a potent immunosuppressive agent. This is a significant observation because a large number of compounds are known to have in vitro immunosuppressive properties, but relatively few exhibit in vivo activity [205, 206]. Work continues to appear in this area [207].

Despite the other biochemical properties of the didemnins, it is as antitumor agents that they show the greatest promise. The cytotoxicity of the crude extracts against monkey kidney cells was confirmed by assays using L1210 leukemia cells. The pure didemnins A (20) and B (21) were highly cytotoxic toward L1210 cells in culture (ED_{50} 0.031 µg/ml and 0.0022 µg/ml, respectively). Their potential antitumor activity was established by in vivo tests against P388 leukemia in mice. Didemnin B provided a maximum life-extension of 100%; didemnin A was not as effective and higher dose rates were required. The efficacy of the didemnins was dependent on whether the mice were inoculated intraperitoneally or intravenously. Antitumour activity was also recorded against the B16 melanoma in vivo (maximum life extension was 57%) [83].

In more selective testing using the human tumor stem cell assay (see Sect. 5.3), two independent studies established that didemnin B was active against a variety of human neoplastic cell types [208, 209]. Significant activity was found against ovarian, breast and renal carcinomas as well as mesothelioma and sarcoma [209].

Didemnin B has been tested by the NCI (NSC 325319) in their in vivo panel of murine and human tumor lines. It was active in the P388 leukemia, M5076 sarcoma and the B16 melanoma systems. Significant biological activity, using the NCI criteria, was observed for the B16 melanoma with increases in lifespan of 50–70%. On this basis didemnin B was selected for advanced preclinical development. An adequate supply of didemnin B for the formulation, analytical, toxicological and clinical trials was obtained as a result of a 250 kg collection of the *Trididemnum* tunicate. Following the filing of an Investigational New Drug Application in June, 1984, didemnin B became the first marine natural product to enter Phase I clinical trials [94].

The didemnins A and B are both very potent agents, but have yet to be proven useful and acceptable as anticancer, immunosuppressive, or, less likely, antiviral drugs. Should they ultimately prove to be unacceptable, compounds with similar structures may fare better. Although the didemnins A and B are structurally very similar, there is a diversity in their biological properties. This particular feature is most encouraging for the future production of analogs which may be tailored to suit specific therapeutic areas.

4. Cytotoxic Compounds

Cytotoxic effects of chemical agents on cells may include altered cellular morphology, failure of the cell to attach to surfaces, or changes in the rates of cell processes such as growth, death, and disintegration [210]. Since the ultimate goal of a selection process is to find compounds which are selectively toxic to tumor cells (or cells infected by a virus), with little toxicity to normal cells, a screening procedure using renegade cells such as P388, L1210 or the KB cell lines is usually the best model for cytotoxicity studies. Nonetheless, it is sometimes more convenient to determine whether a compound is cytotoxic by observing the effects on other types of cells. Examples of such alternate cell systems are the sea urchin embryo and potato crown-gall assay (see Sect. 3.2.1).

The natural products listed below have been shown to be cytotoxic by one or more of these systems. Some of these compounds have also been tested in vivo and found to be inactive or too toxic. Most, however have only been tested in vitro, and because of their cytotoxic nature are clear candidates for further testing as potential antineoplastic agents.

The compounds reviewed have been grouped along biogenetic lines.

4.1 Cytotoxic Terpenoids

4.1.1 Monoterpenoids

Gonzalez and co-workers investigated the effects of four polyhalogenated monoterpenes from the red alga *Plocamium cartilagineum* on HeLa 229 human carci-

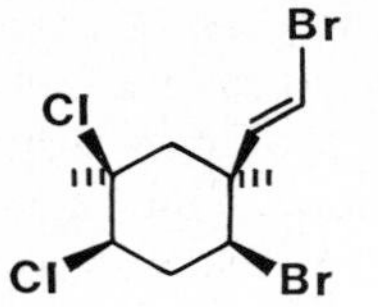

71 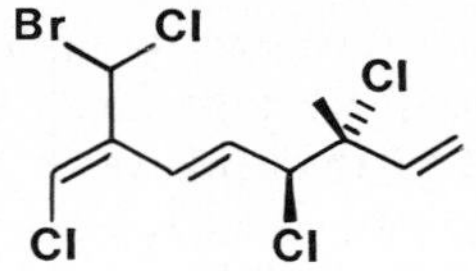   72 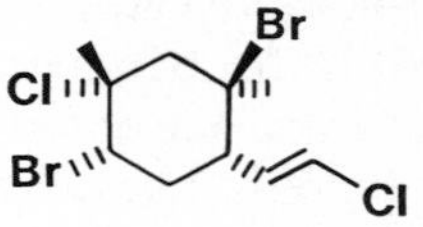   73

noma cells [77, 78, 211]. The two tetrahalogen compounds, (71) and (72), each had an ED_{50} of 1 µg/ml. Two related bromodichloro monoterpenes showed ED_{50} values of 10 µg/ml. Protocols of the NCI recommend that ED_{50} values of 1 µg/ml or less are required for a compound to be considered active in this assay [93].

Crews and co-workers reported that *Plocamium* polyhalogenated monoterpenes, such as (73), inhibit cell division in sea urchin embryos; no concentrations were reported [76].

4.1.2 Sesquiterpenoids

A series of polyhalogenated sesquiterpenes from the red alga *Laurencia obtusa* have shown cytotoxicity in various tests. Isoobtusol acetate (74) and isoobtusol (75) inhibited the growth of HeLa 229 cells with ED_{50} values of 4.5 and 10 µg/ml, respectively [77, 78], while a related compound, elatol (76) and its synthetic oxidation product elatone (77) were both active in the sea urchin egg assay. The effective dose which inhibits cleavage in 100% of the eggs (ED_{100}) is 16 µg/ml for each. Jacobs has suggested that compounds which inhibit cell division in fertilized sea urchin eggs at a concentration < 16 µg/ml are considered active in this assay [74]. Even though elatol (76) was cytotoxic to sea urchin embryos, and isoobtusol acetate (74) showed some activity against HeLa cells, both compounds were inactive in the P388 in vitro assay, with ED_{50} values of 26 and 40 µg/ml, respectively [212].

Elatol has been isolated from the Australian alga *L. elata* [213] and from Atlantic and Caribbean collections of *L. obtusa* [19, 214]. The Caribbean collection yielded the enantiomorph of the Australian and Atlantic collections. Although isoobtusol is present in *L. obtusa*, it is isoobtusol acetate that is found in the digestive gland of the sea hare *Aplysia dactylomela* [78]. Comparative biochemical studies have shown that algal metabolites are often concentrated in the digestive glands of sea hares [215, 216].

Another example of this concentration of algal metabolites by sea hares is the cytotoxic sesquiterpenoid deodactol (78). The same precursor proposed for the

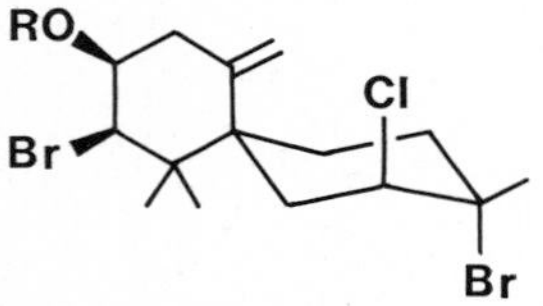

74   R = Ac

75   R = H

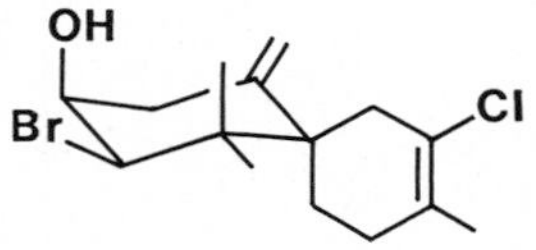

76 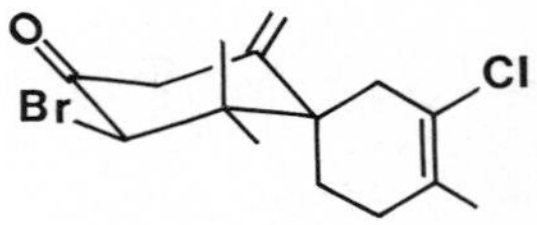   77

biosynthesis of the chamigrenes in *Laurencia* species can be used to rationalize the formation of deodactol, which has been isolated from the sea hare *Aplysia dactylomela* [217]. Schmitz initiated the investigation of this mollusc because of a report that its extracts had in vivo tumor-inhibitory activity [3]. Deodactol was moderately cytotoxic to the L1210 cell line with an ED_{50} of 12 µg/ml. In this test, active materials display an $ED_{50} \leqq 20$ µg/ml [93]. Insufficient material was available for in vivo testing. Deodactol is closely related to isocaespitol acetate (**79**), an *Aplysia dactylomela* metabolite, which was weakly cytotoxic toward HeLa cells (ED_{50} 10 µg/ml). The *Laurencia caespitosa* metabolite, isocaespitol, was much less active than its acetate in this assay (ED_{50} 50 µg/ml) [78].

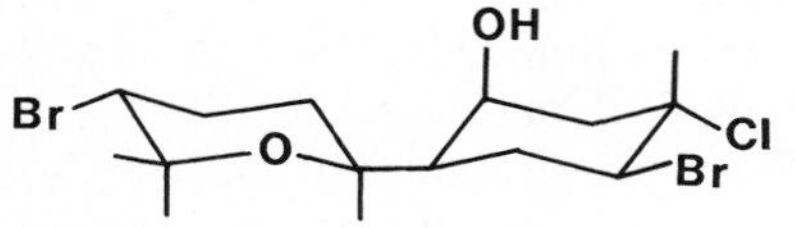

78

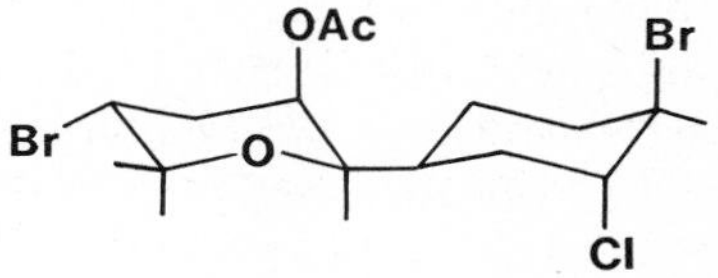

79

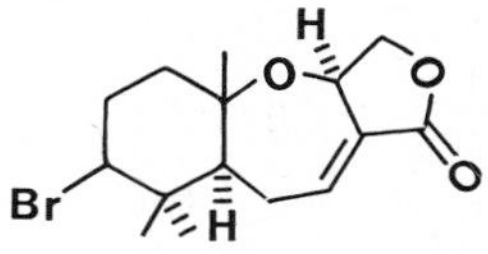

80

A related sea hare, *Aplysia angasi*, also provided a cytotoxic halogenated sesquiterpene, aplysistatin (**80**) [218]. A 2-propanol extract of this mollusc inhibited the P388 cell line both in vitro and in vivo (T/C 175% at 400 mg/kg). P388 was used as a bioassay to guide isolation of the active substance. The ED_{50} of aplysistatin was 2.7 µg/ml for the P388 and 2.4 µg/ml for the KB cell lines. Three syntheses of racemic aplysistatin [219, 220] and a chiral biomimetic synthesis [221] have been reported.

From the gorgonian *Pseudopterogorgia rigida*, a Caribbean sea plume, Fenical obtained two compounds which inhibited cell division in fertilized sea urchin eggs [222]. Curcuhydroquinone (**81**) was more active than curcuquinone (**82**), with ED_{100} values of 8 µg/ml and 16 µg/ml, respectively. Curcuphenol (**83**), also present in the extract, was inactive in the sea urchin embryo assay.

The Mediterranean sponge *Axinella cannabina* has been the source of several sets of novel sesquiterpenes with isonitrile, isothiocyanate, and/or formamide functional groups. Two of the compounds, axisonitrile-1 (**84**) [223] and axisothiocyanate-3 (**85**) [224] have shown cytotoxic activity in vitro on KB and/or P388

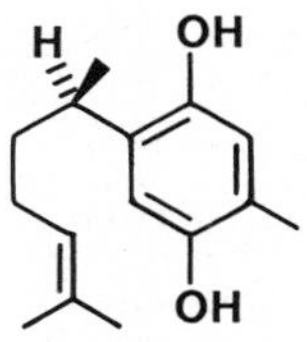

81

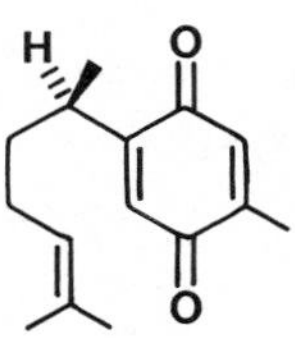

82

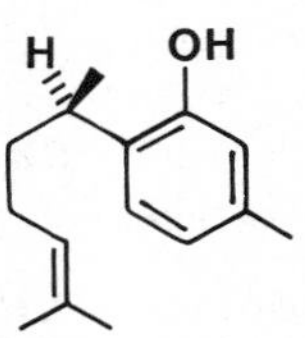

83

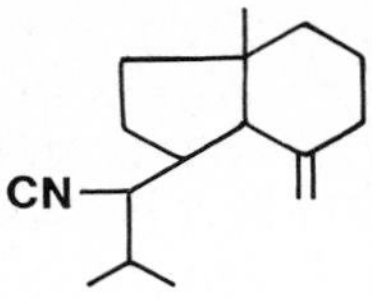

84

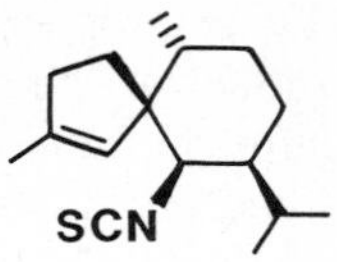

85

cells [225] (ED_{50} values were not reported). Two sesquiterpene isothiocyanate-thiourea pairs were reported from a Japanese sponge *Epipolasis kushimotoensis* [226]. Epipolasinthiourea-A (**86**) and epipolasinthiourea-B (**87**) both showed moderate cytotoxic activities in the L1210 cell line (ED_{50} 4.1 µg/ml and 3.7 µg/ml, respectively), whereas the corresponding isothiocyanates did not show any significant bioactivities (antibacterial, antiviral, cytotoxic).

The most characteristic feature of metabolites from green algae (Chlorophyta) is the 1,4-diacetoxybutadiene moiety that is found in over half of the reported compounds [227]. The conjugated bis-enol acetate functionality represents a

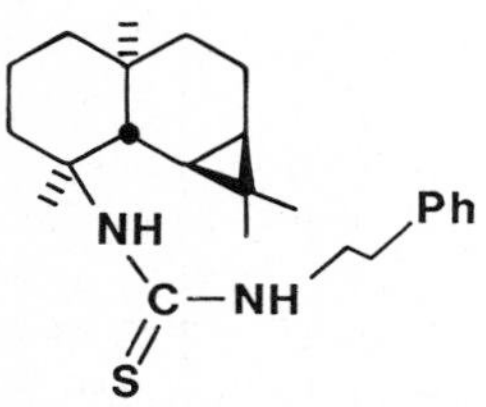

86 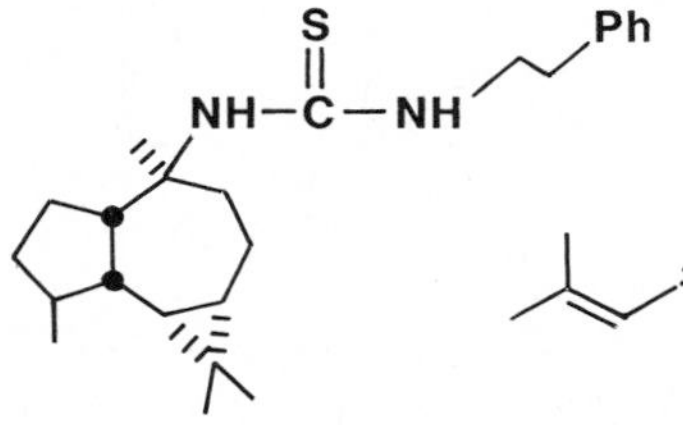   87 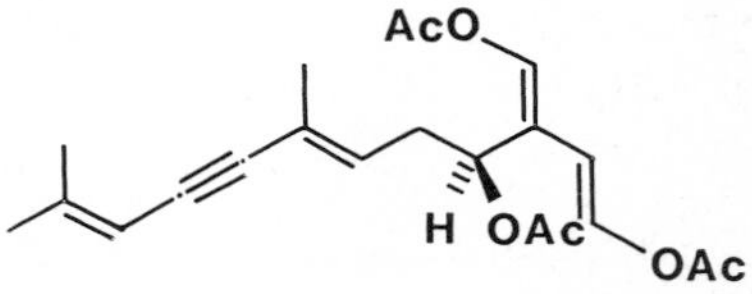   88

"masked" dialdehyde, often associated with high biological activity [228]. Many, but not all of these terpenoids are cytotoxic. Caulerpenyne (**88**), an acetylenic sesquiterpenoid with this substructure, was isolated from *Caulerpa prolifera* [229]. This substance was later reported to have "strong" cytotoxicity toward the KB cell line [13]; the reported ED_{50} of 2.8 g/ml probably has a misprint associated with the units. Fenical found that a series of four related sesquiterpenoids (**89–92**) from the family Udoteaceae [228, 230] inhibited cell division in the sea urchin egg assay. Their structures, sources, and ED_{100} are shown below. Flexilin (**91**) had been isolated previously from *Caulerpa flexilis* [231].

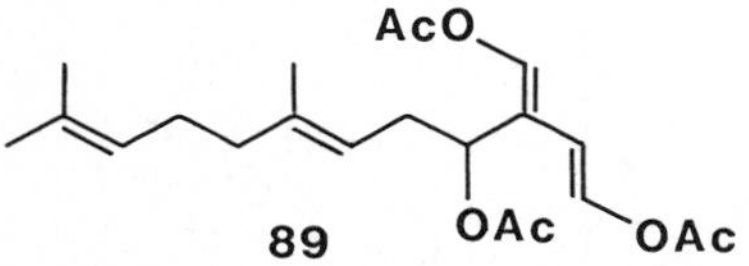

89   90

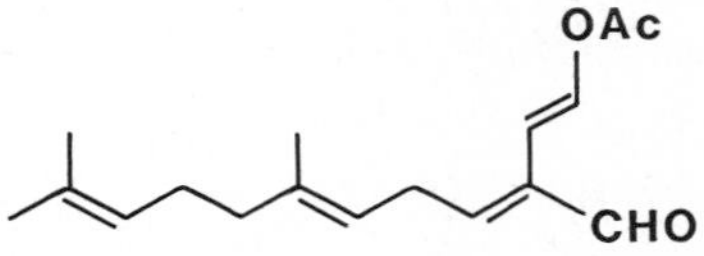

Penicillus   capitatus   **Penicillus   capitatus**

Udotea   flabellum   **Udotea   cyathiformis**

ED_{100} 2 µg/ml ED_{100} 2 µg/ml

91   **92**

Caulerpa flexilis	Rhipocephalus phoenix
Udotea conglutinata	
ED_{100} 16 µg/ml	ED_{50} 8 µg/ml

4.1.3 Diterpenoids

In addition to the sesquiterpenoids mentioned above, the green algae in the family Udoteaceae also produce related cytotoxic diterpenoids (**93–99**) [228, 232]. Structures, sources, and ED_{100} values for the sea urchin egg assay are shown below. Halimedatrial (**98**) with an ED_{100} of 1 µg/ml is strongly cytotoxic. In addition to the cytotoxic effect on fertilized sea urchin eggs, it was toxic toward reef fishes, significantly reduced feeding in herbivorous fishes, and was antimicrobial [233].

Brown algae of the family Dictyotaceae have provided several interesting cytotoxic diterpenoids. Piatelli obtained the dolabellane 3-epoxide (**100**) from *Dictyota dichotoma* [234]. This substance was reported to be significantly cytotoxic, and noted to have in vivo activity against *Adenovirus* and *Influenza* virus [39], but see Sect. 2.3. A similar, but distinguishable *Dictyota* species contained a dolabellane 3,16-diol derivative (**101**), which exhibited significant cytotoxic activity against KB cells (no ED_{50} value was reported) [234a].

From *Dictyota dentata* collected in Barbados, Gerwick obtained a complex mixture of known and unknown diterpenoids [235]. One of the components, dictyol H (**102**) was reported to be weakly cytotoxic in the KB9 assay (ED_{50} 22 µg/ml). A more potent cytotoxin, spatol (**103**), was obtained from *Spatoglossum schmittii* (Dictyotaceae) from the Galapagos Islands [236]. This tricyclic diterpenoid was cytotoxic in the sea urchin egg assay with a reported ED_{50} of 1.2 µg/ml; it also inhibited cell division in human T242 melanoma and 224C astrocytoma neoplastic cell lines at a cell culture testing concentration of 1–5 µg/ml.

Two additional spatane diterpenoids, (**104, 105**), inhibited cell cleavage of sea urchin eggs at 16 µg/ml [237]. These compounds have been isolated from an Indian Ocean collection of the brown alga *Stoechospermum marginatum* [238] as well as the related *Spatoglossum howleii* from the Galapagos Islands. The diol (**104**), but not the tetrol (**105**), was also found in the extract which provided spatol (*S. schmitti*).

Cytotoxic diterpenoids have also been obtained from sea hares. 14-Bromoobtus-1-ene-3,11-diol (**106**), isolated by Schmitz and co-workers from *Aplysia dactylomela* is probably a *Laurencia* metabolite [239]. The relatively uncommon skeleton of this compound was encountered first in a metabolite from *L. obtusa*. The halogenated diol (**106**) showed marginal activity against the KB (ED_{50} 4.5 µg/ml) and P388 (ED_{50} 10 µg/ml) cell lines.

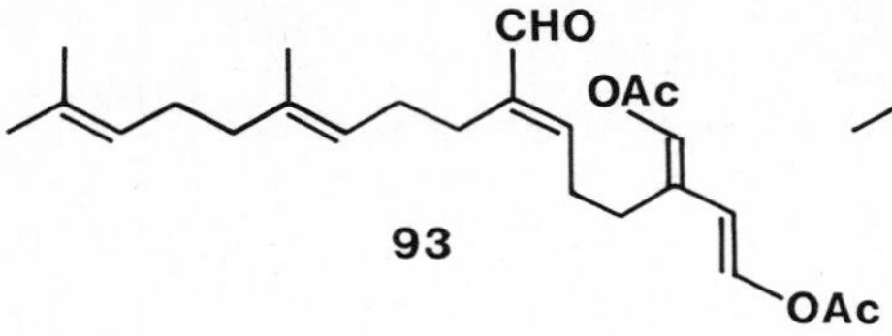

93

Udotea   flabellum
ED$_{100}$ 2 µg / ml

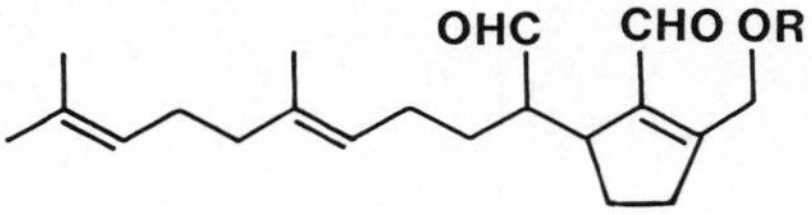

Udotea   flabellum
                   ED$_{100}$

94   R=H   1 µg / ml
95   R=Ac   1 µg / ml

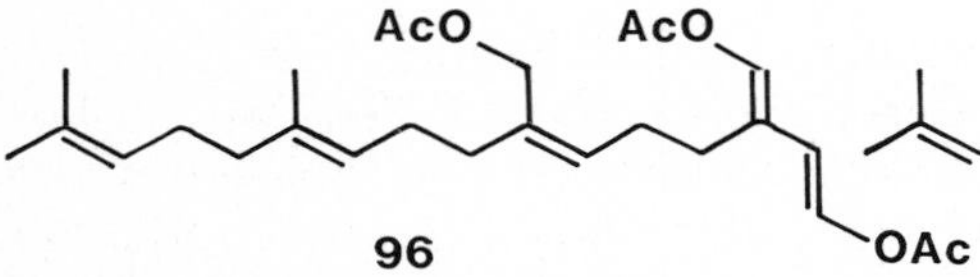

96

Penicillus   dumetosus
ED$_{100}$ 16 µg / ml

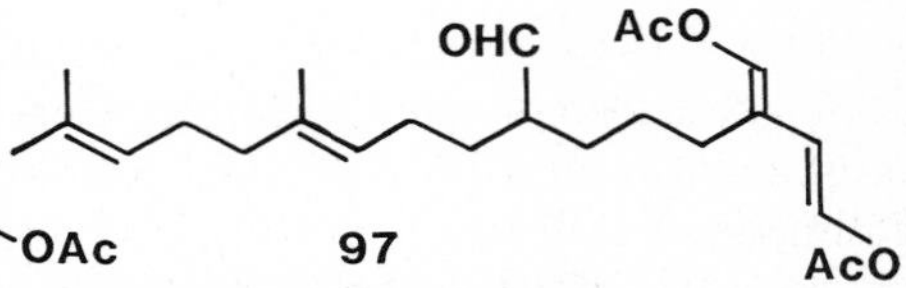

97

Penicillus   dumetosus
ED$_{100}$ 16 µg / ml

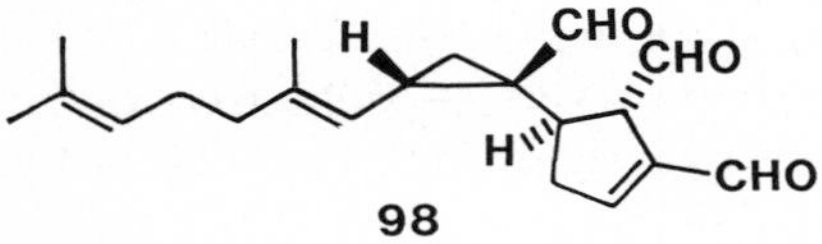

98

99

Halimeda   tuna
ED$_{100}$ 1 µg / ml

H.   opuntia

H.   incrassata

H.   copiosa   H.   simulans

H.   scabra   H.   cylindracea

H.   monile   H.   gigas

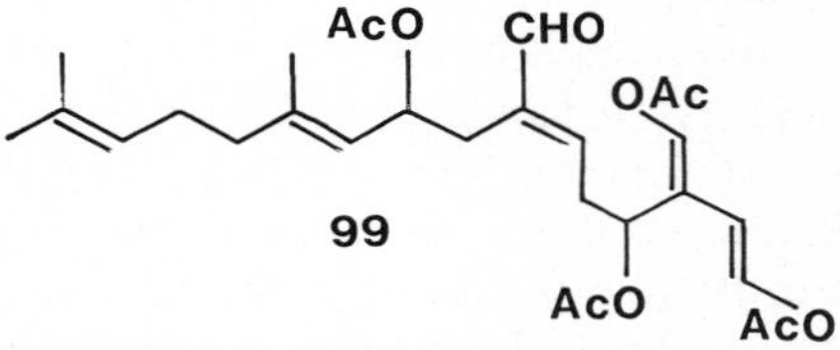

Halimeda   discoidea
ED$_{100}$ 16 µg / ml

H.   macroloba

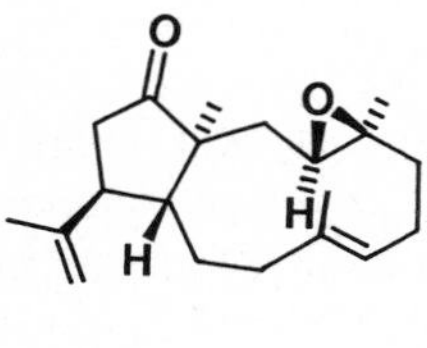

100

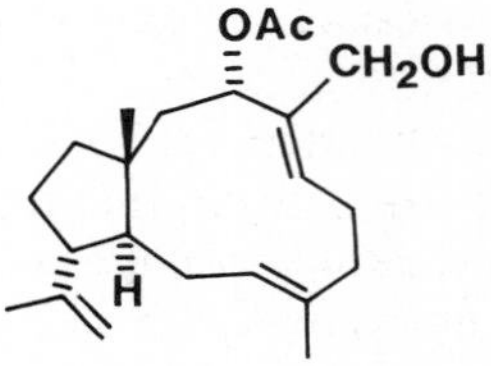

101

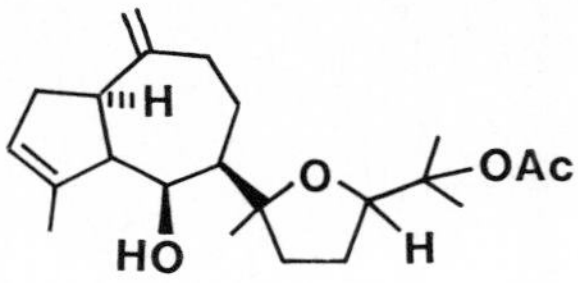

102 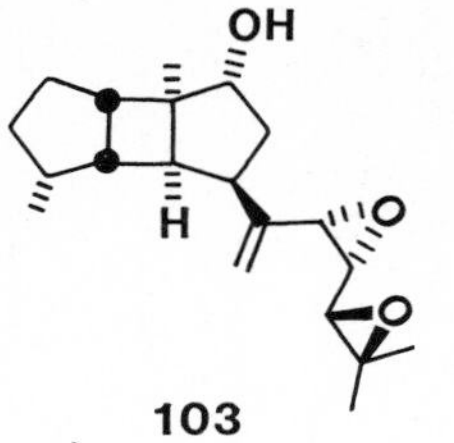   103 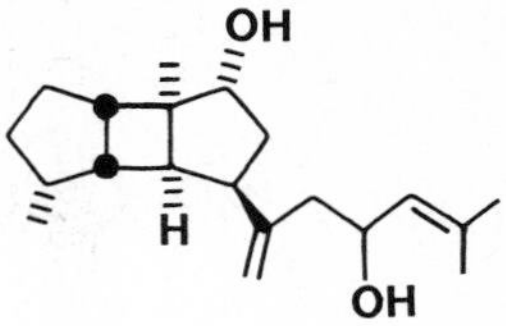   104

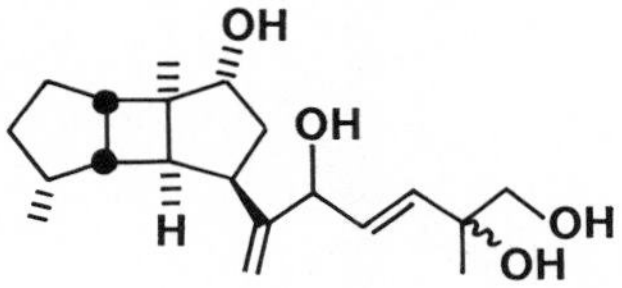

105 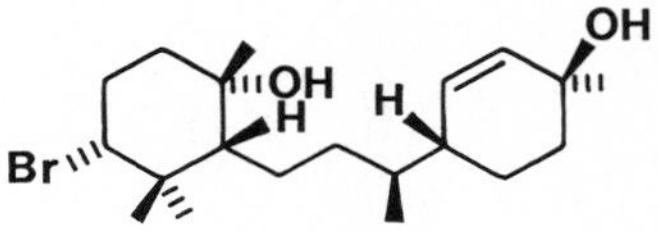   106

Because of their success with *A. dactylomela* collected in the Bahamas, and the dietary dependence of the contents of the sea hare digestive glands, the Oklahoma group analysed specimens collected in Puerto Rico. The complex mixtures from these Caribbean locations did not have any common components [212]. Five cytotoxic brominated diterpenes (**107–111**) with modified pimarane skeletons were found in the Puerto Rican sea hares. The ED_{50} values in the P388 in vitro assay are shown below the structures.

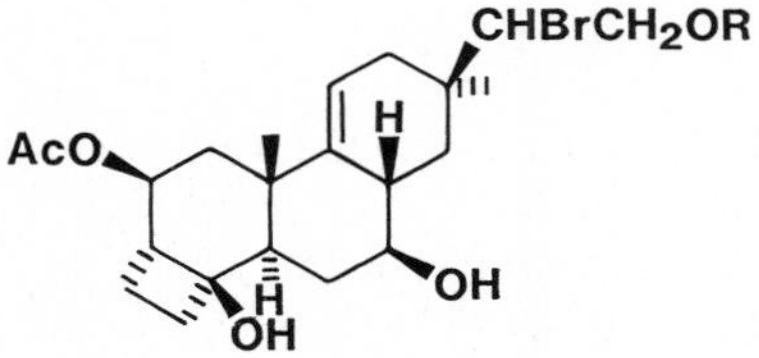

		ED_{50} (µg/ml)
107	R = H	4.6
108	R = Ac	0.52

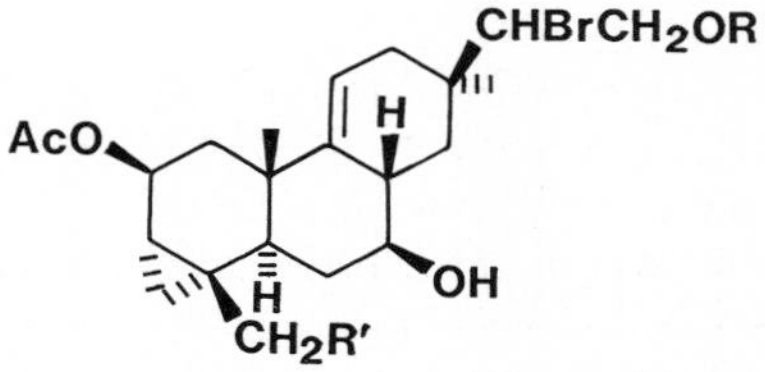

		ED_{50} (µg/ml)
109	R = H , R′ = OH	3.8
110	R = Ac , R′ = OH	4.3
111	R = H , R′ = H	0.38

Bioasssay-guided isolation of active components of a cytotoxic extract from the sea hare *Dolabella auricularia* resulted in the isolation of a pair of compounds with the new dolastane skeleton [240]. Dolatriol (**112**) and dolatriol 6-acetate (**113**) were cytotoxic toward the P388 cell line with ED_{50} values of 13 µg/ml and 10 µg/ml, respectively [95].

Several tetracyclic diterpenoids with the spongian skeleton have been isolated from the sponge *Spongia officinalis* [116]. Isoagatholactone (**114**) and three related lactones in this series have been reported to be cytotoxic. Cimino and co-workers were the first to find isoagatholactone from a Mediterranean collection [241]; it was re-isolated from the Canary Islands sponge along with the 11-hy-

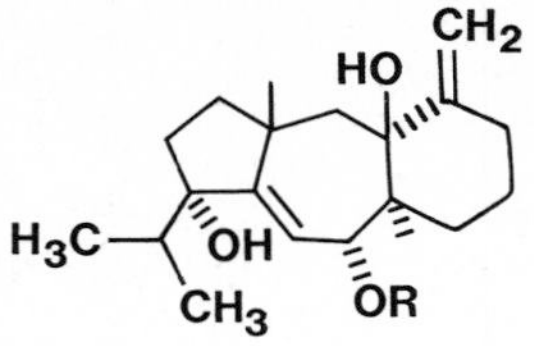

		HeLa cell			P388
		ED_{50} µg/ml			ED_{50} µg/ml
114	R=H	10	117	R=COCH$_2$CH$_2$CH$_3$	6.5
115	R=OH	5	118	R=COCH$_3$	—
116	R=OAc	1	119	R=H	—

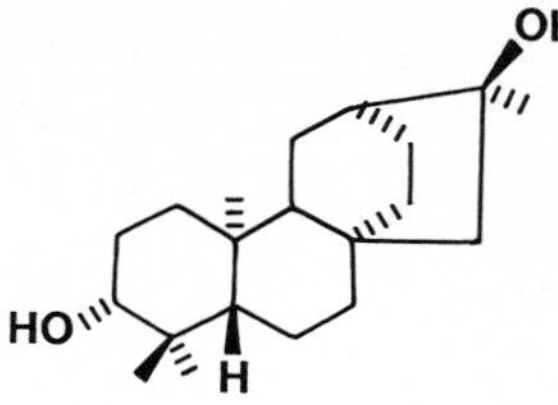

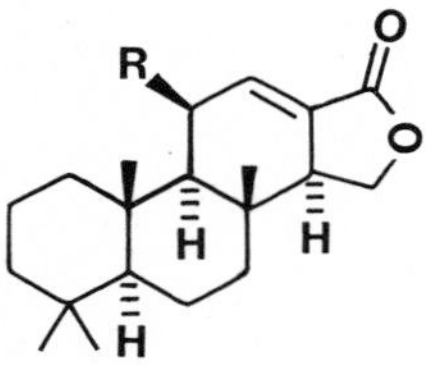

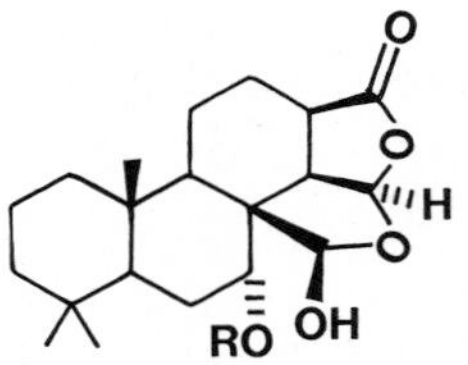

droxy and 11-acetoxy derivatives, 11β-hydroxyspongi-12-en-16-one (115) and 11β-acetoxyspongi-12-en-16-one (116) [242]. ED_{50} values for cytostatic activity toward HeLa cells are shown next to the structures [243]. In the Canary Island sponge, (115) and (116) together made up over 6% of the dry weight of the animal. Schmitz obtained a series of three more highly functionalized spongian lactones (117–119) from the Caribbean sponge *Igernella notabilis*, but only 7α,17β-dihydroxy-15,17-oxidospongian-16-one 7-butyrate (117) was reported to be mildly cytotoxic in the P388 cell line [244]. Stereoselective syntheses of racemic [245] and chiral isoagatholactone (114) [246] have been reported.

A mildly cytotoxic diterpenoid, atisane-3β,16α-diol (120), with a tetracyclic bridged atisane skeleton, was isolated from the Caribbean fire sponge *Tedania ignis* [247]. With an ED_{50} of 21 µg/ml in a test against KB cells, this might be considered almost active by the NCI criterion ($\leq$ 20 µg/ml).

Hoffmann and Rabe have estimated that approximately 10% of the 25,000–30,000 structurally elucidated natural products are α-methylene-γ-butyrolactones, mainly of the sesquiterpene type. Many of these 5-membered ring lactones are cytotoxic [247a]. Diterpenes (114–116) in the isoagatholactone series and lobohedliolide (121) fall into this category. Uchio and co-workers isolated the latter compound from the Japanese soft coral *Lobophytum hedleyi*, and found that it inhibited the growth of HeLa cells at 5 µg/ml [248]. No ED_{50} value was reported. In contrast to the γ-lactone present in these compounds, other cytotoxic cembranolides have larger lactone rings. Crassin acetate (40), sinularin (122), and dihy-

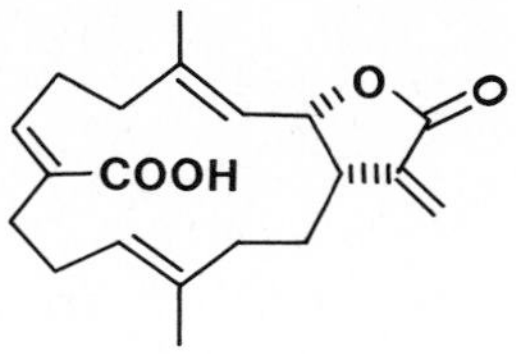

121

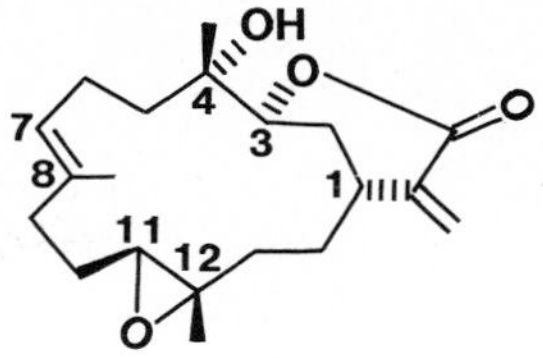

122

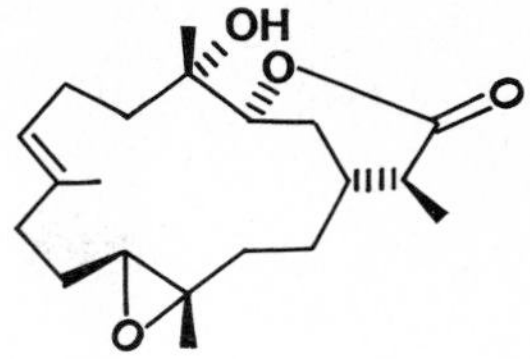

123

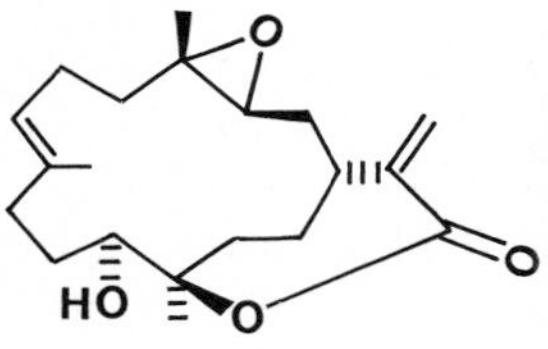

124

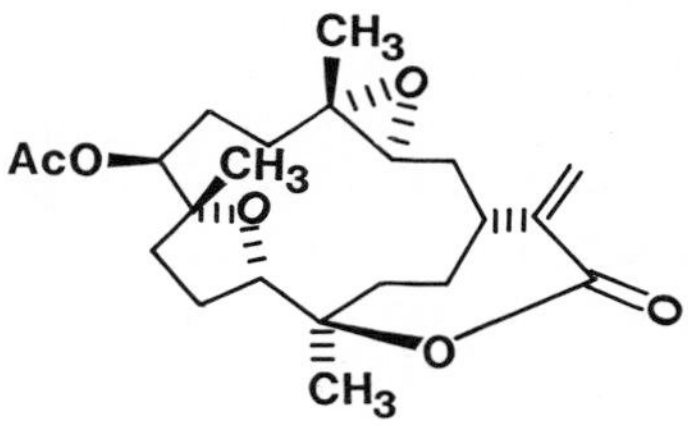

125

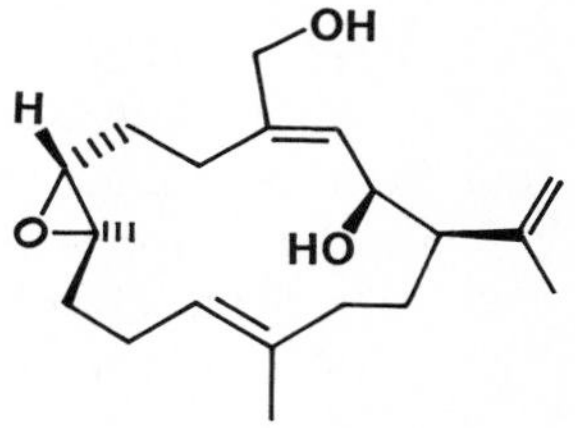

126

drosinularin (**123**) are examples of cembranes with δ-lactones. Both were isolated from the Australian soft coral *Sinularia flexibilis*, by Weinheimer [249], and almost at the same time by Kazlauskas and co-workers, [250] who assigned the redundant names, flexibilide and dihydroflexibilide. As might be expected, sinularin, with the α-methylene lactone function, is far more cytotoxic than dihydrosinularin. ED_{50} values for cytotoxicity toward the KB and P388 cell lines are 0.3 µg/ml and 0.3 µg/ml for sinularin and 16 µg/ml and 1.1 µg/ml for dihydrosinularin; the latter compound does not have the conjugated exocyclic double bond. Sinularin is representative of the *alpha*-series of cembranes isolated from the order Alcyonacea (soft corals), whereas crassin acetate is typical of the *beta*-series of cembranes found in the order Gorgonacea (gorgonians) [5].

Sinularia flexibilis was also the source of two cytotoxic cembranes with a 7-membered ε-lactone ring. Sinulariolide (**124**) from the Indonesian/Australian soft coral has ED_{50} values in the KB and P388 cell lines of 20 µg/ml and 7.0 µg/ml, respectively [249, 251]. The related 8,11-bridged ether (**125**), (in addition to (**122–124**), and four other known cembranoids) was obtained from a Japanese collection; it showed cytotoxic activity against DBA/MC fibrosarcoma at a concentration of 100 µg/ml [252] although an ED_{50} value was not reported.

Asperdiol (**126**) is one of the only non-lactonic cembranes which has significant cytotoxic activity against the KB, P388, and L1210 cell lines (ED_{50}'s 24 µg/ml, 6 µg/ml, and 6 µg/ml, respectively) [253]. The *cis* configuration of the C(3) double bond is the most unusual feature of this substance. Extracts of Caribbean gorgonians *Eunicea asperula* and *E. tourneforti* had in vivo activity in the P388 test. Bioassay-guided purification (in vitro P388 and KB) led to the isolation of **126** from each. Two synthetic routes to asperdiol have been reported [254, 255].

Surprisingly, two cembranes which are devoid of functional groups proved to be cytotoxic in tests carried out by the NCI. The hydrocarbon cembrene-A (**127**) had ED_{50} values of 2.8, 0.31 and 0.22 µg/ml in the KB, P388, and L1210 assays, respectively. The corresponding values for the hydrocarbon flexibilene (**128**) were

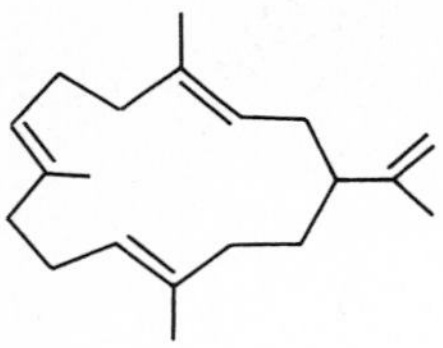

CO2CH3

127 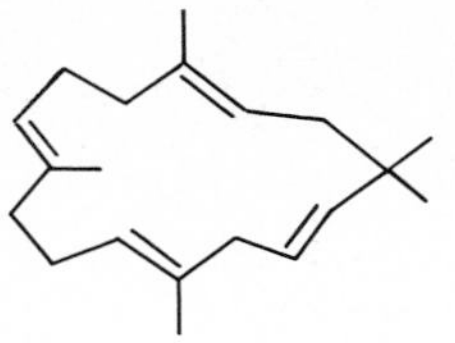   **128** 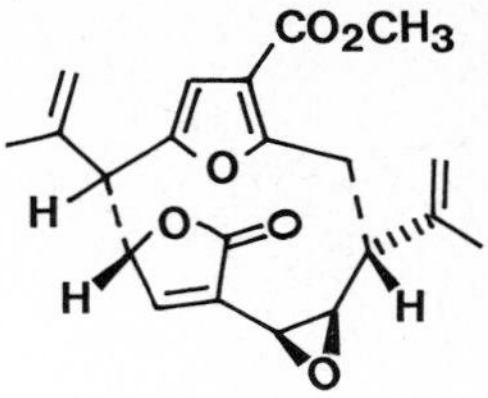   **129**

24, 1.2 and 0.33 µg/ml. Results of in vivo testing against P388 leukemia cells showed that both compounds were inactive [144]. Schmitz and co-workers found that these compounds were the major hydrocarbon metabolites from the Pacific soft coral *Sinularia conferta* [256]. The compounds were isolated first from a marine organism by Tursch and co-workers in their examination of *S. flexibilis* [257, 258]. Cembrene-A has been found in many soft corals [116] and terrestrial sources [259].

An irregular 12-membered ring diterpenoid was obtained from another Caribbean gorgonian, the sea whip *Pseudopterogorgia acerosa* [260]. The isolation of the major metabolite pseudopterolide (**129**) was guided by the sea urchin egg assay. The concentrations required for inhibition were not indicated.

4.1.4 Sesterterpenoids

Yasuda and Tada collected samples of the marine sponge *Cacospongia scalaris* from two different locations in Japan [261]. Only one of the collections provided the very potent cytotoxic agent desacetylscalaradial (**130**), which had an ED_{50} of 0.58 µg/ml in the L1210 assay. Scalaradial (**131**) which had been obtained earlier from *C. mollior*, was also isolated from the same sponge; there was no report of it being cytotoxic. Braekman and co-workers later reported that some closely related sesterterpenes are ichthyotoxic [262].

Another collection of the Japanese sponge *C. scalaris* by Fusetani and co-workers provided two new examples of furanosesterterpenes, (**132**) and (**133**) [263]. Both of these compounds inhibited the cell division of fertilized starfish eggs at a concentration of 1.0 µg/ml; this assay is a variation on the test with sea urchin embryos. The structures of these compounds are reminiscent of the sponge metabolite variabilin (**14**), which was discussed previously.

The same research group obtained spongionellin (**134**) and dehydrospongionellin (**135**) from the Japanese sponge *Spongionella* sp. [264]. Both inhibited cell division of fertilized starfish eggs at 2.0 µg/ml. This was the first report of finding furanosesterterpenes from a sponge of the genus *Spongionella*.

Muqubilin (**136**) is a norsesterterpenoid carboxylic acid which contains a cyclic peroxide linkage. Kashman first isolated muqubilin from a brown sponge *Prianos* sp. from the Red Sea [265]. Crews and co-workers subsequently reisolated it from an undescribed Tongan *Prianos* sp., whose crude extract inhibited sea urchin egg cell division. Muqubilin was responsible for the biological activity (ED_{100} 16 µg/ml) [266].

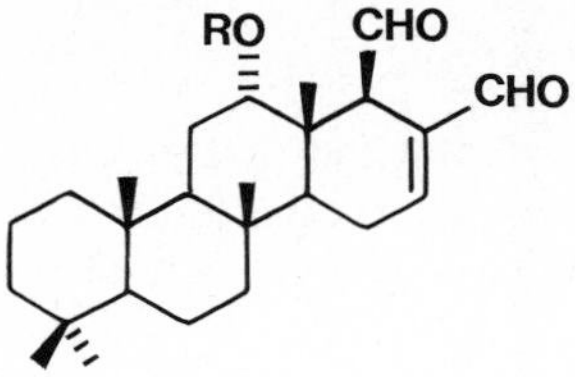

130   R = H

131   R = Ac

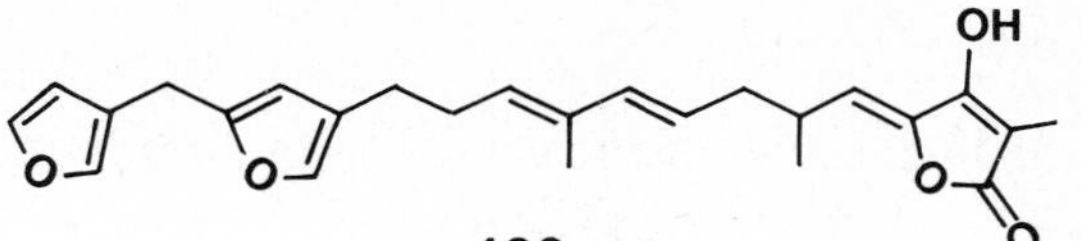

132

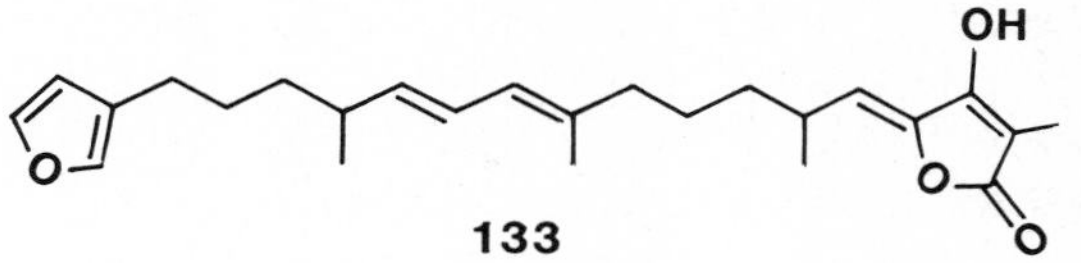

133

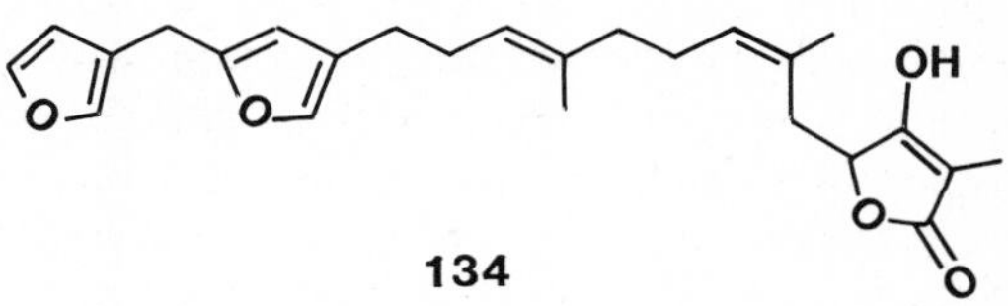

134

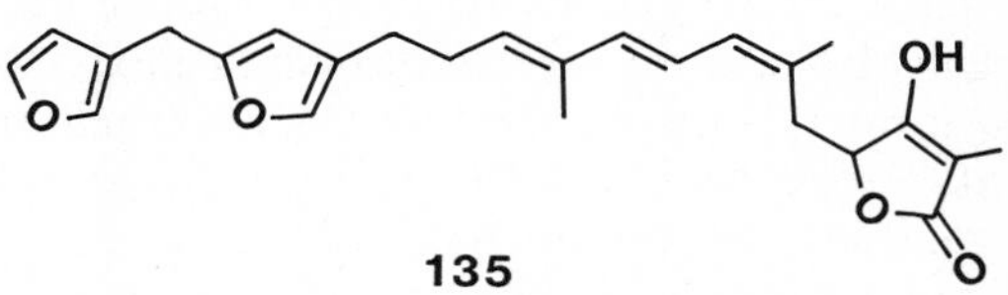

135

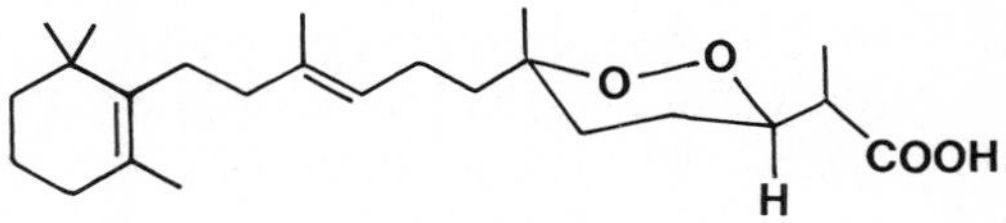

136

4.1.5 Triterpenoids

The Japanese red alga *Laurencia obtusa* was the source of two cyclic bromoethers, thyrsiferyl 23-acetate (**137**) and thyrsiferol (**138**), which have the squalene skeleton [267]. Thyrsiferyl 23-acetate was remarkably cytotoxic toward the P388 cell line with an ED_{50} of 0.0003 µg/ml. Thyrsiferol, which had been isolated previously from *L. thyrsifera* (Hook) by the New Zealand group [268] was strongly cytotoxic with ED_{50} 0.010 µg/ml. Many other remarkably cytotoxic constituents, whose structural elucidation is in progress, are also present in the Japanese *L. obtusa* whose crude methanol extract had an ED_{50} of 0.18 µg/ml.

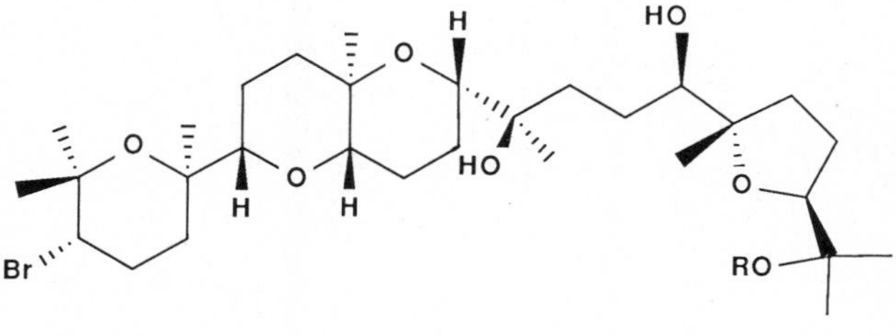

137   R = Ac

138   R = H

4.1.6 Echinoderm Saponins

Saponins obtained from sea cucumbers (holothurins) and starfish (asterosaponins) are water-soluble natural products composed of a sugar moiety attached to a triterpenoid or steroid aglycone. Echinoderm saponins have been reviewed recently [269] and will be covered only briefly here.

4.1.7 Holothurins – Triterpenoid Saponins

Sea cucumbers were the first source of saponins isolated from an animal. Holothurins from *Actinopyga agassizi* increased the survival times of white mice with sarcoma 180 [270, 271] and Krebs-2 ascites tumors [272], but not of C57 black mice with B16 melanoma tumors. The crude extract was more potent than the purified preparation [271, 273], but the effective dose required was only slightly lower than a lethal dose. Holothurins were also cytotoxic in vitro. Although Hashimoto attributes the cytotoxicity to their surface active agent properties [127], Cairns and Olmstead have argued that other factors are also implicated [271]. Both crude and purified holothurin showed cytotoxicity toward the KB cell line at 5 μg/ml [15]. Holothurins were also cytotoxic in the sea urchin egg assay (274–277), but these effects were counteracted by cholesterol and cholesterol-containing neutral lipids [278]. Sea cucumber saponins also had a toxic effect on a variety of organisms, including mammals, and were antimicrobial [127].

The structures of several of the holothurins have now been elucidated [269]; those of holothurin A (**139**) [270a, 271a] and holothurin B (**140**) [272a, 273a, 279] are shown. Holothurin A was cytotoxic to sea urchin eggs at 0.78 μg/ml [141,

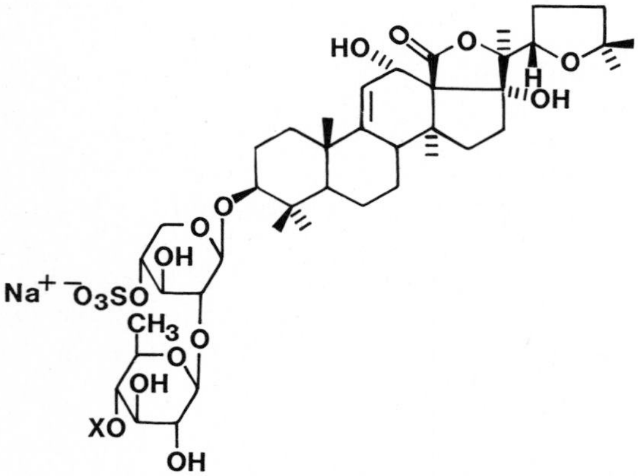

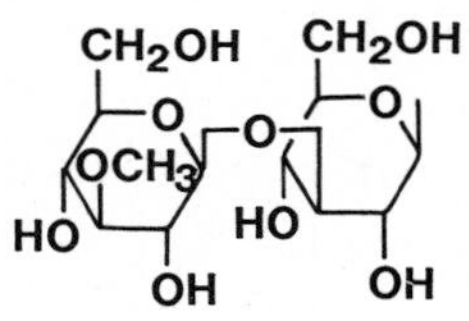

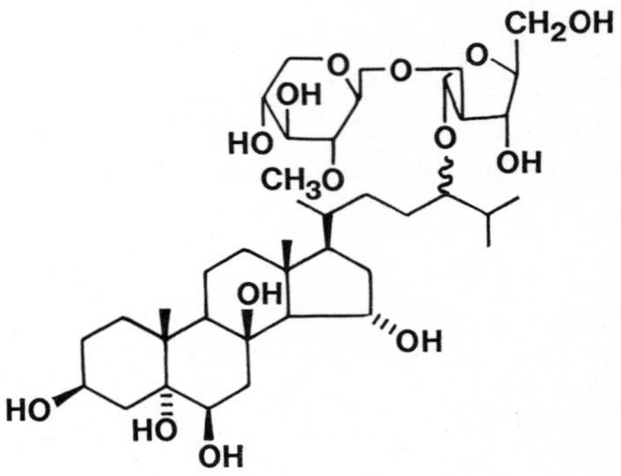

280]. Elyakov and co-workers found that holothurins A and B, or very similar holothurins, are the major saponins in all of the 25 species of the genera *Holothuria* and *Actinopyga* that they examined [269, 281, 282].
ponins stichostatin 1 (from *Stichopus chloronatus*, Australia), P388 ED_{50} 2.9 µg/ml; thelenostatin 1 (from *Thelenota ananas*, Taiwan and Marshall Islands), P388 ED_{50} 1.5 µg/ml; and actinostatin 1 (from *Actinopyga mauritiana*, Hawaii) KB ED_{50} 2.6 µg/ml, L1210 ED_{50} 2.1 µg/ml [197, 283].

4.1.8 Asterosaponins – Steroidal Saponins

Like the holothurins, the asterosaponins are toxic to many types of organism [269]. The extract of *Acanthaster planci* caused a variety of cytological changes in tissue cell cultures of KB cells and Gray Seal Kidney cells at concentrations of 50–100 µg/ml [15, 284] but an *Asterias forbesi* extract showed no activity against P388 in vivo [285]. Asterosaponins from several sea stars induced abnormalities in the early development of sea urchin larvae [274, 284, 286–288], but higher concentrations were necessary to observe the effects caused by holothurin [274, 280]. Asterosaponins are less cytotoxic to urchin embryos than holothurins.

Cytotoxicity has been reported also for characterized asterosaponins. Nodososide (**141**) is a moderately cytotoxic steroidal glycoside, which has been isolated from *Protoreaster nodosus*, and in small amounts from *Acanthaster planci* and *Linkia laevigata* [269, 289–291]. The bioassay and effective dose were not reported.

Fusetani and co-workers carried out a systematic study of seventeen asterosaponins (**142–158**) from Japanese *Acanthaster planci, Luidia maculata* and *Asterias amurensis* [c.f.] *versicolor* [292]. From the ED_{50} values against fertilized sea urchin eggs and starfish eggs, which are listed below the structures, only **153**, **155**, **157**, and perhaps **143**, **146**, **154**, and **156** would be considered active. Compound (**153**) was the most active, and **158** the least, which suggests the importance of the side chain for activity. The sulfate moiety at C-3 is not important in affecting activity.

Testing by the National Cancer Institute has led to the conclusion that asterosaponins, which are cytotoxic in the KB assay, have no useful in vivo activity [92].

4.1.9 Steroids

Cytotoxic marine steroids have been isolated from a variety of marine invertebrates. Faulkner investigated the sponge *Toxadocia zumi* because, compared to other sponges in the same environment, it remained free of epiphytic growth [293]. Three steroidal sulfates (**159–161**) with antimicrobial, antifeedant and cytotoxic properties were obtained from this sponge. The mixture of the three, in the same proportions as found in the sponge, inhibited cell division in the fertilized sea urchin egg assay at 5 µg/ml.

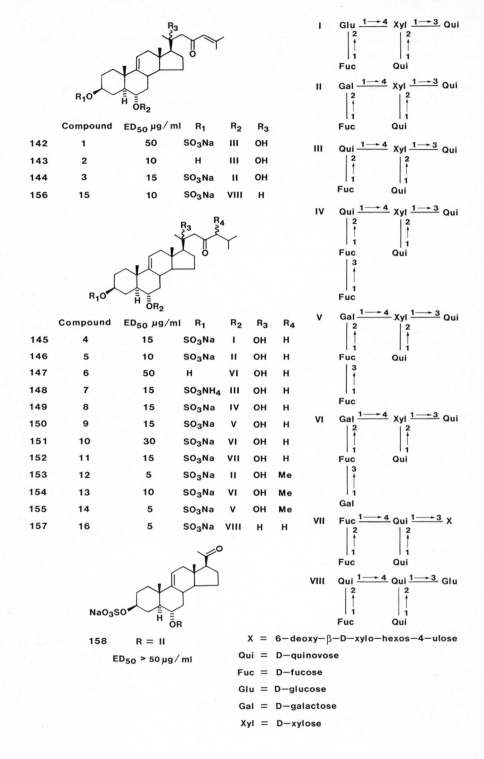

Compound		ED$_{50}$ µg/ml	R$_1$	R$_2$	R$_3$
142	1	50	SO$_3$Na	III	OH
143	2	10	H	III	OH
144	3	15	SO$_3$Na	II	OH
156	15	10	SO$_3$Na	VIII	H

Compound		ED$_{50}$ µg/ml	R$_1$	R$_2$	R$_3$	R$_4$
145	4	15	SO$_3$Na	I	OH	H
146	5	10	SO$_3$Na	II	OH	H
147	6	50	H	VI	OH	H
148	7	15	SO$_3$NH$_4$	III	OH	H
149	8	15	SO$_3$Na	IV	OH	H
150	9	15	SO$_3$Na	V	OH	H
151	10	30	SO$_3$Na	VI	OH	H
152	11	15	SO$_3$Na	VII	OH	H
153	12	5	SO$_3$Na	II	OH	Me
154	13	10	SO$_3$Na	VI	OH	Me
155	14	5	SO$_3$Na	V	OH	Me
157	16	5	SO$_3$Na	VIII	H	H

158   R = II

ED$_{50}$ > 50 µg/ml

X   = 6−deoxy−β−D−xylo−hexos−4−ulose

Qui = D−quinovose

Fuc = D−fucose

Glu = D−glucose

Gal = D−galactose

Xyl = D−xylose

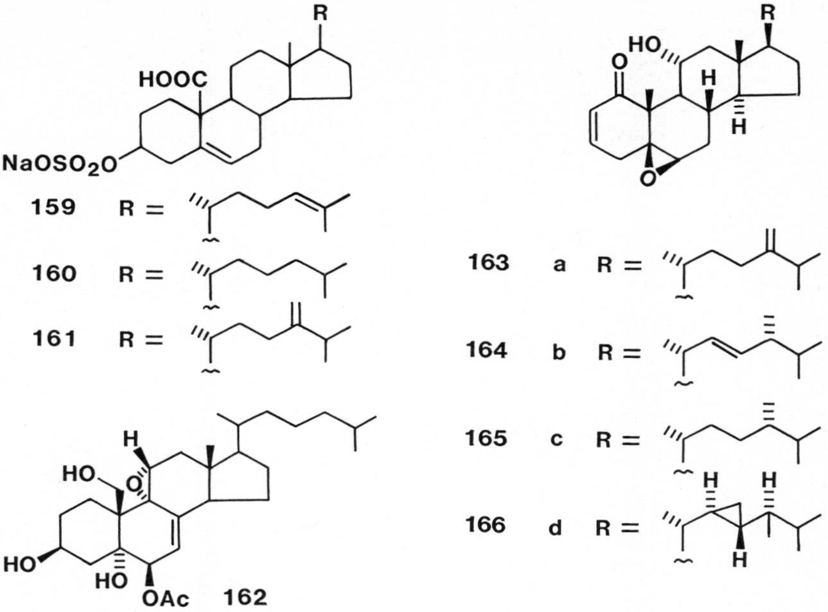

159   R =

160   R =

161   R =

163   a   R =

164   b   R =

165   c   R =

166   d   R =

162

Several of the cytotoxic sterols contained an epoxide ring. A *Dysidea* sp. sponge collected in Guam was the source of a polyhydroxylated steroidal 9,11-epoxide (162), which was reported to be slightly cytotoxic (P388 ED$_{50}$ 4.9 µg/ml [294]). Four epoxy, polyoxygenated sterols stoloniferones a–d (163–166) were isolated from the stoloniferan soft coral *Clavularia viridis*, the same organism that produced the antitumor clavulone eicosanoids discussed earlier. At a concentration of 1 µg/ml the stoliniferones inhibited growth in 69% of the P388 leukemia cells [295].

The gorgonian *Isis hippuris*, collected in Japan, was the source of two potent polyoxygenated steroids containing a spiroketal function – hippuristanol (167) and 2α-hydroxyhippuristanol (168) [296]. The ED$_{50}$ values for the two compounds against in vitro DBA/MC fibrosarcoma were 0.8 and 0.1 µg/ml, respectively. Hippuristanol was "active" in the P388 assay in mice, and the 2α-hydroxy derivative was "slightly active", but ED$_{50}$ values were not given.

Minale and co-workers reported that a series of highly hydroxylated sterols, exemplified by 169–171 from the starfish *Protoreaster nodosus* and *Hacelia attenuata*, were moderately cytotoxic, but no details of the method or concentrations were provided [297]. A common feature of all but one of the compounds is an hydroxyl group at the unusual C-8 position, and 3β,6(α or β),15α,16β,26-pentahydroxy substitution.

The observation of cytotoxicity in a methylene chloride extract of the tunicate *Ciona intestinalis* led Guyot and co-workers to the isolation of a cytotoxic peroxide (172) [298, 299]. It was present also in the tunicate *Phallusia mamillata*, and was not considered to be an artifact. This hydroperoxide (172) was synthesized by reacting fucosterol with singlet oxygen. Three other derivatives of fucosterol (173–175), all cytotoxic toward L1210 cells (ED$_{50}$ 2–4 µg/ml), were also obtained

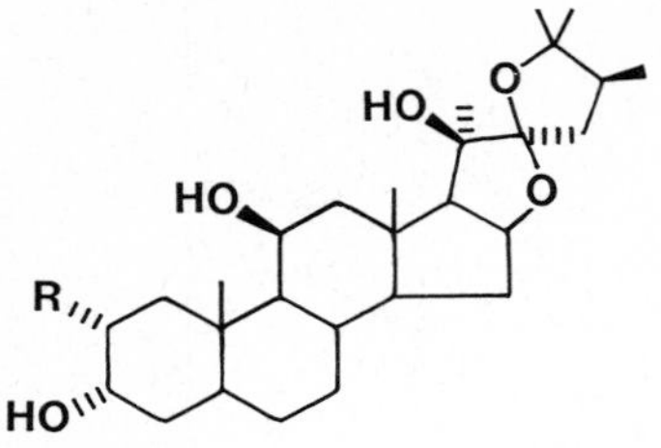

167   R=H

168   R=OH

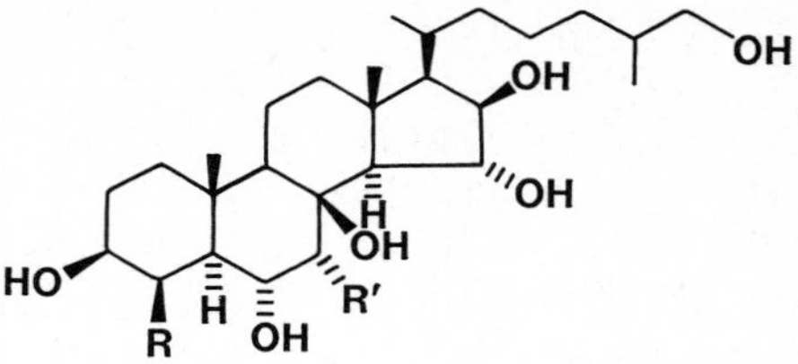

169   R = H, R′ = H

170   R = H, R′ = OH

171   R = R′ = OH

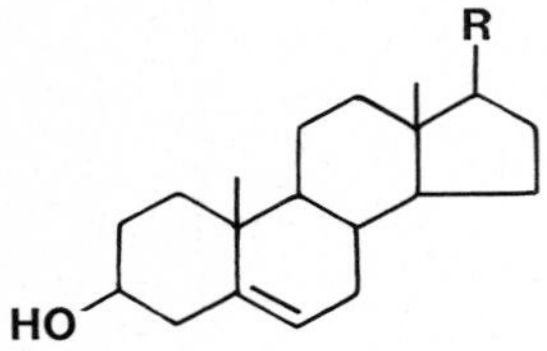

172
R =

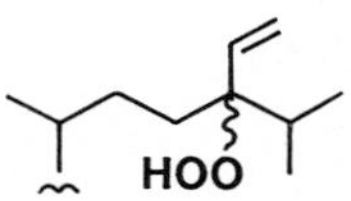

173
R =

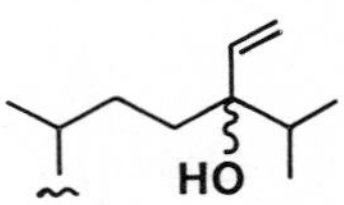

175
R =

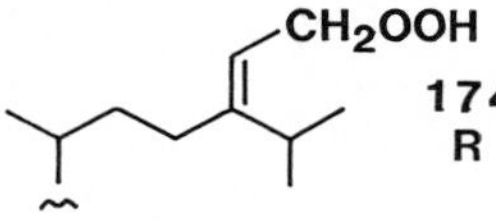

174
R =

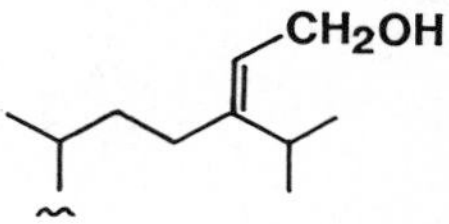

by synthetic transformations [299]. The most interesting of the series was **175** which was obtained from **172** in chloroform solution by an allylic rearrangement. Only **175** was selectively cytotoxic toward cultured spleen lymphocytes stimulated by the mitogen concanavalin A (ED_{50} <1 µg/ml). Cytotoxicity toward normal lymphocytes was much lower (ED_{50} >16 µg/ml). Other examples of steroids which are cytotoxic toward hepatoma strains HTC and ZHC, but which are much less toxic toward the non-tumorous 3TC mouse fibroblasts have been reported [300]. The diol (**174**) corresponding to hydroperoxide (**175**) was cytotoxic (L1210), but the diol (**173**) corresponding to hydroperoxide (**172**) was not.

4.1.10 Tetraterpenoids

Halocynthiaxanthin (**176**) was one of several carotenoids isolated from the Japanese tunicate *Halocynthia roretzi*. It displayed weak activity toward the P388 leukemia cell line (ED_{50} 12.5 µg/ml), but was inactive in the L1210 leukemia assay (ED_{50} 25.0 µg/ml) [301].

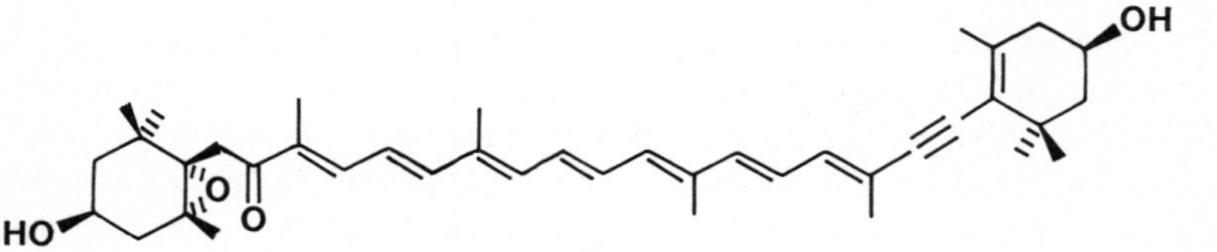

176

4.2 Degraded Terpenoids

Loliolide (**177**) has been isolated from numerous plants, and is thought to be an oxidation product of certain plant carotenoids. Pettit and co-workers isolated (**177**) from the Indian Ocean opisthobranch mollusc *Dolabella ecaudata* along with the diterpene dolatriol (**112**) [302]. The (−)-loliolide was cytotoxic against the KB and P388 cell lines, with ED_{50} values of 10 µg/ml and 3.5–22 µg/ml, respectively. It was tested in vivo against P388 leukemia cells at a dose of 2.5–10 mg/kg and was found to be inactive. Since sea hares are known to concentrate algal

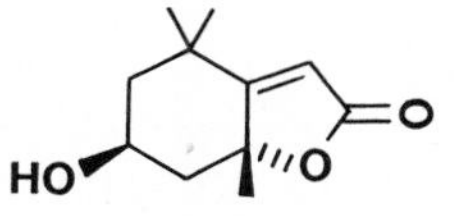

177 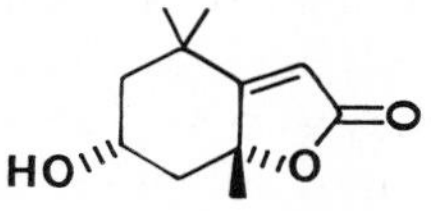   178

metabolites, it seems likely that the loliolide is of dietary origin. Its C-5 epimer, epiloliolide (**178**), was obtained from the fire sponge *Tedania ignis* and was almost as cytotoxic as loliolide (ED_{50} 21 µg/ml against KB cells) [247]. Several syntheses of (−)-loliolide have appeared [303].

4.3 Terpenoid-Shikimates

Many marine natural products of mixed terpenoid-shikimate biogenetic origin with interesting biological properties have been isolated [304]. Some have been discussed earlier [see the antitumor compounds prenylhydroquinone (**41**) and stypoldione (**47**). Avarol (**179**) was the first sesquiterpene-substituted hydroquinone which contains a rearranged drimane skeleton [305, 306]. The Mediterranean sponge *Dysidea avara* produces large amounts (6%, dry weight) of avarol, accompanied by smaller amounts (0.8%) of the corresponding quinone, avarone (**180**). Cariello and co-workers observed that avarol induced developmental aberrations in the eggs of the sea urchin *Sphaerechinus granularis*, but only at concentrations greater than 30 µg/ml (100 µM) [307]. Using the in vitro mouse L5178y lymphoma cell system, Müller and co-workers observed that, within a limited concentration range, avarol showed a pronounced cytostatic activity toward this cell line; when the compound was washed out the cells grew normally [ED_{50} 0.3 µg/ml (0.9 µM)] [308]. At concentrations greater than 0.6 µg/ml the compound was cytotoxic for L5178y cells.

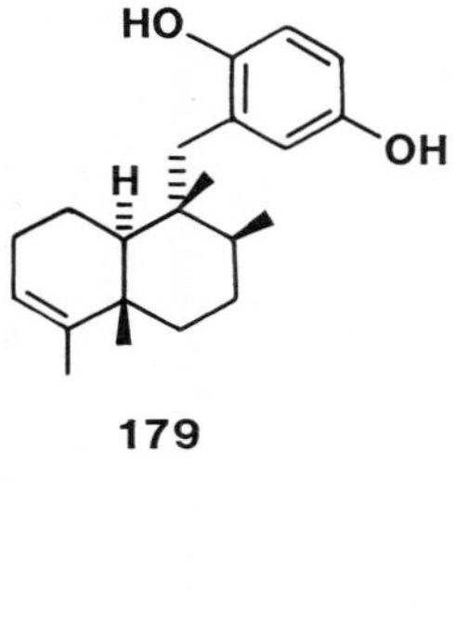

179 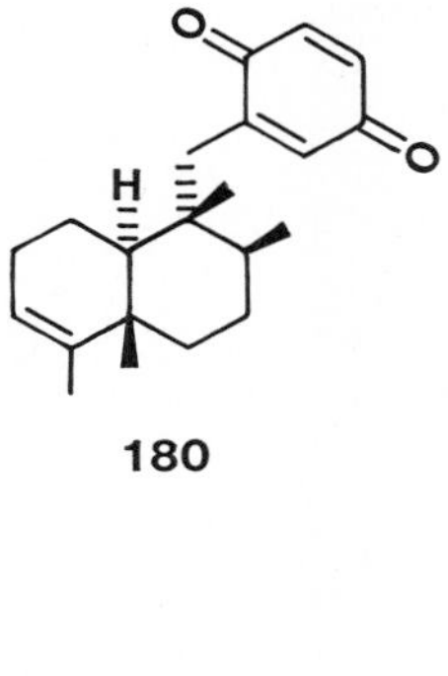   180

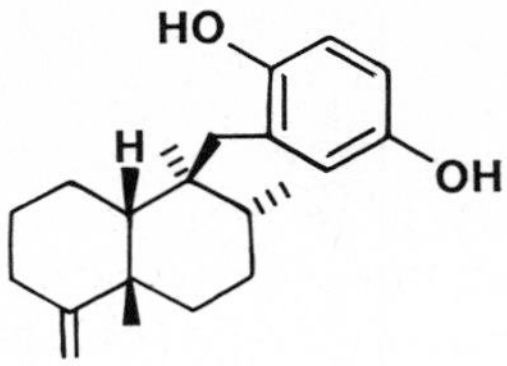

181 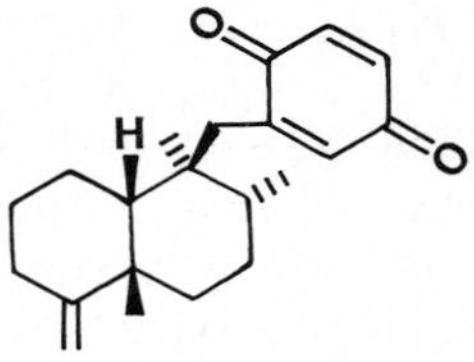   **182**

The Pacific sponge *Dysidea arenaria* was the source of the cytotoxic hydro-quinone-quinone pair of compounds, arenarol (**181**) and arenarone (**182**) [309]. The quinone (**182**) was much more cytotoxic than the hydroquinone; ED_{50} values for the P388 assay were 17.5 and 1.7 µg/ml, respectively. These compounds have the same rearranged drimane skeleton of avarol and avarone, but with *cis*- rather than *trans*-decalin stereochemistry. The absolute configuration was not deter-mined.

Isospongiaquinone (**183**) was isolated from the Australian sponge *Fascio-spongia rimosa* (= *Stelospongia conulata*) [116, 310, 311]. It also has the rear-ranged drimane skeleton, with a *trans*-decalin stereochemistry, as found in avarol (**179**). (For comments on the stereochemical assignment see Footnote 5 in Ref. [312].) The absolute configuration of isospongiaquinone (**183**) has not been deter-mined. Morita and Endo found that isospongiaquinone was cytotoxic towards PV_4 cultured cells transformed with polyoma virus. The Roche group determined that isospongiaquinone and its dihydroderivative were the most potent inhibitors of oxidative phosphorylation that they had encountered in their routine screening of marine natural products [313].

Ilimaquinone (**184**) from the Hawaiian sponge *Hippiospongia metachromia*, shares the rearranged drimane skeleton, with *trans*-decalin stereochemistry, but its absolute configuration is enantiomeric with avarol [314]. Ilimaquinone was cy-totoxic towards monkey kidney cells (BSC) at 20 µg/disc [315].

Faulkner obtained an inseparable mixture of ilimaquinone (**184**) and 5-*epi*-ili-maquinone (**185**) (3:2) from a *Fenestraspongia* sp. sponge collected from Palau [312]. Like arenarone (**182**), 5-*epi*-ilimaquinone has the *cis*-decalin ring fusion in the rearranged drimane skeleton. The mixture was active in the sea urchin assay (concentration not indicated) and, when applied to fish pellets at 5 µg/ml, signifi-cantly inhibited the feeding of goldfish.

An unusual farnesyl hydroquinone glucoside was isolated from a Japanese gorgonian *Euplexaura* sp., whose lipophilic extract inhibited cell division in the

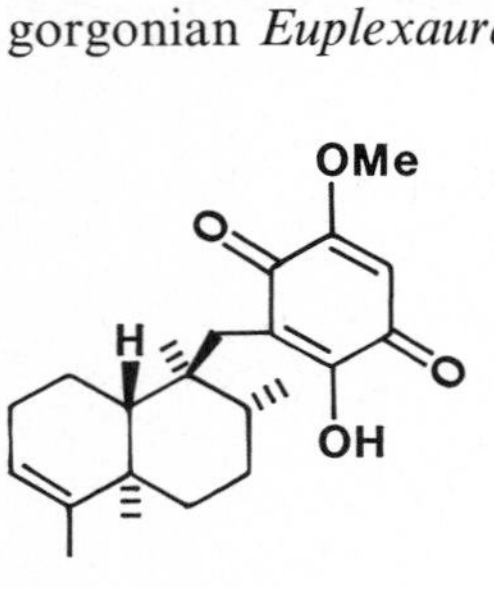

183 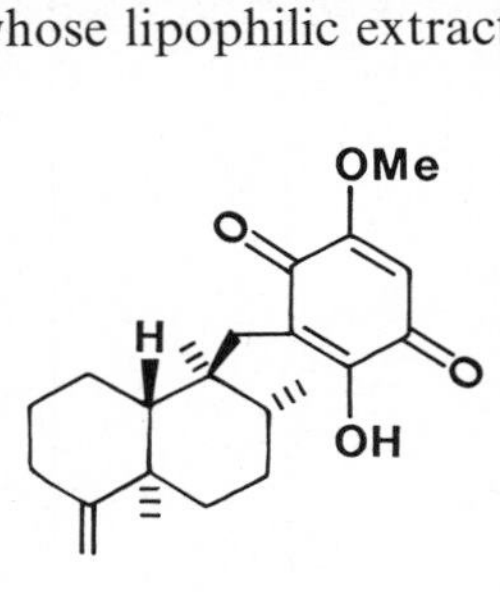   **184** 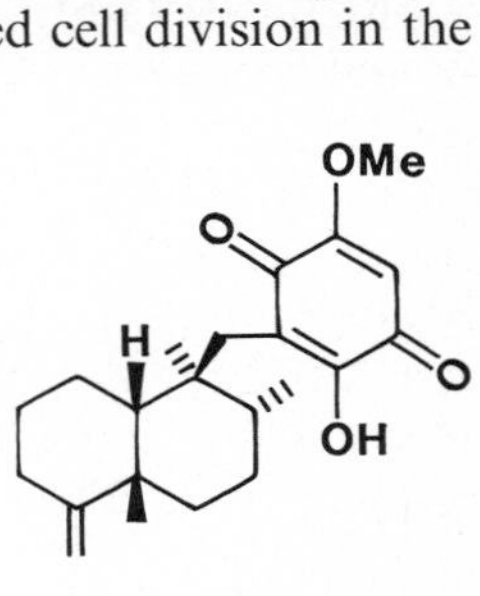   **185**

fertilized starfish egg assay. Moritoside (186) was the active compound, and was effective at 1 µg/ml [316]. This was the first example of the occurrence of D-altrose in a natural product.

Brown seaweeds were an additional source of cytotoxic terpenoid-shikimate natural products. Fenical and co-workers obtained the diterpenoid hydroquinone, bifurcarenone (187), from a Galapagos Islands collection of the brown alga *Bifurcaria galapagensis* [317]. Bifurcarenone inhibited cell cleavage of sea urchin embryos, ED_{50} 4.0 µg/ml.

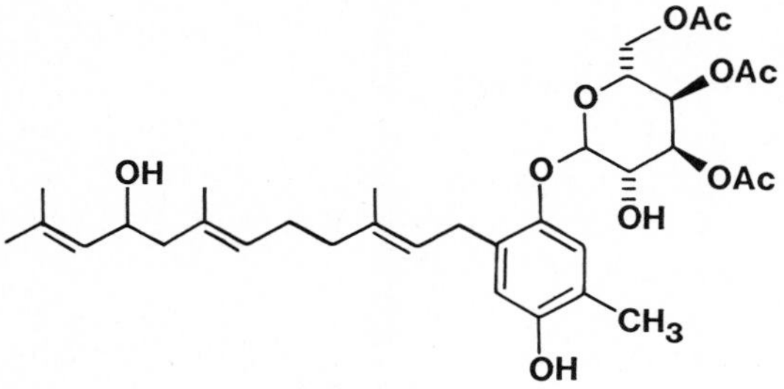

186 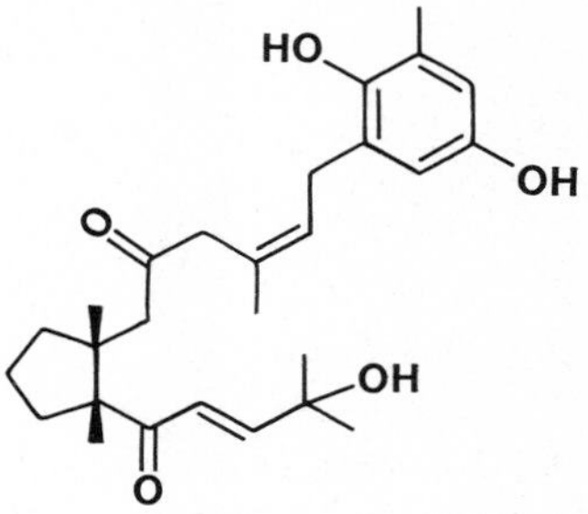   187

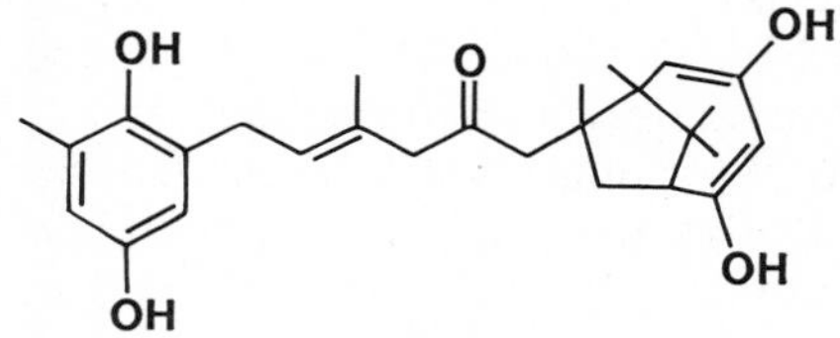

188

Another brown seaweed of the Cystoseiraceae family, *Cystoseira mediterranea*, collected from the French Mediterranean coast by Francisco and co-workers, provided mediterraneol A (188) [318] (ED_{50} in the sea urchin assay 2 µg/ml). Especially striking in the rearranged diterpenoid part of the molecule is the bicyclo[4.2.1]nonane skeleton. The stereochemistry and exact nature of the tautomers which comprise mediterraneol A were not reported in the preliminary publication.

4.4 Shikimates

The viscera of the hard-shelled snail *Kelletia kelletii* was the source of two moderately antimicrobial tetraesters of erythritol and 2-deoxyribose, kelletinin I and II (189 and 190, respectively) [319]. Both compounds were cytotoxic toward monkey kidney (CV-1) cells and L1210 leukemia cells, at the low concentration of 0.4 µg/ml for the L1210 assay. The kelletinins I and II appear to be produced *de novo* by the mollusc. Since most animals cannot biosynthesize aromatic amino acids by the shikimate pathway, the ultimate source of the *p*-hydroxybenzoic acid is presumably a plant species.

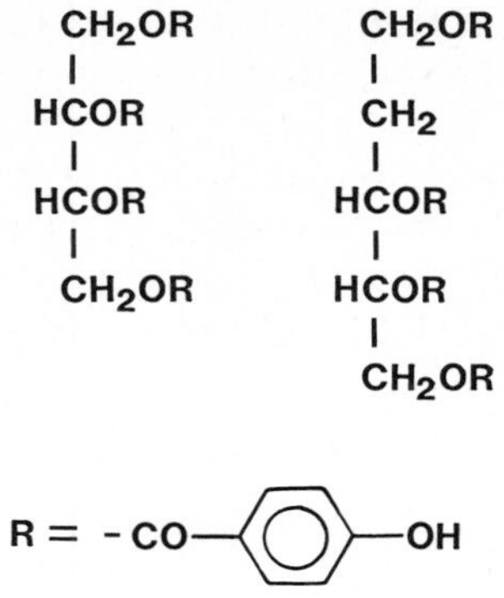

$$CH_2OR$$
$$|$$
$$HCOR$$
$$|$$
$$HCOR$$
$$|$$
$$CH_2OR$$

$$CH_2OR$$
$$|$$
$$CH_2$$
$$|$$
$$HCOR$$
$$|$$
$$HCOR$$
$$|$$
$$CH_2OR$$

R = -CO—⟨○⟩—OH

189   190

4.5 Polyketides

4.5.1 Acetogenins

Clavularin A (**191**) and clavularin B (**192**) are two simple epimeric metabolites from the Japanese soft coral *Clavularia koellikeri* [320]. Both compounds have a strong lethal effect on PV_1, cultured cells transformed with polyoma virus. The structures originally proposed were later modified.

The crude extract of the sponge *Xestospongia muta* showed in vivo tumor-inhibiting ability [321]. A dibromoacetylenic acid (**193**) was isolated from this extract. It showed slight cytotoxicity in the P388 and L1210 assays (ED_{50} 24 µg/ml and 34 µg/ml, respectively), and would be considered inactive by the NCI standards. Neither the acid (**193**) nor its methyl ester showed in vivo P388 tumor inhibitory activity.

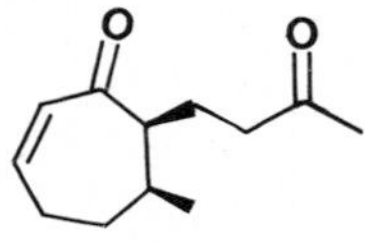

191 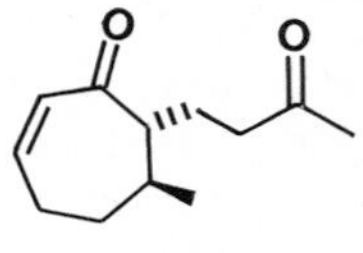   192

$$BrCH=CH-CBr=CH-(CH_2)_4-CH=CH-C\equiv C-(CH_2)_3-CO_2H$$

193

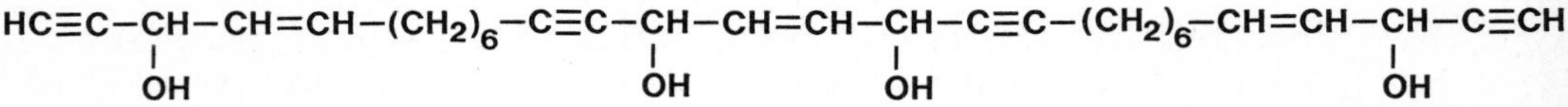

$$HC\equiv C-CH-CH=CH-(CH_2)_6-C\equiv C-CH-CH=CH-CH-C\equiv C-(CH_2)_6-CH=CH-CH-C\equiv CH$$
$$\quad\quad |\quad\quad\quad\quad\quad\quad\quad\quad\quad\quad\quad\quad |\quad\quad\quad\quad\quad |\quad\quad\quad\quad\quad\quad\quad\quad\quad\quad\quad\quad\quad |$$
$$\quad\quad OH\quad\quad\quad\quad\quad\quad\quad\quad\quad\quad\quad\quad OH\quad\quad\quad\quad OH\quad\quad\quad\quad\quad\quad\quad\quad\quad\quad\quad\quad\quad OH$$

194

Fusetani obtained a symmetrical linear polyacetylene alcohol triaconta-4,15,26-triene-1,23,28,29-tetrayne-3,14,17,28-tetrol, (**194**) from a new *Petrosia* sp. sponge. This compound inhibited the cell division of sea urchin eggs at a concentration of 1 µg/ml [322].

Patil and Rinehart obtained acetogenins from a new sponge (Order: Lithistida; Family: Theonellidae) collected at a depth of 800 m using the Johnson-Sea-Link submersible. This sponge provided three examples from a series of homologous alkyl substituted 1,2-dioxolane-3-acetic acids (195–199), which were strongly antimicrobial and strongly cytotoxic (ED_{50} <0.5 μg/ml) toward the P388 cell line [323].

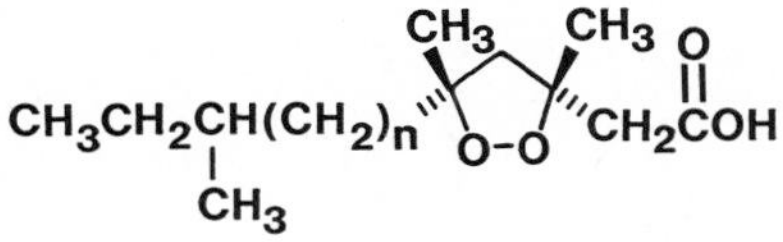

195	n = 9	198	n = 12
196	n = 10	199	n = 13
197	n = 11		

Acanthifolicin (200) and okadaic acid (201), two related sponge polyether derivatives of a C_{38} fatty acid, were reported simultaneously [324, 325]. Both were remarkably cytotoxic. Acanthifolicin was a trace component of the Caribbean sponge *Pandaros acanthifolium*. Strong cytotoxicity in the bioassay-guided isolation did not arouse great interest during fractionation, since it is known that oleic and palmitoleic acids exhibit ED_{50} values of 0.67 and 0.96 μg/ml against the P388 cell line [256]. However, repeated chromatography rewarded the group with pure acanthifolicin (200), a class of compound previously obtained from bacteria, but which contains a unique episulfide group which was unprecedented among the polyether antibiotics. Acanthifolicin was remarkably cytotoxic, with ED_{50} values of 0.000208 μg/ml, 0.0021 μg/ml and 0.0039 μg/ml, respectively against P388, KB, and L1210 cell cultures. Unfortunately, it was also toxic to mammals. The in vivo tests showed toxicity down to 0.14 mg/kg doses. At sub-toxic doses acanthifolicin did not meet the NCI Tumor Panel criteria for activity [144].

Okadaic acid (201) was obtained independently from the black Japanese sponge *Halichondria* (syn *Reniera*) *okadai* and *H. melanodocia* collected in the Caribbean. Schmitz and Gopichand used cytotoxicity to monitor the separation of the *H. melanodocia* extract; Tsukitani and Kikuchi measured toxicity to mice as a bioassay. The pure compound exhibited ED_{50} values of 0.0017 μg/ml and 0.017 μg/ml respectively against P388 and L1210 cell lines and KB cells were inhibited by more than 30% at 0.0025 μg/ml and more than 80% at 0.005 μg/ml.

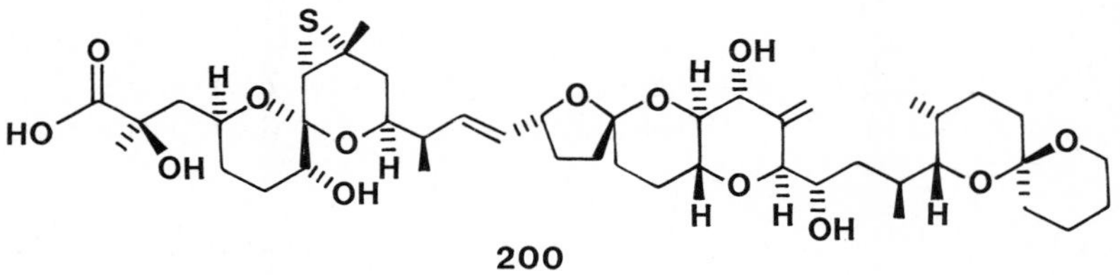

200

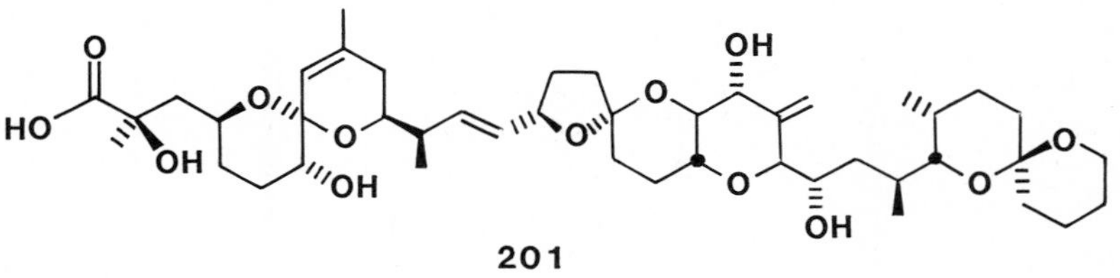

201

Testing in mice showed that okadaic acid was toxic at doses >0.12 mg/kg (intra-peritoneal) (LD$_{50}$ 0.192 mg/kg) and showed no inhibition against P388 at sub-toxic doses. Although okadaic acid proved to be ineffective as an antitumor agent, this ionophoric polyether causes contraction of smooth muscle both in the presence and absence of calcium [326]. Okadaic acid was identified as a toxic com-ponent of the widely distributed marine dinoflagellate *Prorocentrum lima*, and was isolated from scallops and mussels as one of the causative agents of diarrhetic shellfish poisoning. The dinoflagellate *Dinophysis acuminata* is the probable source of okadaic acid in the shellfish [327, 328].

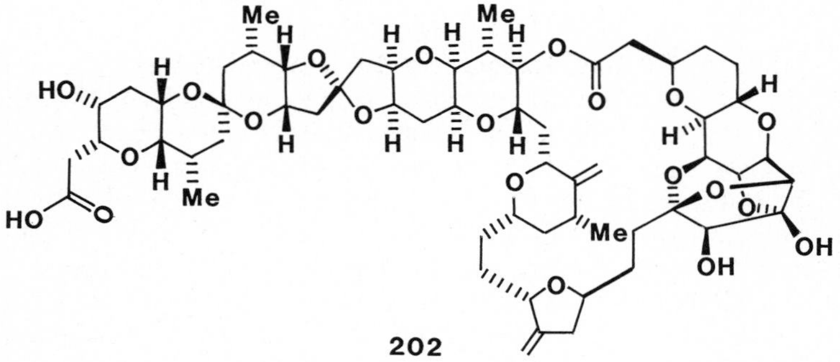

202

Halichondria okadai, the sponge from which okadaic acid was obtained, was also the source of norhalichondrin A (**202**) (EC$_{50}$ 0.005 µg/ml). Attracted by the in vivo antitumor activity of the crude extract, Uemura and co-workers used in vitro activity against B16 cells as a bioassay for isolating the active constituents [329]. An even more potent analog, halichondrin B, with high in vivo activity against L1210, P388, and B-16 cell lines has been announced in a footnote [329]. A total of 8 halichondrins have now been reported [330]. At the time of this review no quantitative antitumor data had been published for this series.

4.5.2 Acetate-Propionates

A macrolide of mixed acetate-propionate biogenesis was obtained from the Caribbean fire sponge *Tedania ignis* [331]. Tedanolide (**203**) differs from other macrolides in that the site of lactonization is not close to the terminus of the car-

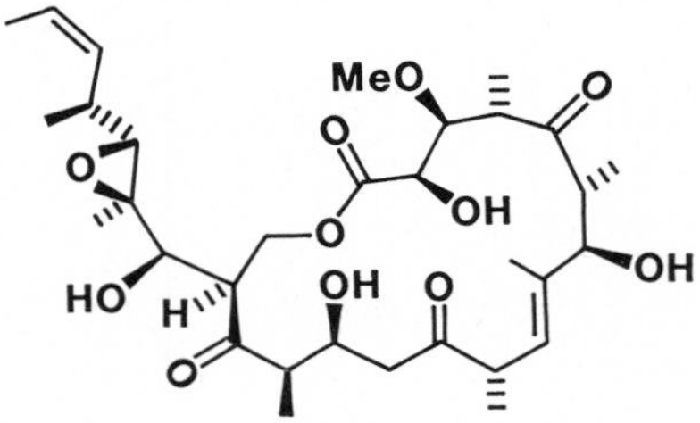

203

bon skeleton. The 18-membered ring substance was highly cytotoxic, exhibiting an ED_{50} of 0.00025 µg/ml in KB and 0.000016 µg/ml in P388 cell lines. Results of in vivo evaluation of the antineoplastic activity were disappointing. Tedanolide was toxic to mice at 12–40 µg/kg. In the P388 assay a T/C value of 123% was obtained at 1.56 µg/kg; tedanolide was inactive against M5 (ovarian cancer) cells [144].

4.5.3 Propionates

Ksebati and Schmitz obtained ten metabolites from the sacoglossan mollusc *Tridachia crispata* collected in Jamaica. Five of the eight compounds tested for cytotoxicity were active against P388 cells at the indicated ED_{50} (µg/ml): crispatone (**204**), crispatene (**205**), tridachiapyrone-A (**206**), tridachiapyrone-B (**207**), and tridachiapyrone-D (**208**) [332]. Crispatene and crispatone had been isolated previously by Ireland from *T. crispata* collected in Belize and Panama [333]. Although tridachiapyrone-A and -B were cytotoxic, their stereoisomers, with the opposite configuration at carbon 14, were inactive in the bioassay. The stereochemistry for tridachiapyrone-D was not established completely.

Siphonaria diemenensis is an air-breathing gastropod mollusc with a limpet-like shell. An Australian collection provided diemenensin-A (**209**), an α-pyrone of polypropionate origin which was antimicrobial and inhibited cell division in the fertilized sea urchin egg assay at 1 µg/ml [334].

Biskupiak and Ireland obtained a crude extract of the shell-less mollusc *Peronia peronii* (Guam) which inhibited the growth of L1210 cells in vitro (ED_{50} 0.5 µg/ml) [335]. Chromatography yielded an intractable mixture of related esters

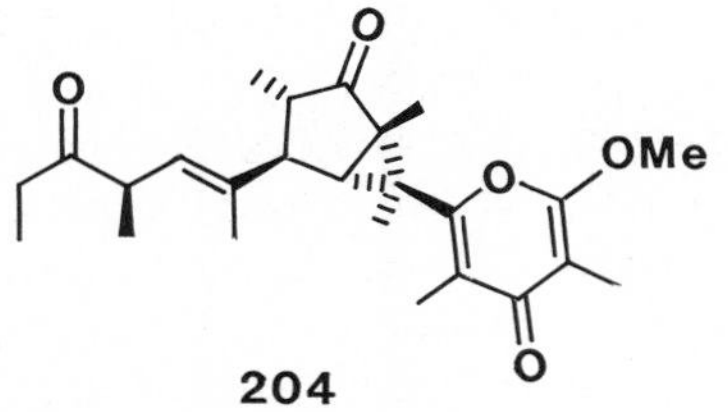

204

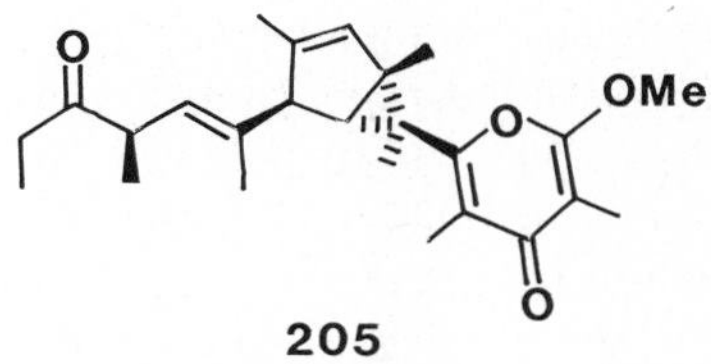

ED$_{50}$ (µg/ml)   7.2

205

3.7

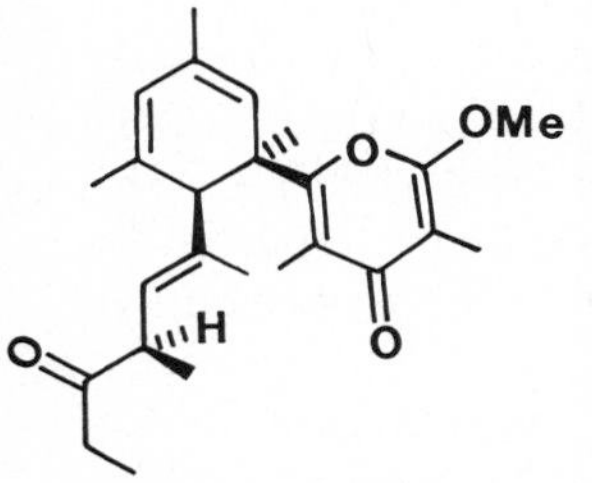

206

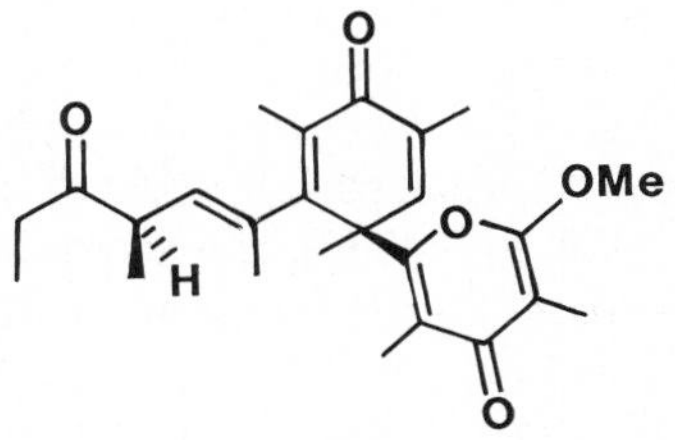

ED$_{50}$(µg/ml) 5

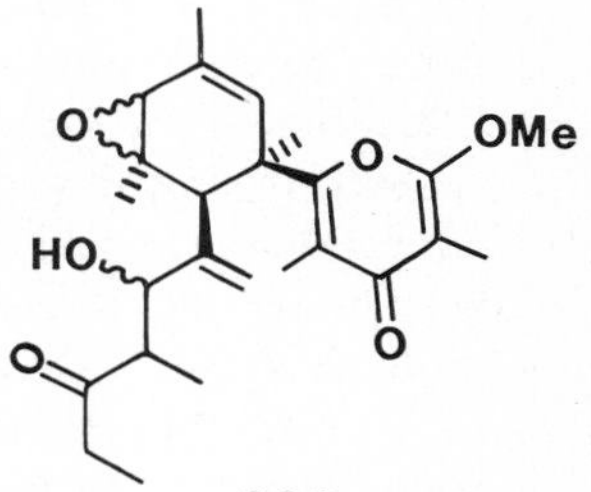

207

6

208

3.1

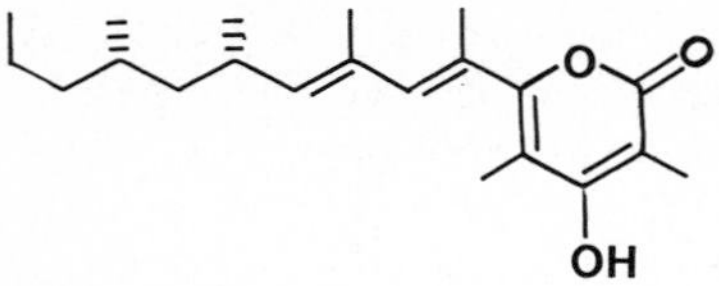

209 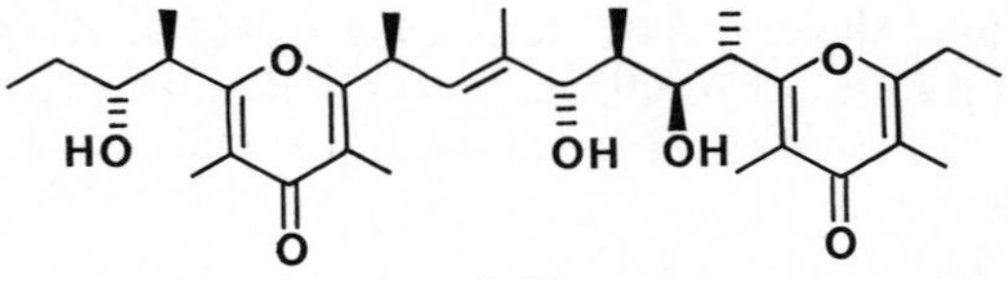   **210**

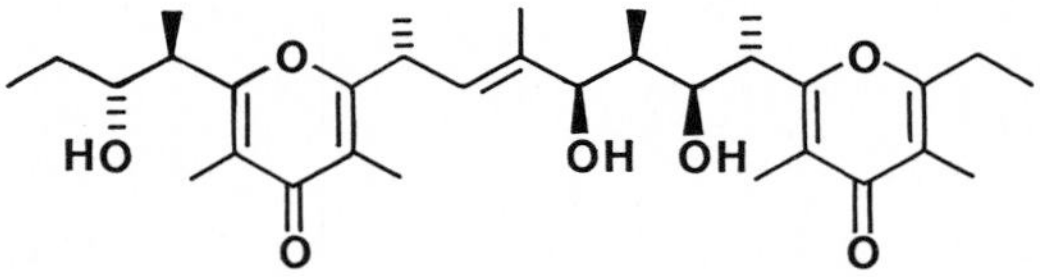

211

(ED$_{50}$ 0.07 µg/ml). Saponification of the mixture gave equal quantities of peroniatriol I (**210**) and II (**211**), which were less active than the esters (ED$_{50}$ 5.5 µg/ml and 3.1 µg/ml), respectively. Structures for the esters have not been proposed.

4.6 N-Containing Compounds

4.6.1 Tyrosine Derived Bromo Compounds

Almost all sponges in the Families Aplysinidae and Aplysinellidae (Order Verongida) contain brominated tyrosine derivatives [116, 336, 336a, 337]. The major exception is found in the Brazilian *Aplysina* spp. which are strikingly devoid of brominated metabolites [338]. Several bromotyrosine derivatives are cytotoxic.

Aeroplysinin-1 (**212**) has been isolated in three forms. The dextrorotatory enantiomer was obtained from Mediterranean collections of *Aplysina* (formerly *Verongia*) [339] *aerophoba* and *A. cavernicola* [340–342]. The Caribbean sponges *A. archeri* [343, 344], *Pseudoceratina* (reported as *Ailochroia*) *crassa*, and *Verongula rigida* [344, 345], and the Pacific sponge *Psammaplysilla purpurea* also contained the same (+)-aeroplysinin-1 [346]. The levorotatory enantiomer was found in the Caribbean sponges *Pseudoceratina crassa* (reported as *Ianthella ardis*) [336, 347] and *Verongula gigantea* [344]. Racemic aeroplysinin-1 was found in other Caribbean collections of *Verongula gigantea*, and *Pseudoceratina* (reported as *Ailochroia*) *crassa* [344]. It is unusual to obtain a natural product in racemic and both of the enantiomeric forms. It is even more striking that different collections of the same sponge, *Pseudoceratina crassa* produced all three forms.

Aeroplysinin-1 was cytotoxic toward KB and HL-5 or HS-5 cells [348, 349] and active in the L1210 assay in vivo [337]; concentrations were not reported for these bioassays. Racemic aeroplysinin-1 has been synthesized by Andersen and Faulkner [350].

Another dibromotyrosine derivative **213** was active in the KB test at an unspecified concentration [337]. This dienone has been assigned at least four different systematic names. The trivial name, dibromoverongiaquinol, is used in this review [351]. This strongly antimicrobial substance has been obtained from sev-

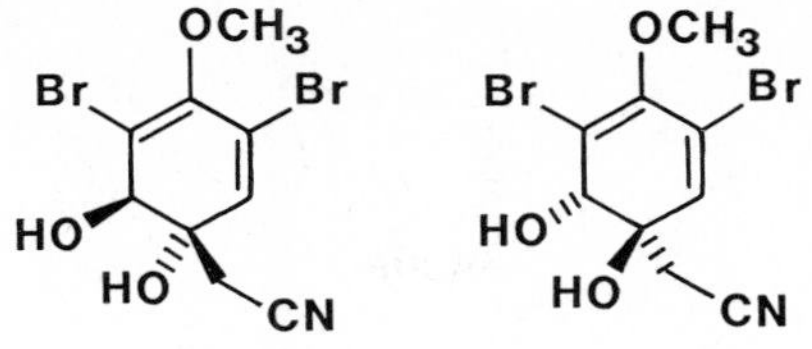

(+) 212   (−)

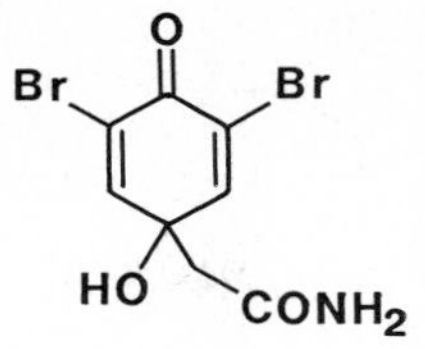

213

eral *Aplysina* sponges: *A. archeri, A. fistularis* (fulva) [344], *A. fistularis* [352, 353], *A. cauliformis* [352, 354], *A. aerophoba, A. cavernicola* [336a, 340, 342, 355], *A. hiona,* and *A.* sp. [356]. From a series of labelling experiments with *A. fistularis,* Tymiak and Rinehart concluded that both phenylalanine and tyrosine are biosynthetic precursors of dibromoverongiaquinol [353]. A synthesis of **213** has been reported [357].

Gopichand and Schmitz observed cytotoxicity in a high molecular weight dibromotyrosine-related metabolite, fistularin-3 (**214**), which they obtained from

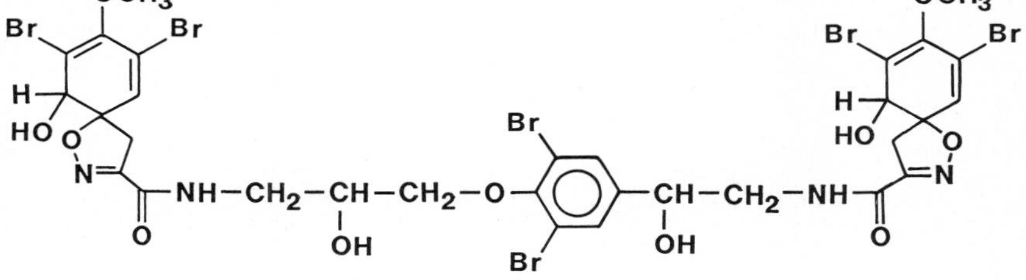

214

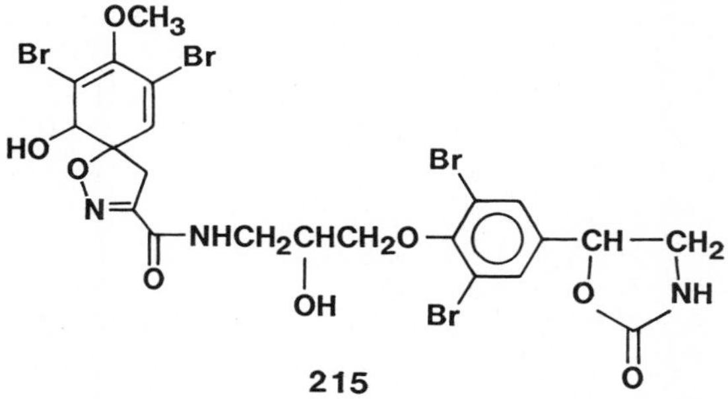

215

the Caribbean sponge *Aplysina fistularis* forma *fulva* [358]. Fistularin-3 inhibited cell growth in the KB, P388, and L1210 in vitro assays, with ED_{50} values of 4.1, 4.3, and 1.3 µg/ml, respectively. More recent bioassays showed ED_{50} values of 23, 7.1, and 3.3 µg/ml, respectively, in the same tests [144]. Its synthetic tetraacetate, originally reported to be active in the P388 system (ED_{50} 14 µg/ml) was later found to be less active (ED_{50} 24 µg/ml). A related compound, fistularin-1 (**215**), which contains an oxazolidone ring, was originally reported to be inactive against the KB, P388, and L1210 cell cultures (ED_{50} 21–35 µg/ml), but later tests showed activity in these cell lines with ED_{50} values of 14, 4.1, and 10 µg/ml, respectively.

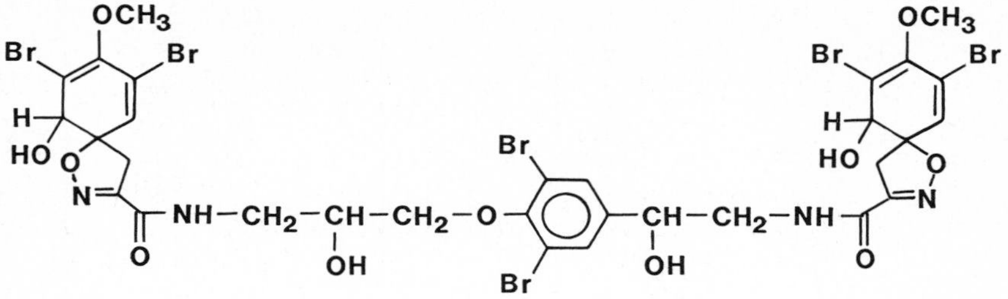

216

Fistularin-3 was tested in the in vivo P388 assay, but was not active [144]. An isomer of fistularin-3 (**214**), isofistularin-3 (**216**), which differed from fistularin-3 in the stereochemistry of one or more of the chiral centers, was also cytotoxic in the KB cell line. An "effective dose" was reported at 4 µg/ml [340]. Isofistularin-3 was isolated from an extract of the Mediterranean sponge *A. aerophoba*.

A simple ammonium salt **217**, which is a structural and pharmacologic hybrid of epinephrine and acetylcholine, was obtained from the Caribbean sponge *A. fistularis* [359]. Like epinephrine, the dibromotyramine derivative induced dual adrenergic activity in dogs i.e. moderate increase in blood pressure, followed by a small short-lived decrease. The salt (**217**) was weakly cytotoxic in the P388 screen

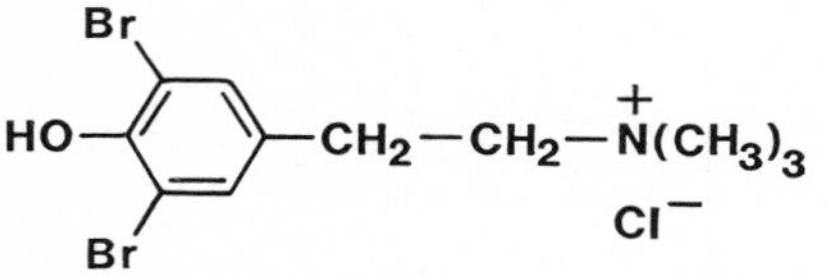

217

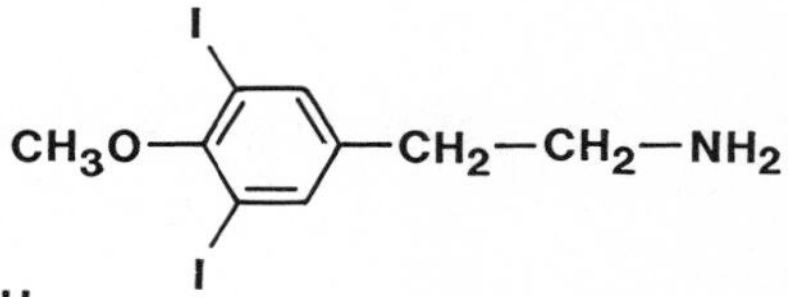

218

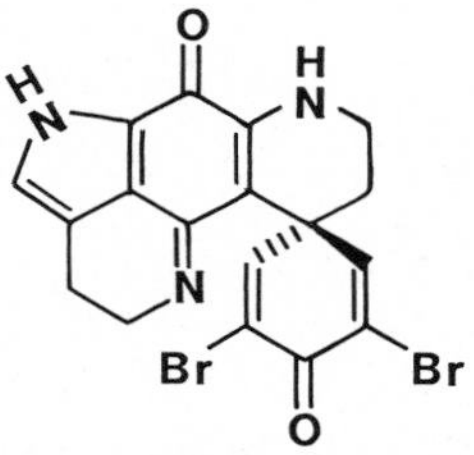

219

(ED$_{50}$ 20 µg/ml). An iodinated analog (**218**) obtained from a tunicate (*Didemnum* sp.) from Guam showed similar cytotoxicity in the L1210 assay (ED$_{50}$ 20 µg/ml) [360].

A highly cytotoxic (ED$_{50}$ in P388 0.03 µg/ml), and structurally very interesting dibromotyrosine derivative, is discorhabdin C (**219**) [361]. This pigment was isolated in a bioassay directed analysis of a sponge of the genus *Latrunculia* du Bocage as part of a wide-scale screening of New Zealand's marine invertebrates for antiviral and antitumor activity. The structure contains a new skeleton with a tetracyclic iminoquinone chromophore and a spiro 2,6-dibromocyclohexadienone ring, which is presumably derived from tyrosine and tryptophan [353]. This compound is an exception to the generalization that bromotyrosine-derived metabolites occur only in sponges in the order Verongida [336, 362].

4.6.2 Indole Alkaloids

Rinehart and co-workers obtained a series of brominated indoles from the red alga *Laurencia brongniartii*. Only one, 2,3,5,6-tetrabromoindole (**220**), was reported to be cytotoxic toward L1210 cells (ED_{50} 3.6 μg/ml), and antimicrobial. The other compounds all had N-methyl groups [363]. Citorellamine (**221**) from the tunicate *Polycitorella mariae* had the same level of activity in the L1210 assay (ED_{50} 3.7 μg/ml), and was strongly antimicrobial [364]. The tunicate *Dendrodoa grossularia* provided an indole with a rare 1,2,4-thiadiazole ring incorporated into the side chain [365]. Dendrodoine (**222**) was reported to have cytotoxic activity toward the L1210 cell line, but the concentration was not specified. Hogan and Sainsbury have reported a "one pot" synthesis of dendrodoine [366].

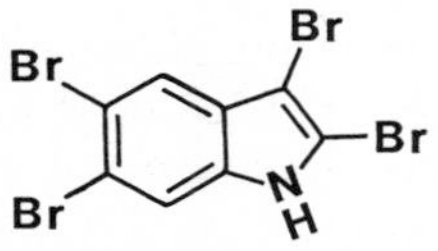

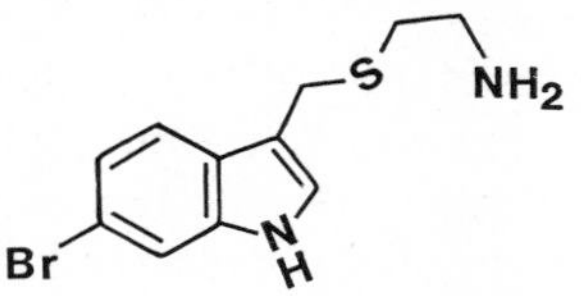

Aplysinopsin (**223**) is a tryptophan derivative first isolated from the sponges *Aplysinopsis* sp. and *Smenospongia echina* (reported as *Verongia spengelii*) [336, 367, 368]. The Australian group confirmed the structure by synthesis. Hollenbeak and Schmitz pursued the isolation of aplysinopsin because of in vivo inhibitory activity of their sponge extract against P388 in mice (T/C 135% at 0.2 mg/kg). Bioassay (KB) guided fractionation led to its characterization. Aplysinopsin was cytotoxic against the cancer cell lines KB, P388, and L1210 (ED_{50} 0.87, 3.8, and 3.7 μg/ml, respectively), but was not active in the in vivo P388 or B16 (melanoma) assays [144].

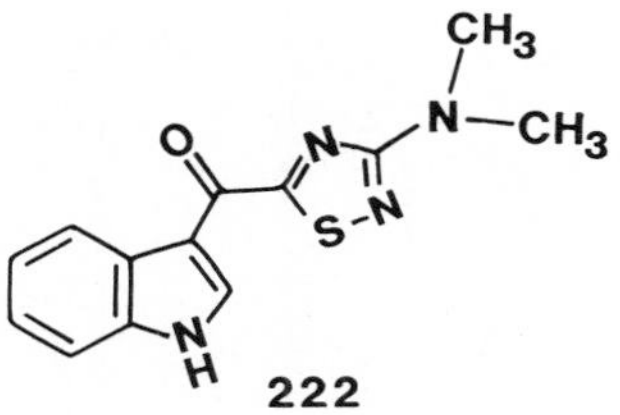

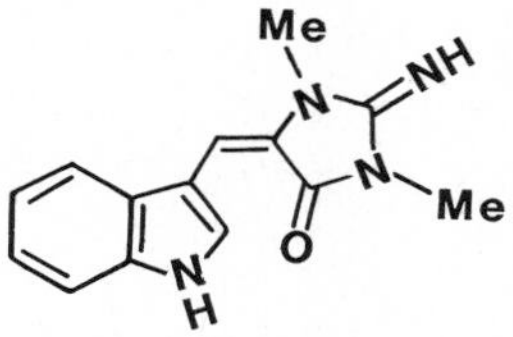

Subsequent reports of isolation of aplysinopsin from sponges of the family Thorectidae have appeared: *Thorectandra choanoides* and *Thorectandra* sp. [336]. It has also been isolated from a sponge identified as *Dercitus* sp., which differs from the Thorectidae sponges even at the subclass level [369]. The occurrence of brominated and methyl aplysinopsin derivatives has been surveyed [336, 362, 370].

4.6.3 Other Alkaloids

Although the cytotoxicity of soft corals is often due to the terpenoid (especially cembranolide) components, the activity of the extract of the Fijian alcyonacean

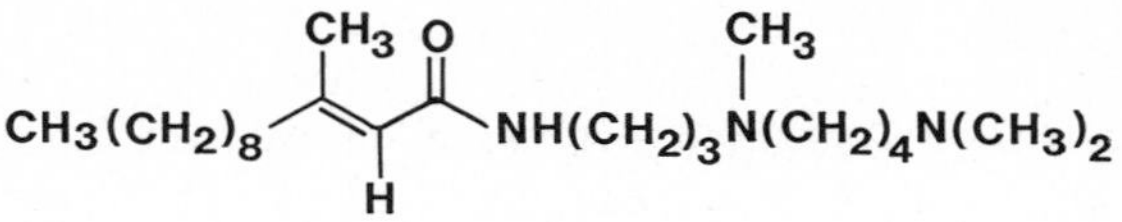

$$CH_3(CH_2)_8 \overset{CH_3}{\underset{H}{\diagup}} \overset{O}{\diagdown} NH(CH_2)_3N(CH_2)_4N(CH_3)_2$$

224

$$CH_3(CH_2)_8\overset{CH_3}{CH}CH_2CONH(CH_2)_3\overset{CH_3}{N}(CH_2)_4N(CH_3)_2$$

225

Sinularia brongersmai was caused by the first two spermidine derivatives, **224** and **225**, found in a marine organism [371]. The inseparable 9:1 mixture of **224** and **225** was converted to **225** by hydrogenation. Comparable cytotoxicity was observed for the 9:1 mixture of **224** and **225** and for the pure dihydro derivative **225**. The ED_{50} values (µg/ml) for the mixture were KB 1.0, P388 0.04, and L1210 0.30. Results of in vivo testing (P388) showed that the compounds were not active [144]. Ganem and co-workers have reported the synthesis of these compounds from the parent polyamine, spermidine [372].

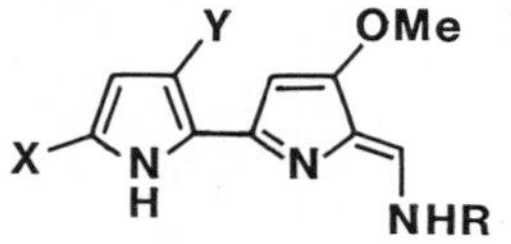

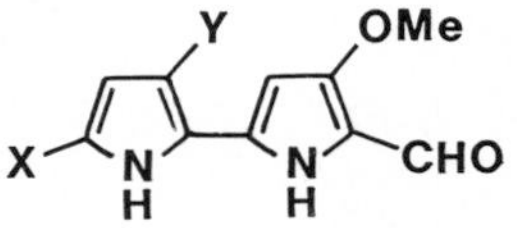

226 X = H, Y = H, R = H	**230** X = H, Y = H
227 X = Br, Y = H, R = H	**231** X = Br, Y = H
228 X = H, Y = H, R = i-Bu	**232** X = H, Y = Br
229 X = H, Y = Br, R = i-Bu	

The bryozoan *Sessibugula translucens* was the dietary source of the tambjamines A–D (**226–229**), which Carté and Faulkner isolated from three nembrothid nudibranchs, *Roboastra tigris* (large carnivorous), and the smaller prey *Tambje eliora* and *Tambje abdere* [373]. Mixtures of tambjamines A–D and a mixture of the three artifacts of the isolation procedure, aldehydes **230–232**, all inhibited cell division in the sea urchin egg assay at 1 µg/ml. The tambjamines A–D were antimicrobial, but the aldehydes were not. When attacked by *Roboastra tigris, T. abdere* produced a yellow mucus, a defensive secretion which contained relatively large amounts of the tambjamines and often caused the predator to break off its attack. Curiously, the tambjamines are also present in low concentrations in the slime trails of the *Tambje* species. Although they might be used to repel most potential predators, *R. tigris* can detect the chemicals and use them to track its preferred prey.

Renierone (**233**) is the major antimicrobial metabolite of a bright blue sponge *Reniera* sp. [374]. Both renierone and minor component N-formyl-1,2-dihydrorenierone (**234**) inhibited cell division in the fertilized sea urchin egg assay at unspecified concentration. Many other minor isoquinoline quinones were also characterized, including renieramycins A–D (compounds similar in structure to the sa-framycin antitumor antibiotics) and mimosamycin. Saframycins and mimosamy-cin have been isolated from *Streptomyces lavendulea* and it is possible that the sponge metabolites were produced by a symbiotic microorganism. Insufficient quantities of the renieramycins were obtained to determine whether they had anti-tumor properties similar to those reported for the saframycins [375]; cytotoxicity evaluations were not reported for the renieramycins.

Schmitz and co-workers found the cytotoxic fused pentacyclic alkaloid am-phimedine (**235**) with some structural similarity to renierone in extracts of a Pa-cific *Amphimedon* sp. sponge [376]. Although an ED_{50} value of 2.8 µg/ml was ob-tained from in vitro testing, the compound was inactive in vivo in the P388 as-say [144].

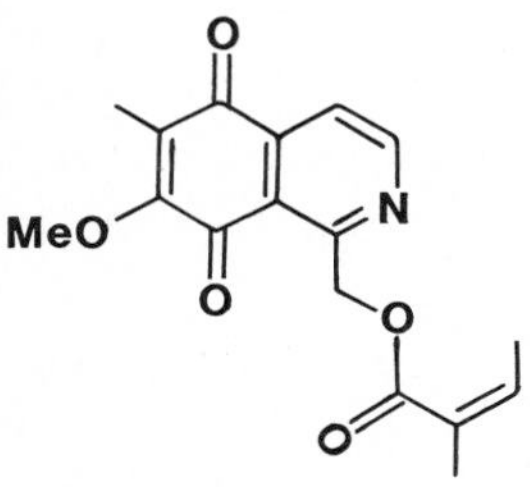

233 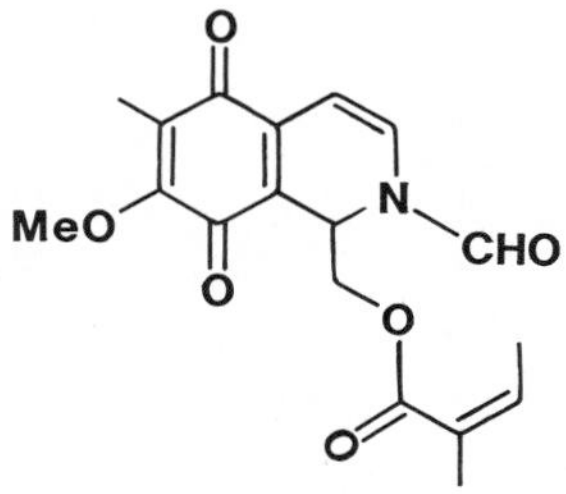   **234** 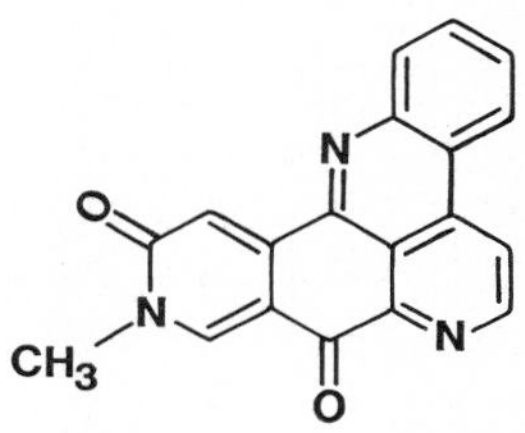   **235**

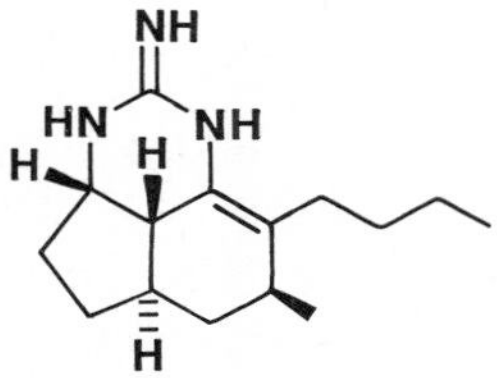

236 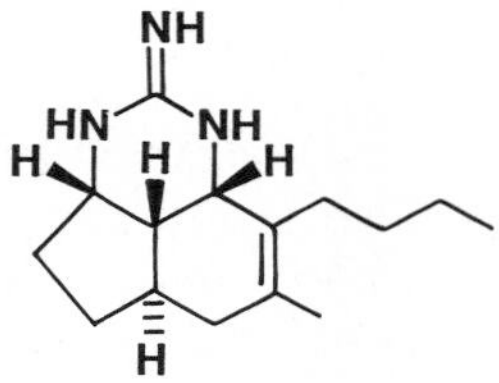   **237**

Two guanidine-containing isomers, ptilocaulin (**236**) and isoptilocaulin (**237**) were isolated from the Caribbean sponge *Ptilocaulis* aff. *P.spiculifer* by Rine-hart's group. These antimicrobial compounds were isolated as their nitrate salts and both inhibited L1210 cells (ED_{50} 0.39 µg/ml and 1.4 µg/ml, respectively) [377]. Rinehart speculated that the biosynthetic pathway leading to ptilocaulin and isoptilocaulin involves the addition of guanidine to a polyketonide chain. A similar retrosynthetic analysis led to a synthesis of chiral ptilocaulin, which estab-lished the absolute configuration of the natural product [378, 379]. An alternate synthesis led to the same conclusion [380].

A brominated guanidine derivative (**238**) was isolated from both the Mediter-ranean sponge *Axinella verrucosa* and the Red Sea sponge *Acanthella aurantiaca* [381]. The yellow solid was "moderately cytotoxic" against KB cells in vitro (ED_{50} not reported), but was inactive in the P388 in vivo assay.

Suvanine (**239**) appears to be a guanidine-containing sesterterpene isolated from an undescribed *Ircinia* sp. sponge from Suva Harbor, Fiji. It inhibited cell

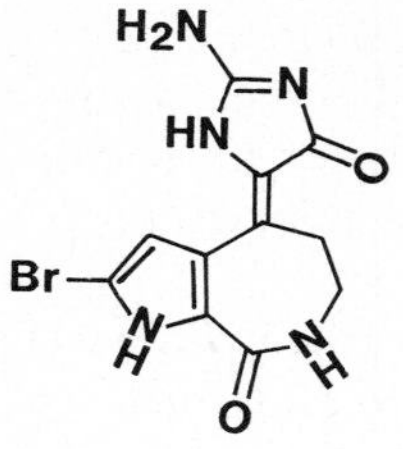

238 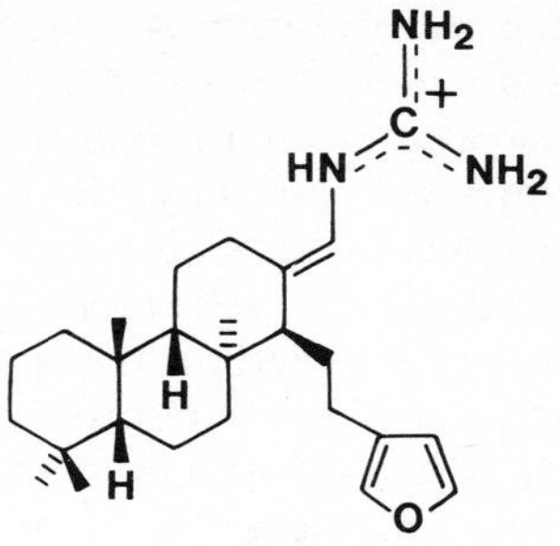   **239**

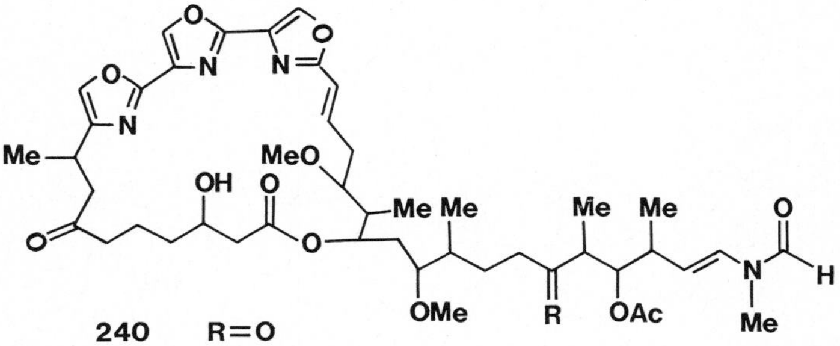

240   R=O

241   R=

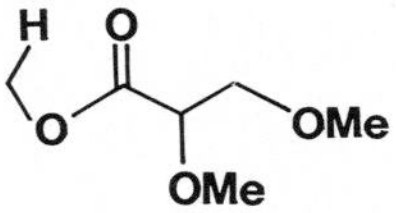

division of fertilized sea urchin eggs with ED_{100} 16 µg/ml [382]. The published structure of suvanine is now known to be incorrect [383].

Ulapualide A (**240**) and B (**241**) are nitrogen-containing macrolides which Roesener and Scheuer obtained from the egg-masses of the nudibranch *Hexabranchus sanguineus* and in lower concentration, from the nudibranch itself. They were strongly antimicrobial, and cytotoxic toward L1210 cells (ED_{50} 0.01–0.03 µg/ml) [384]. Fusetani and co-workers simultaneously announced a similar macrolide, kabiramide C, with marked antifungal activity from *Dendodoris nigra* egg-masses [385]. Cytotoxicity bioassays were not carried out for kabiramide C.

4.6.4 Peptides

Several cytotoxic cyclic peptides have been isolated from marine organisms. Those with reported antiviral and/or antitumor properties, like the didemnins (**20**–**22**) and the dolastatins (**68**), have been discussed earlier (see Sect. 3.2.2).

Lipophilic cytotoxic peptides were unknown until 1980 when Ireland and Scheuer discovered the thiazole-containing ulicyclamide (**242**) and ulithiacyclamide (**243**) in the tunicate *Lissoclinum patella* [386]. Subsequent investigation of the same organism, collected at a different location, disclosed three additional analogs, patellamides A–C, and resulted in clarification of the stereochemistry of ulicyclamide and ulithiacyclamide [387]. The structures proposed originally for

Table 3. Cytotoxicity of peptides from the tunicate *Lissoclinum patella*

Compound	L1210 ED_{50} (µg/ml)
Ulicyclamide	7.2
Ulithiacyclamide	0.35
Patellamide A	3.9
Patellamide B	2.0
Patellamide C	3.0
Three unnamed	> 10

Refs.: [387, 389]

the patellamides A–C (**244**–**246**) are now known to be incorrect (*vide infra*). The absolute configuration of the thiazole amino acids in these compounds was determined by a newly developed method [388]. A third collection of the same organism revealed three more unnamed cyclic peptides **250**–**252**, and resulted in a revised structure (**242**) for ulicyclamide [389]. One of these, **252**, differed from ulicyclamide only in two alkyl groups. The other two, **250** and **251**, had the same alkyl groups as ulicyclamide but with differing stereochemistry, and with one of the thiazole rings as a dihydro analog. All of these metabolites were tested for cytotoxicity toward L1210 leukemia cells; their ED_{50} values are shown in Table 3. Ulithiacyclamide and patellamide A also inhibited the human ALL cell line (T cell acute leukemia) CEM with ED_{50} values of 0.01 µg/ml and 0.028 µg/ml, respectively. The structure of ulicyclamide (**242**) has been confirmed by synthesis [390].

Two groups completed independent syntheses of the compound with the "wrong" patellamide B structure (**245**) [391, 392]; Hamada and co-workers also prepared the "wrong" patellamide C (**246**). Both compounds were different from the natural products, but remarkably the synthetic compounds were as potent in

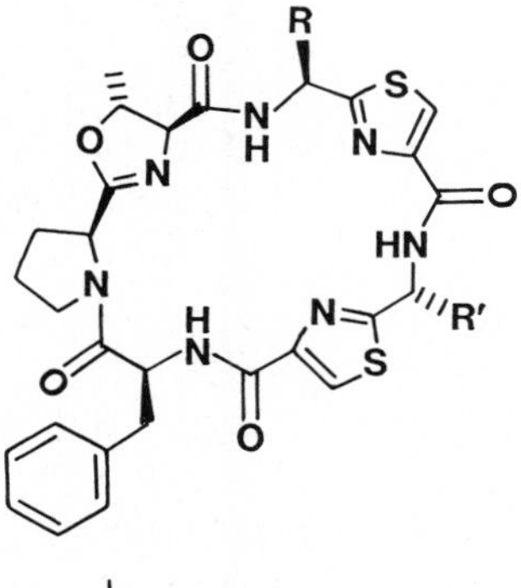

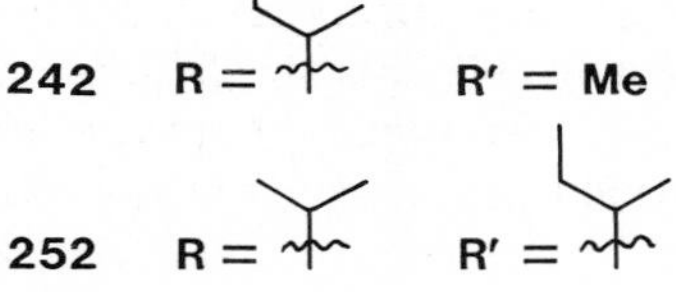

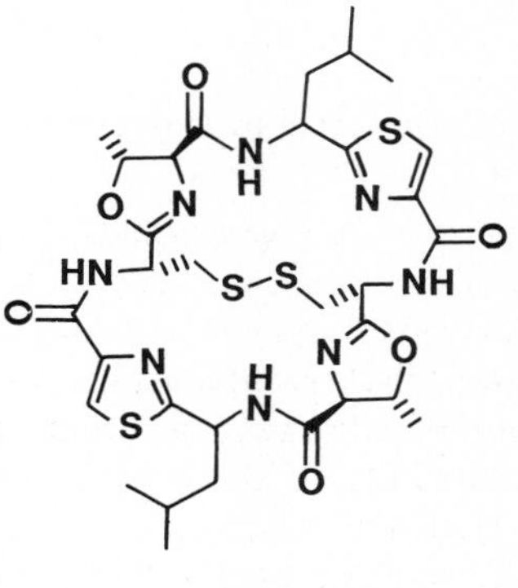

242   R = ⌇⌇   R' = Me

252   R = ⌇⌇   R' = ⌇⌇

243

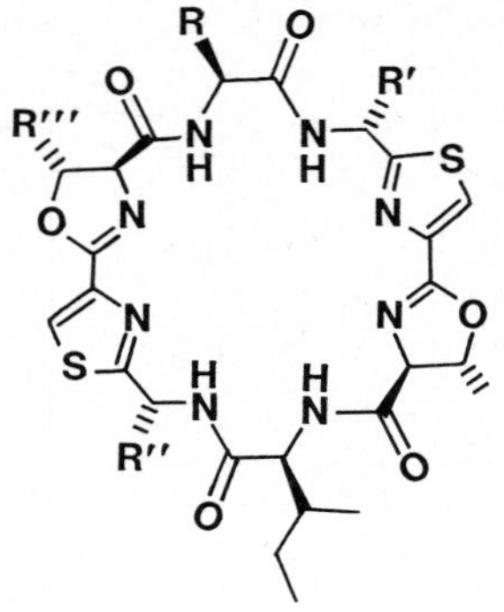

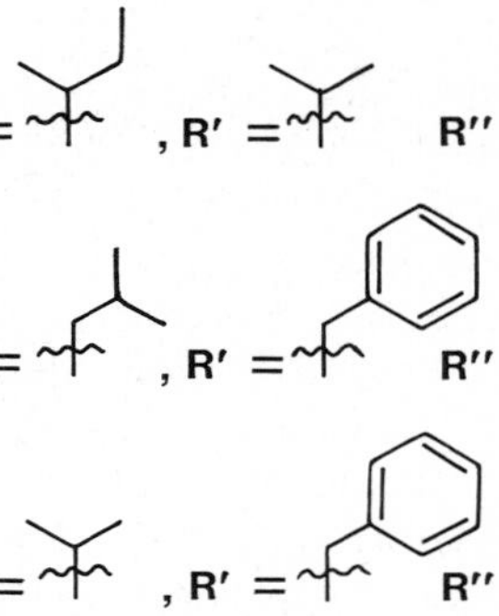

Patellamide A (incorrect) 244 R = ... , R' = ... R'' = ... , R''' = H

B (incorrect) 245 R = ... , R' = ... R'' = Me , R''' = Me

C (incorrect) 246 R = ... , R' = ... R'' = Me , R''' = Me

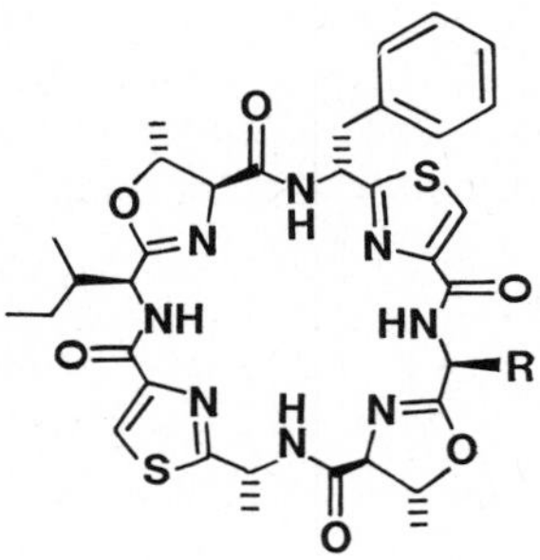

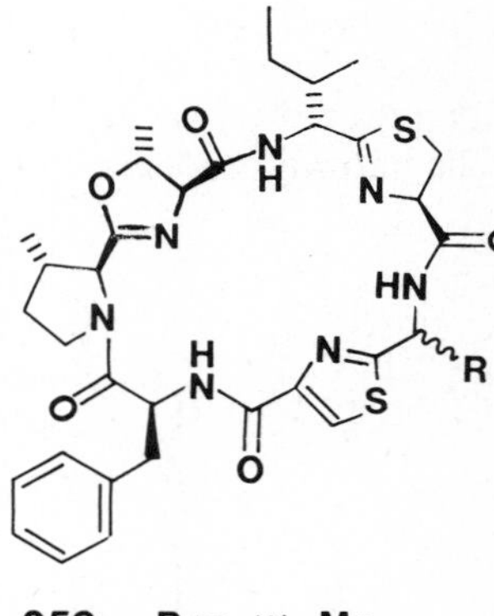

Patellamide B (corrected)   248   R = ...
                               250   R = ''' Me

           C (corrected)   249   R = ...
                               251   R = ◄ Me

their cytotoxic activity toward L1210 cells as the natural compounds from the tunicates [392]. Hamada and co-workers proposed corrected structures for the patellamides B and C, and in an accompanying communication they announced the completion of the synthesis of patellamides B (**248**) and C (**249**) which are identical to the natural products; the patellamide A synthesis is in progress [393]. Shortly thereafter, Schmidt and Griesser confirmed this result with their synthesis of patellamide B [394].

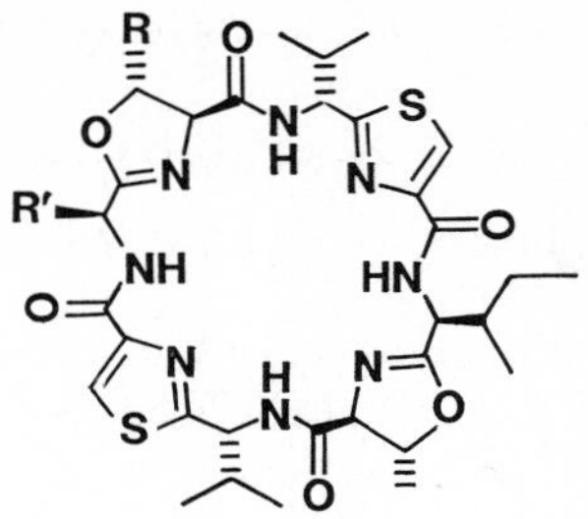

Patellamide A (Probable) 247 R = H , R' =

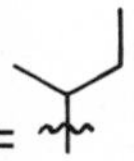

253 R = Me, R' =

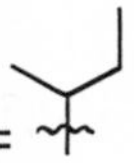

254 R = Me , R' =

A related cyclic peptide, ascidiacyclamide (**253**), along with ulithiacyclamide (**243**), was obtained from an unidentified Australian ascidian [395]. The absolute configuration was determined by synthesis [396]. Ascidiacyclamide differs from the corrected structure of patellamide A (**247**) proposed by Schmidt by a methyl group [391]. A third compound **254** related to these appeared in a Japanese patent as "an anticancer agent" [sic] [397]. This substance, also obtained from a tunicate, differs from ascidiacyclamide by the substitution of an ethyl in place of a *sec*-butyl group. Ascidiacyclamide had a strong lethal effect on PV_1 cultured cells transformed with polyoma virus, and on L1210 cells. The unnamed peptide **254** was 100% effective in inhibiting the growth of PV_4 tumor cells in cultures.

With this group of didemnid tunicates (Trididemnum, Didemnum, Lissoclinum) a point of contention has been the origins of the cytotoxic/antitumor metabolites [125]. Do they originate from the animal or the symbiont, a prochlorphyte of the genus *Prochloron*? Müller and co-workers have observed that extracts of shallow water (<6 m) *Prochloron-Didemnum molle*, and deep water (>25 m) *D. molle* lacking the symbiont, had identical ED_{50} values in the L5178y in vitro assay, and concluded that the cytotoxic compounds in the extracts were produced by the tunicate, not the symbiont [398].

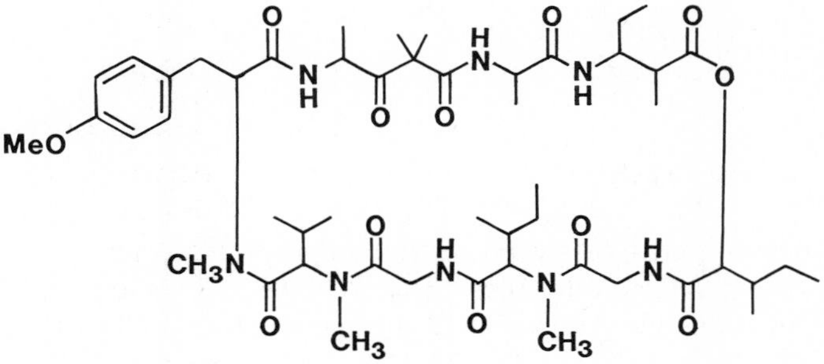

255

HCO-D-Ala-L-Phe-L-Pro-X-D-Trp-L-Arg-D-Cys(O₃H)-L-Thr-L-MeGln-D-Leu-L-Asn-L-Thr-Sar

256   X = D-t-Leu-L-t-Leu

257   X = D-t-Leu-L-Val

Blue-green algae and sponges are also known to contain cytotoxic cyclic peptides. Majusculamide C (**255**) is a depsipeptide which has been isolated from a deep water variety of the blue-green alga *Lyngbia majuscula* by Moore and coworkers [399]. Majusculamide C had marginal to nil activity against P388 leukemia, 6C3HED lymphoma and 755 carcinoma cells, and showed only 35% inhibition against X-5563 myeloma in vivo at 0.5 mg/kg [125]. Fusetani and coworkers have obtained four antimicrobial tetradecapeptides from the sponge *Discodermia kiiensis*; two of them, discodermin A (**256**) and C (**257**), inhibited the development of starfish embryos at 5 µg/ml [400]. The structure of discodermin A was revised from earlier reports [401, 402].

4.6.5 Porphyrin Derivatives

The female echurian worm *Bonellia viridis* produces a green pigment with striking biological activity. It is believed that the pigment is associated with the masculinization of the sexually undifferentiated larvae [403]. If the larvae settle away from an adult female, most develop into females [404].

Extracts of the green pigment inhibited the growth of KB cells in tissue culture and inhibited cell cleavage of sea urchin embryos. However, selective growth inhibition was shown by the fact that they were not toxic to fish, even when injected intraperitoneally, and to rabbits when injected intravenously [15].

Pelter and co-workers have elucidated the structure of the proboscis pigment, bonellin (**258**) [403, 405]. Examination of the pigments of the body integument showed that specimens from Naples consisted mainly of the isoleucine amino acid conjugate (**259**) [406], whereas the Maltese organisms showed a mixture which

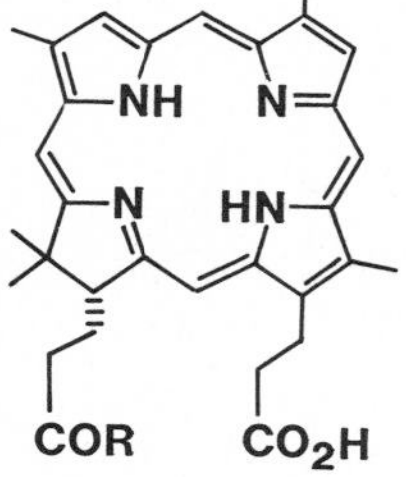

258   R = OH

259   R = −NHCHCHCH₂CH₃ (with CH₃ and CO₂H substituents)

was half bonellin, and half amino acid conjugates of valine (63%), isoleucine (23%), leucine (6%), and alloisoleucine (4%) [405a]. Purified bonellin and the isoleucine conjugate, but not their methyl esters inhibited cell cleavage of sea urchin embryos at 10^{-6} M (0.5 µg/ml) [406], but cytotoxicity was observed only if light was present; the same effect was observed for *B. viridis* embryos [407].

4.6.6 Nucleosides

The mycalisines A (**260**) and B (**261**), which were isolated by Fusetani from a *Mycale* species of sponge, are modified ribosides [408]. Mycalisine A effectively inhibited cell cleavage of fertilized starfish eggs (ED_{50} 0.5 µg/ml). Surprisingly, mycalisine B was considerably less active (ED_{50} 200 µg/ml).

Regular ribosides and 2′-deoxyribosides have been found in marine organisms [41, 409–413].

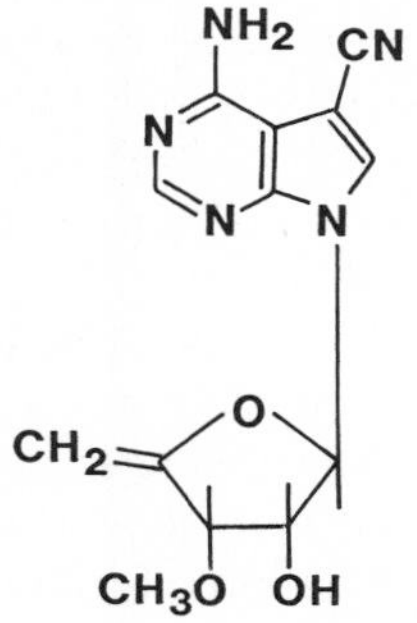

260

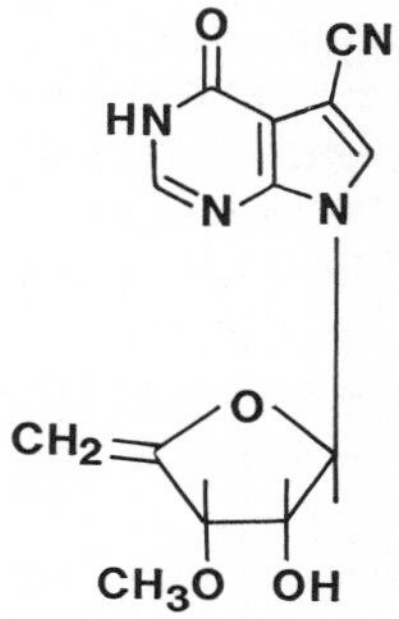

261

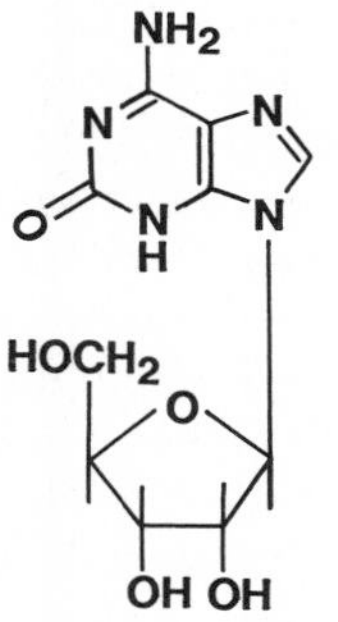

262

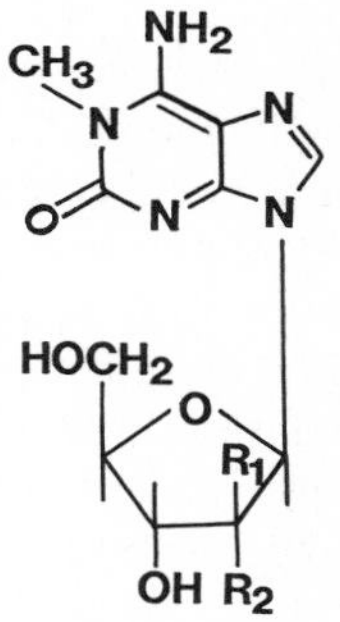

263    $R_1 = H$, $R_2 = OH$

264    $R_1 = OH$, $R_2 = H$

The less common ribosides isoguanosine (**262**) [414] and 1-methylisoguanosine (doridosine) (**263**) [415–420] have featured prominently as marine compounds with strong biological activity. Although many nucleosides have been shown to possess antiviral and/or cytotoxic activity, there were no reports of testing for this type of activity from these compounds.

5. Future Developments

Clearly, the last decade has witnessed substantial progress in the search for antiviral and antitumor drugs. If the pace is to continue to quicken, chemists might examine methods which will increase the proportion of "leads" from the extracts examined, simplify the isolation procedures, or use the newer screening methods that more accurately reflect the human situation.

5.1 Sample Selection

It has been suggested that cytotoxic compounds are more likely to be found in tropical waters than in temperate or cold water environments [4, 421–423]. This conclusion is in direct contrast to observations on the incidence of antimicrobial activity from temperate water sponges [424, 425]. As there is an excellent correlation between antimicrobial activity and cytotoxicity [6], there appears to be little support for this original suggestion. The results from the temperate waters of New Zealand also run counter to this suggestion [25]. It is our conclusion that latitude should not be considered as a factor in sample selection.

The other sample selection variable, which has now been examined, is that of activitiy as a function of depth. With the intense involvement of SeaPharm Inc./ Harbor Branch Oceanographic Institution in the search for antiviral, antitumor, antifungal and immunomodulatory compounds from marine organisms, the resources of the Harbor Branch Oceanographic Institution have been utilised in deep water collections by using the Johnson-Sea-Link submersibles [426]. Since 1984, SeaPharm has collected over 7000 samples of invertebrates off the Bahama Islands from snorkelling and SCUBA depths, down to 800 m. Although the types and varieties of sponges collected from the deeper water varied from the shallow water species, the incidence of biological activity was similar. This is a significant observation as it effectively adds a further dimension to sample selection. The probability of finding certain types of activity as a function of depth was also examined. There appears to be a trend towards a higher incidence of antiviral activity in species living at depths less than 30 m. Conversely, there appears to be a greater likelihood of finding cytotoxicity at depths greater than 30 m [427].

5.2 Isolation Techniques

The overwhelming majority of compounds reported from marine organisms have been lipophilic in nature. In contrast, the majority of the biologically active compounds described are polar molecules, often at the interface of water/organic solvent solubility. Such polar molecules are difficult to isolate by the traditional means of partitioning the crude extract between an organic solvent and water, followed by chromatography of the organic soluble material. With such molecules even desalting can be a difficult task. Shimizu has reviewed the problems associated with handling polar marine extracts. Although reverse-phase chromatography is mentioned favorably its use was not recommended, except for final purification steps, because of low sample capacity [148]. Since then a low cost, high capacity method of partitioning has been described which is based on reverse-phase flash chromatography [428].

5.3 Screening

Few chemists have access to laboratories that offer more sophisticated and dedicated biological screening facilities than the NCI [429]. Since the inception of the National Cancer Chemotherapy Program in 1955, many different tumor systems have been used as primary screens for anticancer compounds. A major consideration in the screening process is whether the models used accurately reflect the cancers which are the target of the drug discovery effort. Over the years the NCI has regularly reviewed the panel of screens used and instituted alternatives as appropriate (see for eg. Refs [430, 431]). The criteria used for selecting these screens have been the validity, selectivity and predictability of the various models in recognizing compounds with potential against human cancers. Murine models, such as the P388, are limiting because they are not necessarily predictors of activity against **human** cancers, and because an animal model is unlikely to identify all **human** antitumor agents. Fortunately, recent developments in assays in which human tumors are grown as xenografts in athymic mice and in cell culture will increase the validity and predictability of the screens [432–435].

The discovery of genetically athymic mice (sometimes called "nude" mice because they have no fur) has revolutionized screening methodology in the cancer field. These mice lack a thymus gland, and hence are immunodeficient and cannot readily reject a transplant of foreign tissue (xenograft). It is possible, therefore, to use human tumors in xenograft experiments in these mice, and actually test for antitumor activity using human tumors [94, 433, 434].

The fairly recent human tumor stem cell assay, also called the human tumor colony forming assay, is an in vitro test which utilizes fresh human tumor surgical or biopsy specimens. Single cell suspensions are plated out in a soft agar feeding layer in the presence of test compound, and the number of tumor colonies formed is measured after incubation. Inhibition of colony formation to 30% or less of control values is considered to be an active test in this system. With the introduction of models which more closely resemble the human situation there will be a greater likelihood that compounds active in this assay will also have clinical activity [94, 436–443].

The use of tissue culture studies can greatly reduce the need for more costly and time-consuming in vivo testing. For example, using cytotoxicity to P388 lymphocytic leukemia (PS) cells would correlate with P388 activity in mice, and could probably reduce the in vivo testing in P388 by 70–80 percent [94]. The disadvantage of this approach is that compounds which require metabolic activation are missed. Nevertheless there are many advantages to in vitro testing. Tissue culture techniques are very sensitive and can be rapidly carried out. They are cost effective and use very little sample. Their sensitivity is 10–1000 times greater than in vivo tests, which can detect only the most potent compound, since active compounds are generally present in the crude extract at low concentrations ($10^{-2}\%$ to $10^{-6}\%$). Even though tissue culture techniques have many advantages in identifying the most interesting extracts, in vivo assessment is then necessary to determine their potential. Most cytotoxic extracts will be inactive in vivo, either because they lack selectivity against tumor cells, or because the active component cannot reach the cell at an effective concentration [92, 94, 444].

The NCI intends to implement a new policy of assessing crude extracts which is based on tissue culture techniques [445]. The Institute will establish a panel of up to 100 human cancer cell lines representing slow-growing refractory tumor types by developing multiple stable cell lines of each selected tumor type. The cell lines selected will be those capable of being transplanted into athymic mice, so that interesting in vitro data can be followed up by in vivo studies. The tumor types to be developed will include lung, colorectal, breast and ovarian. The goal of this approach is to discover agents that have selective activity in one or more of these cell lines and to quickly develop these "leads". This is a disease-orientated approach, and again highlights the disadvantages of a single primary screen like P388. Positive in vitro assays would be followed up by the appropriate in vivo assessment [446].

The NCI has also initiated a series of new projects with a heavy emphasis on marine organisms. Contracts were let in mid-1986 for the extensive collection of shallow water (< 30 m) marine invertebrates from the Indo-Pacific region and for the collection of deep water (> 30 m) marine organisms. The many thousands of samples resulting from these contracts will be screened against the new battery of human tumor cell lines [445, 447].

6. Conclusions

Many of the amazing array of compounds already elicited from marine organisms have never been tested for biological activity or pharmaceutical potential. Too often, the only testing has been against microorganisms. How many interesting antiviral or antitumor "leads" have already been lost? With the promise of free, rapid, more extensive in vitro screening by the NCI, and the evolution of simple systems of testing for cytotoxicity, the opportunity for marine natural products to be screened for antiviral and anticancer potential is greatly increased.

The often novel and interesting compounds from the sea can be manipulated in many ways. One fascinating example was the recent synthesis of ara-doridosine (**264**) [448]. The inspiration for the synthesis of this compound was the report of the resistance of the marine ribonucleoside 1-methylisoguanosine (doridosine) (**263**) to adenosine deaminase [417]. The effectiveness of ara-A as an antiviral agent is limited by its rapid deamination. By synthesizing ara-doridosine it was hoped that the coupling of 1-methylisoguanine with arabinose would create a new nucleoside with prolonged antiviral activity. The implementation of such ideas, even if unsuccessful, support and enhance the search for antiviral and anticancer drugs from marine organisms.

As the search for biological activity in marine organisms widens and the screening methods become more selective, the probability of finding specific compounds with good pharmaceutical potential will increase. With the compounds already discovered there is sufficient basis for the belief that the catchy conference title "Drugs From The Sea" will one day become a reality. The detection and isolation of these compounds is a stimulus for all chemists.

Acknowledgments

Because of the very generous release of pre-prints and information by Professors DJ Faulkner (Scripps Institution), GR Pettit (University of Arizona), KL Rinehart (University of Illinois), PJ Scheuer (University of Hawaii), FJ Schmitz (University of Oklahoma), Dr J Clement (SeaPharm Inc.) and Dr GM Cragg (NCI), the effect of time-lag, one of the hazards facing any review, has been minimized. We also wish to acknowledge the roles played by the University of Canterbury in the supply of resources, the California State University, Hayward for a sabbatical period for one of us (RTL) and Harbor Branch Oceanographic Institution for research funding; the commitment of Victoria Luibrand in assembling data and typing; Mr Simon Petrie for bibliographic research; Mr Alistair Duff for art work; Mr Chris Battershill for taxonomic advice; Professor MP Hartshorn and Dr Nigel Perry for proof-reading, and fellow members of the University of Canterbury marine natural products research group for encouragement and assistance.

7. Appendix: Occurrence of Antiviral, Antitumor and Cytotoxic Compounds in Marine Organisms

The distribution of these types of biological activity through the various marine classes is shown in outline in Scheme 1 and then organised according to phyla in the charts that follow.

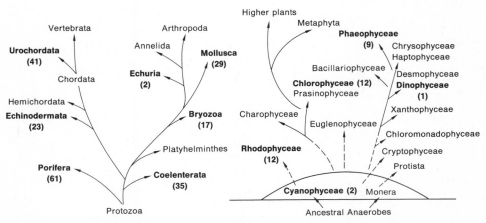

Scheme 1. Distribution of antiviral, antitumor and cytotoxic activity in the marine plant and animal Kingdoms

Genus species	Compound		Refs.
Phylum: Cyanophyta			
Lyngbia majuscula	debromoaplysiatoxin	**44**	[123]
	majusculamide C	**255**	[399]
Oscillatoria nigroviridis	debromoaplysiatoxin	**44**	[124]
Schizothrix calcicola	debromoaplysiatoxin	**44**	[124]
Phylum: Pyrrhophyta			
Prorocentrum lima	okadaic acid	**201**	[327]
Dinophysis acuminata	okadaic acid	**201**	[328]
Phylum: Phaeophyta			
Bifurcaria galapagensis	bifurcarenone	**187**	[317]
Cystoseira mediterranea	mediterraneol A	**188**	[318]
Dictyota dentata	dictyol H	**102**	[235]
Dictyota dichotoma	dolabellane deriv	**100**	[234]
Dictyota sp.	dolabellane deriv	**101**	[234a]
Spatoglossum howleii	spatane derivs	**104–105**	[158]
Spatoglossum schmittii	spatol	**103**	[236]
	spatane deriv.	**104**	[236]
Stoechospermum marginatum	spatane deriv.	**104**	[158]
	spatane deriv.	**105**	[158]
Stypopodium zonale	stypoldione	**47**	[157]
Phylum: Rhodophyta			
Desmia hornemanni	cyclohexadienones	**15–18**	[75]
Laurencia brongniartii	bromoindole deriv	**220**	[363]
Laurencia elata	elatol	**76**	[213]
Laurencia obtusa	elatol	**76**	[214]
	isoobtusol	**75**	[78]
	thyrsiferol	**138**	
	thyrsiferyl deriv.	**137**	[267]
Laurencia thyrsifera	thyrsiferol	**138**	[268]
Plocamium cartilagineum	halomonoterpenes	**71–72**	[77, 78]
	halomonoterpene	**73**	[76]
Phylum: Chlorophyta			
Caulerpa flexilis	flexilin	**91**	[231]
Caulerpa prolifera	caulerpenyne	**88**	[234]
Halimeda spp. (nine)	halimedatrial	**98**	[233]
Halimeda spp. (two)	diterpenoid	**99**	[228]
Penicillus capitatus	sesquiterpenoids	**89–90**	[228]
Penicillus dumetosus	diterpenoids	**96–97**	[228]
Rhipocephalus phoenix	rhipocephanol	**92**	[228]
Udotea conglutinata	flexilin	**91**	[228]
Udotea cyathiformis	sesquiterpenoid	**90**	[228]
Udotea flabellum	sesquiterpenoid	**89**	[228]
	udoteal	**93**	[228]
	diterpenoid	**94–95**	[232]
Phylum: Bryozoa			
Sessibugula translucens	tambjamines A-D	**226–229**	[373]
Bugula neritina	bryostatin 1	**55**	[180]
	bryostatin 2	**56**	[184]
	bryostatin 3	**57**	[183]
	bryostatin 4	**58**	[185]

Genus species	Compound		Refs.
	bryostatin 5–7	**59–61**	[186]
	bryostatin 9	**63**	[191]
	bryostatin 10–11	**64–65**	[189]
	bryostatin 12	**66**	[190]
	bryostatin 13	**67**	[190]
Amathia convoluta	bryostatin 4–6	**58–60**	[187]
	bryostatin 8	**62**	[187]
Phylum: Echinodermata			
a) Actinopyga spp. (several)	holothurin A	**139**	[269]
	holothurin B	**140**	[269]
Holothuria spp. (several)	holothurin A	**139**	[269]
b) Acanthaster planci	asterosaponins	**142–158**	[292]
	nodososide	**141**	[289]
Asterias amurensis [c.f.] versicolor	asterosaponins	**142–158**	[292]
Hacelia attenuata	polyhydroxysterols	**169–171**	[297]
Luidia maculata	asterosaponins	**142–158**	[292]
Linkia laevigata	nodososide	**141**	[289]
Protoreaster nodosus	nodososide	**141**	[289]
	polyhydroxysterols	**169–171**	[297]
a) Class: Holothuroidea			
b) Class: Asteroidea			
Phylum: Porifera			
a) Discodermia kiiensis	discodermin A, C	**256–257**	[400]
b) Dercitus sp.	aplysinopsin	**223**	[369]
c) new	1,2-dioxolane-3-acetic acids	**195–199**	[323]
d) Latrunculia sp.	discorhabdin C	**219**	[361]
e) Axinella aurantiaca	bromoguanidine deriv	**238**	[381]
Axinella cannabina	axisonitrile-1	**84**	[223]
	axisothiocyanate-3	**85**	[224]
Axinella verrucosa	bromoguanidine deriv	**238**	[381]
Pandaros acanthifolium	acanthifolicin	**200**	[324]
Ptilocaulis aff. P. spiculifer	ptilocaulin	**236**	[377]
	isoptilocaulin	**237**	[377]
Phylum: Porifera (continued)			
f) Epipolasis kushimotoensis	epipolasinthiourea-A, B	**86–87**	[226]
g) Acarnus erithacus	acarnidines	**11–13**	[58]
Halichondria melanodocia	okadaic acid	**201**	[325]
Halichondria okadai	okadaic acid	**201**	[325]
	norhalichondrin A	**202**	[329]
h) Prianos sp.	muqubilin	**136**	[265]
i) Mycale sp.	mycalisines A, B	**260–261**	[408]
j) Tedania ignis	atisane-3β,16α-diol	**120**	[247]
	epiloliolide	**178**	[247]
	tedanolide	**203**	[331]
k) Tetrosia sp. (Petrosia?)	polyacetylene alcohol	**194**	[322]
Xestospongia muta	dibromoacetylenic acid	**193**	[321]
l) Amphimedon sp.	amphimedine	**235**	[376]
Reniera sp.	renierone	**233**	[374]
	N-formyl-1,2-dihydrorenierone	**234**	[374]
m) Toxadocia zumi	steroidal sulfates	**159–161**	[293]

Genus species	Compound		Refs.
n) Hippospongia metachromia	ilimaquinone	**184**	[314]
Spongia officinalis	isoagatholactone	**114**	[241]
	11-hydroxyisoagatholactone	**115**	[242]
	11-acetoxyisoagatholactone	**116**	[242]
o) Aplysinopsis sp.	aplysinopsin	**223**	[367]
Cacospongia scalaris	furanosesterterpenes	**132–133**	[263]
	desacetylscalaradial	**130**	[261]
Ircinia variabilis	variabilin	**14**	[70]
Ircinia strobilina	variabilin	**14**	[71]
Ircinia sp.	variabilin	**14**	[73]
Ircinia sp.	suvanine	**239**	[382]
Fascaplysinopsis reticulata	aplysinopsin	**223**	[336]
Fasciospongia rimosa	isospongiaquinone	**183**	[310]
Fasciospongia sp.	variabilin	**14**	[72]
Fenestraspongia sp.	ilimaquinone	**184**	[312]
	5-*epi*-ilimaquinone	**185**	[312]
Smenospongia echina	aplysinopsin	**223**	[368]
Thorectandra choanoides	aplysinopsin	**223**	[336]
Thorectandra sp.	aplysinopsin	**223**	[336]
p) Dysidea arenaria	arenarol	**181**	[309]
	arenarone	**182**	[309]
Dysidea avara	avarol	**179**	[305]
Dysidea sp.	epoxysterol	**162**	[294]
Spongionella sp.	spongionellin	**134**	[264]
	dehydrospongionellin	**135**	[264]
q) Igernella notabilis	spongian lactone	**117**	[244]
r) Aplysina aerophoba	isofistularin-3	**216**	[340]
	(+)-aeroplysinin-l	**212**	[340]
	dibromoverongiaquinol	**213**	[340]
Aplysina archeri	dibromoverongiaquinol	**213**	[344]
	(+)-aeroplysinin-l	**212**	[343]
Aplysina cavernicola	dibromoverongiaquinol	**213**	[342]
	(+)-aeroplysinin-l	**212**	[342]
Aplysina fistularis	dibromotyramine deriv	**217**	[359]
	dibromoverongiaquinol	**213**	[352]
Aplysina fistularis (fulva)	fistularin-1	**215**	[358]
	fistularin-3	**214**	[358]
	dibromoverongiaquinol	**213**	[344]

Phylum: Porifera (continued)

Aplysina sp.	dibromoverongiaquinol	**213**	[356]
Aplysina thiona	dibromoverongiaquinol	**213**	[356]
Verongula gigantia	(−)-aeroplysinin-1	**212**	[344]
	(±)-aeroplysinin-1	**212**	[344]
Verongula rigida	(+)-aeroplysinin-1	**212**	[344]
s) Psammaplysilla purpurea	(+)-aeroplysinin-1	**212**	[346]
Pseudoceratina crassa	(+)-aeroplysinin-1	**212**	[344]
	(−)-aeroplysinin-1	**212**	[347]
	(±)-aeroplysinin-1	**212**	[344]

a) Subclass: Tetractinomorpha: Order: Spirophorida
b)   Order: Choristida   Family: Pachastrellidae
c)   Order: Lithistida   Family: Theonellidae
d)   Order: Hadromerida   Family: Latrunculiidae
e)   Order: Axinellida   Family: Axinellidae

Genus species	Compound		Refs.

f)	Order: Epipolasida	Family: Sollasellidae	
g) Subclass: Ceractinomorpha:	Order: Halichondrida	Family: Halichondridae	
h)		Family: Hymeniacidonidae	
i)	Order: Poecilosclerida	Family: Mycalidae	
j)		Family: Tedaniidae	
k)	Order: Nepheliospongida	Family: Nepheliospongiidae	
l)	Order: Haplosclerida	Family: Haliclonidae	
m)		Family: Adociidae	
n)	Order: Dictyoceratida	Family: Spongiidae	
o)		Family: Thorectidae	
p)		Family: Dysideidae	
q)	Order: Dendroceratida	Family: Dictyodendrillidae	
r)	Order: Verongida	Family: Aplysinidae	
s)		Family: Aplysinellidae	

Phylum: Mollusca

a) Aplysia angasi	aplysistatin	**80**	[218]
Aplysia dactylomela	14-bromoobtus-1-ene-3,11-diol	**106**	[239]
	deodactol	**78**	[217]
	elatol	**76**	[78]
	isocaespitol acetate	**79**	[78]
	isoobtusol acetate	**74**	[78]
	parguerol derivs	**107–111**	[212]
Stylocheilus longicauda	debromoaplysiatoxin	**44**	[118]
b) Dolabella auricularia	dolatriol	**112**	[95]
	dolatriol 6-acetate	**113**	[95]
	dolastatin-3	**68**	[199]
Dolabella ecaudata	dolatriol	**112**	[302]
	loliolide	**177**	[302]
c) Hexabranchus sanguineus	ulapualide A, B	**240–241**	[384]
Roboastra tigris	tambjamines A-D	**226–229**	[373]
Tambje eliora	tambjamines A-D	**226–229**	[373]
Tambje abdere	tambjamines A-D	**226–229**	[373]
d)Tridachia crispata	crispatene	**204**	[333]
	crispatone	**205**	[333]
	tridachiapyrones A, B, D	**206–208**	[332]
e) Peronia peronii	peroniatriol I, II	**210–211**	[335]
Siphonaria diemensin	diemenensin A	**209**	[334]
f) Kelletia kelletii	kelletinin I, II	**189–190**	[319]

Phylum: Mollusca (continued)

a) Subclass: Opisthobranchia	Order: Anaspidea	Family: Aplysiidae	
b)		Family: Dolabellidae	
c)	Order: Nudibranchia		
d)	Order: Sacoglossa		
e)	Order: Basommatophora	Suborder: Pulmonata	
f) Subclass: Prosobranchia	Order: Neogastropoda		

Phylum: Echuria

Bonellia viridis	bonellin	**258**	[403]
	bonellin-isoleucine conjugate	**259**	[406]

Genus species	Compound		Refs.
Phylum: Cnidaria			
a) Clavularia koellikeri	clavularin A, B	**191–192**	[320]
Clavularia viridis	clavulone I–III	**48–50**	[161]
	claviridenone-a	**51**	[163]
	chlorovulone I	**53**	[170]
	stoloniferones a–d	**163–166**	[295]
Lemnalia tenuis	lemnalol	**46**	[154]
Lobophytum hedleyi	lobohedliolide	**121**	[248]
b) Telesto riisei	punaglandin-3	**52**	[167]
c) Sinularia brongersmai	spermidine derivs	**224–225**	[371]
Sinularia conferta	cembrene-A	**127**	[256]
	flexibilene	**128**	[256]
Sinularia flexibilis	sinularin	**122**	[253]
	dihydrosinularin	**123**	[253]
	sinulariolide	**124**	[251]
	sinulariolide bridged ether	**125**	[252]
	cembrene-A	**127**	[257]
	flexibilene	**128**	[258]
d) Eunicea asperula	asperdiol	**126**	[253]
Eunicea calyculata	crassin acetate	**40**	[105]
Eunicea tourneforti	asperdiol	**126**	[253]
Eunicella cavolini	ara-A	**6**	[46]
	3′-0-acetylara-A		[46]
Euplexaura sp.	moritoside	**186**	[316]
Isis hippuris	hippuristanol	**167**	[296]
	2α-hydroxyhippuristanol	**168**	[296]
Pseudoplexaura sp. (4)	crassin acetate	**40**	[104]
Pseudopterogorgia acerosa	pseudopterolide	**129**	[260]
Pseudopterogorgia rigida	curcuhydroquinone	**81**	[222]
	curcuquinone	**82**	[222]
e) Palythoa mammilosa	palytoxin	**45**	[127]
Palythoa toxica	palytoxin	**45**	[126]
Palythoa tuberculosa	palytoxin	**45**	[127]

a) Class: Anthozoa	Subclass: Alcyonaria (Octocorallia)	Order: Stolonifera	
b)		Order: Telestacea	
c)		Order: Alcyonacea	
d)		Order: Gorgonacea	
e)	Subclass: Zoantharia	Order: Zoanthinaria	

Phylum: Chordata			
a) Halocynthia roretzi	halocynthiaxanthin	**176**	[301]
b) Dendrodoa grossularia	dendrodoine	**222**	[365]

Phylum: Chordata (continued)			
c) Polyandrocarpa sp.	polyandrocarpidine I, II	**9–10**	[62, 63]
d) Phallusia mamillata	sterol hydroperoxide	**172**	[298]
e) Ciona intestinalis	sterol hydroperoxide	**172**	[298]
f) Didemnum sp.	diiodotyramine deriv	**218**	[360]
Lissoclinum patella	ulicyclamide	**242**	[386]
	ulithiacyclamide	**243**	[386]
	patellamide A–C	**247–249**	[387]
	cyclic peptides	**250–252**	[389]

Genus species	Compound		Ref.
Trididemnum sp.	didemnin A, B, C	**20–22**	[79, 80]
Unidentified	ascidiacyclamide	**253**	[395]
	ulicyclamide	**243**	[395]
Unidentified	peptide	**254**	[397]
g) Aplidium californicum	prenylhydroquinone	**41**	[112]
Aplidium sp.	aplidiasphingosine	**8**	[58]
h) Eudistoma olivaceum	eudistomins A–Q	**23–39**	[87, 88]
Polycitorella mariae	citorellamine	**221**	[364]

a) Order: Pleurogonia   Suborder: Stolidobranchia   Family: Pyuridae
b)   Family: Styelidae
                                                                          Subfamily: Styelinae
                                                                          Subfamily: Polyzoinae
d) Order: Enterogonia   Suborder: Phlebobranchia   Family: Ascidiidae
e)   Suborder: Aplousobranchia   Family: Cionidae
f)   Family: Didemnidae
g)   Family: Polyclinidae
                                                                          Subfamily: Polycliniae
h)   Family: Polycitoridae

References

1. der Marderosian A (1969) J Pharm Sci 58:1
2. Pettit GR, Day JF, Hartwell JL, Wood HB (1970) Nature 227:962
3. Sigel MM, Wellham LL, Lichter W, Dudek LE, Gargus JL, Lucas AH (1970) In: Youngken Jr HW (ed) Food-Drugs from the sea: Proceedings, 1969, Mar Tech Soc, Washington, DC, p 281
4. Weinheimer AJ, Karns TKB (1974) In: Kaul PN, Sindermann CJ (eds) Drugs and food from the sea, University of Oklahoma Press, p 491
5. Weinheimer AJ, Matson JA, Karns TKB, Hossain MB, van der Helm D (1978) In: Kaul PN, Sindermann CJ (eds) Drugs and food from the sea, University of Oklahoma Press, p 117
6. Nemanich JW, Theiler RF, Hager LP (1978) In: Kaul PN, Sindermann CJ (eds) Drugs and food from the sea, University of Oklahoma Press, p 123
7. Faulkner DJ (1979) Oceans 22:44
8. Suffness M, Douros J (1979) In: DeVita Jr VT, Busch H (eds) Methods in cancer research, Vol. XVI, Academic Press, New York, p 73
9. Baker JT, Wells RJ (1981) In: Beal JL, Reinhard E (eds) Natural products as medicinal agents, Hippokrates Verlag, Stuttgart, p 281
10. de Souza NJ, Ganguli BN, Reden J (1982) Ann. Rep. Med. Chem. 17:301
11. Kashiwagi M, Mynderse JS, Moore RE, Norton TR (1980) J Pharm Sci 69:735
12. Patterson GML, Norton TR, Furusawa E, Furusawa S, Kashiwagi M, Moore RE (1984) Botanica Marina 27:485
13. Hodgson LM (1984) Botanica Marina 27:387
14. Ciereszko LS (1977) In: Faulkner DJ, Fenical W (eds) Marine natural products chemistry, Plenum Press, New York, p 1
15. Nigrelli RF, Stempien Jr MF, Ruggieri GD, Liguori VR, Cecil JT (1967) Fed Proc 26:1197
16. Baslow MH (1969) In: Marine pharmacology, Williams and Wilkins, Baltimore; republished with update, Krieger, Huntington, New York, 1977
17. Ruggieri GD (1976) Science 194:491

18. Grant PT, Mackie AM (1977) Nature 267:786
19. Rinehart Jr KL, Shaw PD, Shield LS, Gloer JB, Harbour GC, Koker MES, Samain D, Schwartz RE, Tymiak AA, Weller DL, Carter GT, Munro MHG, Hughes Jr RG, Renis HE, Swynenberg EB, Stringfellow DA, Vavra JJ, Coats JH, Zurenko GE, Kuentzel SL, Li LH, Bakus GJ, Brusca RC, Craft LL, Young DN, Conner JL (1981) Pure and Appl Chem 53:795
20. Naqvi SWA, Solimabi SY, Kamat L, Fernandes L, Reddy CVG, Bhakuni DS, Dhawan BN (1980) Botanica Marina 24:51
21. Caccamese S, Azzolina R, Furnari G, Cormaci M, Grasso S (1981) Botanica Marina 24:365
22. Blunden G, Barwell CJ, Fidgen KJ, Jewers K (1981) Botanica Marina 24:267
23. Andersson L, Lidgren G, Bohlin L, Magni L, Ogren S, Afzelius L (1983) Acta Pharmaceutica Suecica 20:401
24. Clement J (1986) Personal communication
25. Blunt JW, Munro MHG, Barrow CJ, Calder VL, Fenwick GD, Ingham DM, Jennings LC, Lake RJ, Lewis SJ, McCombs JD, Perry NB (1985) FECS Third International Conference on Chemistry of Biologically Active Natural Products, Sophia, Bulgaria, September 16–21
26. Li CP, Goldin A, Hartwell JL (1974) Cancer Chemotherapy Rep 4:97
27. Bergmann W, Feeney RJ (1951) J Org Chem 16:981
28. Bergmann W, Burke DC (1955) J Org Chem 20:1501
29. Bergmann W, Watkins JC, Stempien Jr MF (1957) J Org Chem 22:1308
30. Whitley RJ, Alford CA (1978) Ann Rev Microbiol 32:285
31. Hahn FE (1980) Antibiotics Chemother 27:1
32. Smith RA, Sidwell RW, Robins RK (1980) Ann Rev Pharmacol Toxicol 20:259
33. Robins RK (1986) Chemical and Engineering News 28
34. Shannon WM, Schabel Jr FM (1980) Pharm Ther 11:263
35. Shugar D (1985) Pure Appl Chem 57:423
36. Herrmann Jr EC (1961) Prog Med Virol 3:158
37. Schroeder AC, Hughes Jr RG, Bloch A (1981) J. Med Chem 24:1078
38. Fenical W, Gifkins KB, Clardy J (1974) Tetrahedron Lett 1507
39. Tringali C, Piattelli M, Nicolosi G (1984) Tetrahedron 40:799
40. Shimizu Y (1971) Experientia 27:1188
41. Cohen SS (1963) Perspect Biol Med 6:215
42. Cohen SS (1966) Prog Nucleic Acid Res Mol Biol 5:1
43. Cohen SS (1976) Med Biol 54:299
44. North TW, Cohen SS (1979) Pharmac Ther 4:81
45. Park NH, Pavan-Langston D (1982) In: Came PE, Caliguiri LA (eds) Chemotherapy of viral infections, Springer-Verlag, Berlin, New York, p 117
46. Cimino G, De Rosa S, De Stefano S (1984) Experientia 40:339
47. Lee WW, Benitez A, Goodman L, Baker BR (1960) J Am Chem Soc 82:2448
48. Privat de Garilhe M, de Rudder J (1964) Compt Rend 259:2725
49. Miller FA, Dixon GJ, Ehrlich J, Sloan BJ, McLean Jr IW (1968) In: Antimicrobial agents and chemotherapy, American Society for Microbiology, p 136
50. Whitley RJ, Soong S, Dolin R, Galasso GJ, Ch'ien RT, Alford CA (1977) N Engl J Med 297:289
51. Kazlauskas R, Murphy PT, Wells RJ, Baird-Lambert JA, Jamieson DD (1983) Aust J Chem 36:165
52. Li CP, Prescott B, Eddy B, Caldes G, Green WR, Martino EC, Young AM (1965) Ann NY Acad Sci 130:374
53. Ehresmann DW, Deig EF, Hatch MT (1979) In: Hoppe H, Levring T, Tanaka Y (eds) Marine algae in pharmaceutical science, W. deGruyter, Berlin, New York, p 293
54. Came PE, Steinberg BA (1982) In: Came PE, Caliguiri LA (eds) Chemotherapy of viral infections, Springer-Verlag, Berlin, New York, p 505
55. Shimizu Y, Kamiya H (1983) In: Scheuer PJ (ed) Marine natural products, chemical and biological perspectives, Vol V, Academic Press, New York, p 391
56. Ehresmann DW, Deig EF, Hatch MT, DiSalvo LH, Vedros NA (1977) J Phycol 13:37

57. Hatch MT, Ehresmann DW, Deig EF (1979) In: Hoppe H, Levering T, Tanaka Y (eds) Marine algae in pharmaceutical science, W.deGruyter, Berlin, New York, p 343
58. Carter GT, Rinehart Jr KL (1978) J Am Chem Soc 100:7441
59. Mori K, Umemura T (1981) Tetrahedron Lett 22:4429
60. Mori K, Umemura T (1981) Tetrahedron Lett 22:4433
61. Mori K, Umemura T (1982) Tetrahedron Lett 23:3391
62. Cheng MT, Rinehart Jr KL (1978) J Am Chem Soc 100:7409
62a. Andersen RJ, Faulkner DJ (1973) In: Food-drugs from the sea: Proc 1972, Mar Tech Soc, Washington, DC, p III
63. Carté B, Faulkner DJ (1982) Tetrahedron Lett 23:3863
64. Rinehart Jr KL, Harbour GC, Graves MD, Cheng MT (1983) Tetrahedron Lett 24:1593
65. Carter GT, Rinehart Jr KL (1978) J Am Chem Soc 100:4302
66. Blunt JW, Munro MHG, Yorke SC (1982) Tetrahedron Lett 23:2793
67. Yorke SC, Blunt JW, Munro MHG, Cook JC, Rinehart Jr KL (1986) Aust J Chem 39:447
68. Boukouvalas J, Golding BT, McCabe RW, Slaich PK (1983) Angew Chemie Int Ed Engl 22:618
69. Blunt JW, Munro MHG (1986) Unpublished results
70. Faulkner DJ (1973) Tetrahedron Lett 3821
71. Rothberg I, Shubiak P (1975) Tetrahedron Lett 769
72. Kazlauskas R, Murphy PT, Quinn RJ, Wells RJ (1976) Tetrahedron Lett 2635
73. Gonzalez AG, Rodriguez ML, Barrientos ASM (1983) J Nat Prod 46:256
74. Jacobs RS, White S, Wilson L (1981) Fed Proc 40:26
75. Higa T, Sakai R, Snader KM, Cross SS, Theiss W (1985) Abstracts of the Fifth International Symposium on Marine Natural Products, Paris, September 2–6
76. Naylor S, Hanke FJ, Manes LV, Crews P (1983) Fortschritte Chemie Organischer Naturstoffe 44:189
77. Gonzalez AG, Darias V, Estevez E, Arteaga JM, Martin VS, Gomez MCG (1980) Planta Medica 39:256
78. Gonzalez AG, Darias V, Estevez E (1982) Planta Medica 44:44
79. Rinehart Jr KL, Gloer JB, Cook Jr JC, Mizsak SA, Scahill TA (1981) J Am Chem Soc 103:1857
80. Rinehart Jr KL, Gloer JB, Hughes Jr RG, Renis HE, McGovren JP, Swynenberg EB, Stringfellow DA, Kuentzel SL, Li LH (1981) Science 212:933
81. Rinehart Jr KL, Cook Jr JC, Pandey RC, Gaudioso LA, Meng H, Moore ML, Gloer JB, Wilson GR, Gutowsky RE, Zierath PD, Shield LS, Li LH, Renis HE, McGovren JP, Canonico PG (1982) Pure Appl Chem 54:2409
82. Rinehart Jr KL (1986) Personal communication
83. Rinehart Jr KL, Gloer JB, Wilson GR, Hughes Jr RG, Li LH, Renis HE, McGovren JP (1983) Fed Proc 42:87
84. Canonico PG, Pannier WL, Huggins JW, Rinehart Jr KL (1982) Antimicrobial Agents and Chemotherapy 22:696
85. Weed SD, Stringfellow DA (1983) Antiviral Research 3:269
86. Martin MB, Rinehart Jr KL, Canonico PG (1986) Antimicrobial Agents and Chemotherapy, submitted
87. Rinehart Jr KL, Kobayashi J, Harbour GC, Hughes Jr RG, Mizsak SA, Scahill TA (1984) J Am Chem Soc 106:1524
88. Kobayashi J, Harbour GC, Gilmore J, Rinehart Jr KL (1984) J Am Chem Soc 106:1526
89. Rinehart Jr KL, Kobayashi J, Harbour GC, Gilmore J, Mascal M, Holt TG, Shield LS, Lafargue F (1987) J Am Chem Soc 109:3378
90. Blunt JW, Lake RJ, Munro MHG, Toyokuni T (1987) Tetrahedron Lett 28:1825
91. Han S-Y, Lakshmikantham MV, Cava MP (1985) Heterocycles 23:1671
92. Suffness M, Douros J (1982) J Nat Prod 45:1
93. Gueran RI, Greenberg NH, Macdonald MM, Schumacher AM, Abbott BJ (1972) Cancer Chemother Rep, Part 3, 3 No 2 (Sept.)
94. Suffness M (1985) In: Vlietinck AJ, Domisse RA (eds) Advances in medicinal plant research, Wissenschaftliche Verlagsgesellschaft, Stuttgart, p 101

95. Pettit GR, Cragg GM (1978) In: Biosynthetic products for cancer Chemotherapy, Vol II, Plenum Press, New York and London, p 118
96. Cornman I (1950) J Nat Cancer Inst 10:1123
97. Ikegami S, Kawada K, Kimura Y, Suzuki A (1979) Agric Biol Chem 43:161
98. Galsky AG, Kozimor R, Piotrowski D, Powell RG (1981) J Nat Cancer Inst 67:689
99. Ferrigni NR, Putnam JE, Anderson B, Jacobsen LB, Nichols DE, Moore DS, McLaughlin JL, Powell RG, Smith Jr CR (1982) J Nat Prod 45:679
100. Ferrigni NR, McLaughlin JL, Powell RG, Smith Jr CR (1984) J Nat Prod 47:347
101. McLaughlin JL (1985) Abstracts of the International Symposium on Organic Chemistry of Medicinal Natural Products, IUPAC, Shanghai, November 10–14
102. Ciereszko LS, Sifford DH, Weinheimer AJ (1960) Ann NY Acad Sci 90:917
103. Ciereszko LS (1960) Trans NY Acad Sci 24:502
104. Weinheimer AJ, Matson JA (1975) Lloydia 38:378
105. Look SA, Fenical W, Qui-tai Z, Clardy J (1984) J Org Chem 49:1417
106. Perkins DL, Ciereszko LS (1973) Hydrobiologia 42:77
107. Hadfield MG, Ciereszko LS (1978) In: Kaul PN, Sindermann CJ (eds) Drugs and food from the sea, myth or reality?, University of Oklahoma Press, 1978, p 145
108. Ciereszko LS (1962) Trans NY Acad Sci 24:502
109. Kupchan SM (1974) Fed Proc 33:2288
110. Hanson RL, Lardy HA, Kupchan SM (1970) Science 168:378
111. Kupchan SM, Giacobbe TJ, Krull IS, Thomas AM, Eakin MA, Fessler DC (1970) J Org Chem 35:3539
112. Howard BM, Clarkson K, Bernstein RL (1979) Tetrahedron Lett 4449
113. Fenical W (1976) In: Webber HH, Ruggieri GD (eds) Food-Drugs from the sea, conf proc (4th), Marine Technology Society, p 388
114. Baranger P (1964) Chem Abstr 61:15940e
115. Targett NM, Keeran WS (1984) J Nat Prod 47:556
116. Faulkner DJ (1984) Nat Prod Rep 551
117. Minale L (1978) In: Scheuer PJ (ed) Marine natural products, chemical and biological perspectives Vol I, Academic Press, New York, p 175
118. Kato Y, Scheuer PJ (1974) J Am Chem Soc 96:2245
119. Kato Y, Scheuer PJ (1975) Pure Appl Chem 41:1
120. Kato Y, Scheuer PJ (1976) Pure Appl Chem 48:29
121. Moore RE, Blackman AJ, Cheuk CE, Mynderse JS, Matsumoto GK, Clardy JC, Woodward RW, Craig JC (1984) J Org Chem 49:2484
122. Mynderse JS, Moore RE, Kashiwagi M, Norton TR (1977) Science 196:538
123. Moore RE (1981) In: Scheuer PJ (ed) Marine natural products, chemical and biological perspectives Vol IV, Academic Press, New York, p 1
124. Mynderse JS, Moore RE (1978) J Org Chem 43:2301
125. Moore RE (1982) Pure Appl Chem 54:1919
126. Moore RE, Scheuer PJ (1971) Science 172:495
127. Hashimoto Y (1979) Marine toxins and other bioactive marine metabolites, Japan Scientific Societies Press, Tokyo, p 248
128. Moore RE, Bartolini G (1981) J. Am Chem Soc 103:2491
129. Uemura D, Ueda K, Hirata Y, Naoki H, Iwashita T (1981) Tetrahedron Lett 22:2781
130. Moore RE, Bartolini G, Barchi J, Bothner-By AA, Dadok J, Ford J (1982) J Am Chem Soc 104:3776
131. Klein LL, McWhorter Jr WW, Ko SS, Pfaff KP, Kishi Y, Uemura D, Hirata Y (1982) J Am Chem Soc 104:7362
132. Ko SS, Finan JM, Yonaga M, Kishi Y, Uemura D, Hirata Y (1982) J Am Chem Soc 104:7364
133. Fujioka H, Christ WJ, Cha JK, Leder J, Kishi Y, Uemura D, Hirata Y (1982) J Am Chem Soc 104:7367
134. Cha JK, Christ WJ, Finan JM, Fujioka H, Kishi Y, Klein KL, Ko SS, Leder J, McWhorter Jr WW, Pfaff KP, Yonaga M, Uemura D, Hirata Y (1982) J Am Chem Soc 104:7369
135. Quinn RJ, Kashiwagi M, Moore RE, Norton TR (1974) J Pharm Sci 63:257
136. Pettit GR, Herald CL, Vanell LD (1979) J Nat Prod 42:407

137. Pettit GR, Hasler JA, Paull KD, Herald CL (1981) J Nat Prod 44:701
138. Pettit GR, Rideout JA, Hasler JA (1981) J Nat Prod 44:588
139. Pettit GR, Rideout JA, Hasler JA, Doubek DC, Reucroft PR (1981) J Nat Prod 44:713
140. Pettit GR, Fujii Y, Hasler JA, Schmidt JM, Michel C (1982) J Nat Prod 45:263
141. Pettit GR, Cragg GM, Herald CL (1985) Biosynthetic products for cancer chemotherapy, Vol V, Elsevier, New York
142. Schmitz FJ, Hollenbeak KH, Campbell DC (1978) J Org Chem 43:3916
143. Lad PJ, Brown JW, Shier WT (1978) Biochem Biophys Res Commun 85:1472
144. Schmitz FJ (1986) Personal communication
145. Yamamoto I, Nagumo T, Takahashi M, Fujihara M, Suzuki Y, Iizima N (1981) Japan J Exp Med 51:187
146. Yamamoto I, Takahashi M, Tamura E, Maruyama H (1982) Botanica Marina 25:455
147. Yamamoto I, Takahashi M, Tamura E, Maruyama H, Mori H (1984) Hydrobiologia 116/117:145
148. Shimizu Y (1985) J Nat Prod 48:223
149. Lichter W, Wellham LL, Van der Werf BA, Middlebrook RE, Sigel MM (1972) In: Worthen LR (ed) Food-Drugs from the sea: Proc., Mar Tech Soc, Washington, DC, p 117
150. Sigel MM, Lichter W, McCumber LJ, Ghaffar A, Wellham LL, Hightower JA (1984) Mononucl. Phagocyte Biol 26:451
151. Lichter W, Ghaffar A, Wellham LL, Sigel MM (1978) In: Kaul PN, Sindermann CJ (eds) Drugs and food from the sea: myth or reality?, University of Oklahoma Press, p 137
152. Lichter W, Lopes DM, Wellham LL, Sigel MM (1975) Proc Soc Exp Biol Med 150:475
153. Dunn WC, Carrier WL, Regan JD (1982) Toxicon 20:703
154. Kikuchi H, Tsukitani Y, Yamada Y, Iguchi K, Drexler SA, Clardy J (1982) Tetrahedron Lett. 23:1063
155. Schultz RM, Papamatheakis JD, Chirigos MA (1977) Science 197:674
156. Kikuchi H, Manda T, Kobayashi K, Yamada Y, Iguchi K (1983) Chem Pharm Bull 31:1086
157. Gerwick WH, Fenical W, Fritsch N, Clardy J (1979) Tetrahedron Lett 145
158. Gerwick WH, Fenical W (1981) J Org Chem 46:22
159. Jacobs RS, Culver P, Langdon R, O'Brien T, White S (1985) Tetrahedron 41:981
160. O'Brien ET, White S, Jacobs RS, Boder GB, Wilson L (1984) Hydrobiologia 116/117:141
161. Kikuchi H, Tsukitani Y, Iguchi K, Yamada Y (1982) Tetrahedron Lett 23:5171
162. Kikuchi H, Tsukitani Y, Iguchi K, Yamada Y (1983) Tetrahedron Lett 24:1549
163. Kobayashi M, Yasuzawa T, Yoshihara M, Akatsu H, Kyogoku Y, Kitagawa I (1982) Tetrahedron Lett 23:5331
164. Kobayashi M, Yasuzawa T, Yoshihara M, Son BW, Kyogoku Y, Kitagawa I (1983) Chem Pharm Bull 31:1440
165. Iguchi K, Yamada Y (1983) Tetrahedron Lett 24:4433
166. Kitagawa I, Kobayashi M, Yasuzawa T, Son BW, Yoshihara M, Kyogoku Y (1985) Tetrahedron 41:995
167. Baker BJ, Okuda RK, Yu PTK, Scheuer PJ (1985) J Am Chem Soc 107:2976
168. Fukushima M, Kato T, Yamada Y, Kitagawa I, Kurozumi S, Scheuer PJ (1985) Proc Am Assoc Cancer Res 26:249
169. Fukushima M, Kato T (1985) In: Havaishi O, Yamamoto S (eds) Advances in prostaglandin, thromboxane, and leukotriene research, Vol XV, Raven Press, New York, p 415
170. Iguchi K, Kaneta S, Mori K, Yamada Y, Honda A, Mori Y (1985) Tetrahedron Lett 26:5787
171. Nagaoka H, Iguchi K, Miyakoshi T, Yamada N, Yamada Y (1986) Tetrahedron Lett 27:223
172. Honda A, Yamamoto Y, Mori Y, Yamada Y, Kikuchi H (1985) Biochem Biophys Res Commun 130:515
173. Corey EJ, Mehrotra MM (1984) J Am Chem Soc 106:3384
174. Nagaoka H, Miyakoshi T, Yamada Y (1984) Tetrahedron Lett 25:3621
175. Hashimoto S, Arai Y, Hamanaka N (1985) Tetrahedron Lett 26:2679
176. Shibasaki M, Ogawa Y (1985) Tetrahedron Lett 26:3841

177. Pettit GR, Hartwell JL, Wood HB (1968) Cancer Res 28:2168
178. Christophersen C (1985) Acta Chemica Scandinavica B39:517
179. Pettit GR, Kamano Y, Herald CL, Schmidt JM, Zubrod CG (1986) Pure Appl Chem 58:415
180. Pettit GR, Herald CL, Doubek DL, Herald DL, Arnold E, Clardy J (1982) J Am Chem Soc 104:6846
181. Pettit GR, Holzapfel CW, Cragg GM, Herald CL, Williams P (1983) J Nat Prod 46:917
182. Pettit GR, Holzapfel CW, Cragg GM (1984) J Nat Prod 47:941
183. Pettit GR, Herald CL, Kamano Y (1983) J Org Chem 48:5354
184. Pettit GR, Herald CL, Kamano Y, Gust D, Aoyagi R (1983) J Nat Prod 46:528
185. Pettit GR, Kamano Y, Herald CL, Tozawa M (1984) J Am Chem Soc 106:6768
186. Pettit GR, Kamano Y, Herald CL, Tozawa M (1985) Can J Chem 63:1204
187. Pettit GR, Kamano Y, Aoyagi R, Herald CL, Doubek DL, Schmidt JM, Rudloe JJ (1985) Tetrahedron 41:985
188. Pettit GR, Kamano Y, Herald CL (1986) J Nat Prod 49:661
189. Pettit GR, Kamano Y, Herald CL (1987) J Org Chem 52:2848
190. Pettit GR, Leet JE, Herald CL, Yoshiaki K, Boettner FE, Baczynskyj L, Nieman RA (1987) J Org Chem 52:2854
191. Pettit GR (1986) Personal communication
192. Berkow RL, Kraft AS (1985) Biochem Biophys Res Commun 131:1109
193. Smith JB, Smith L, Pettit GR (1985) Biochem Biophys Res Commun 132:939
194. Fox JL (1982) Chem Eng News, Dec 6, p 29
195. Masamune S, Choy W, Petersen JS, Sita LR (1985) Angew Chem Int Ed Eng 24:1
196. Plinius GS (ca 60) Historia Naturalis, Lib IX, Lib. XXII
197. Pettit GR, Herald CL, Herald DL (1976) J Pharm Sci 65:1558
198. Pettit GR, Kamano Y, Fujii Y, Herald CL, Inoue M, Brown P, Gust D, Kitahara K, Schmidt JM, Doubek DC, Michel C (1981) J Nat Prod 44:482
199. Pettit GR, Kamano Y, Brown P, Gust D, Inoue M, Herald CL (1982) J Am Chem Soc 104:905
200. Schmidt U, Utz R (1984) Angew Chem Int Ed Engl 23:725
201. Hamada Y, Kohda K, Shioiri T (1984) Tetrahedron Lett 25:5303
202. Pettit GR, Nelson PS, Holzapfel CW (1985) J Org Chem 50:2654
203. Kashman Y, Groweiss A, Lidor R, Blasberger D, Carmely S (1985) Tetrahedron 41:1905
204. Montgomery DW, Zukoski CF (1983) Fed Proc 42:374
205. Montgomery DW, Zukoski CF (1985) Transplantation 40:49
206. Montgomery DW, Zukoski CF (1984) Fed Proc 43:368
207. Russell DH, Kibler R, Montgomery DW, Gout PW, Beer CT, Zukoski CF (1985) J Leuk Biol 38:187
208. Rossof AH, Johnson PA, Kimmell BD, Graham JE, Roseman DL (1983) Cancer Res. 315
209. Jiang TL, Lui RH, Salmon SE (1983) Cancer Chemother Pharmacol 11:1
210. Horvath S (1980) Toxicology 16:59
211. Gonzalez AG, Arteaga JM, Martin JD, Rodriguez ML, Fayos J, Martinez-Ripolls M (1978) Phytochem 17:947
212. Schmitz FJ, Michaud DP, Schmidt PG (1982) J Am Chem Soc 104:6415
213. Sims JJ, Lin GHY, Wing RM (1974) Tetrahedron Lett 3487
214. Gonzalez AG, Darias J, Diaz A, Fourneron JD, Martin JD, Perez C (1976) Tetrahedron Lett 3051
215. Stallard MO, Faulkner DJ (1974) Comp Biochem Physiol 49B:25
216. Stallard MO, Faulkner DJ (1974) Comp Biochem Physiol 49B:37
217. Hollenbeak KH, Schmitz FJ, Hossain MB, van der Helm D (1979) Tetrahedron 35:541
218. Pettit GR, Herald CL, Allen MS, Von Dreele RB, Vanell LD, Kao JPY, Blake W (1977) J Am Chem Soc 99:262
219. Hoye TR, Caruso AJ, Dellaria Jr JF, Kurth MJ (1982) J Am Chem Soc 104:6704
220. White JD, Nishiguchi T, Skeean RW (1982) J Am Chem Soc 104:3923
221. Shieh HM, Prestwich GD (1982) Tetrahedron Lett 23:4643
222. McEnroe FJ, Fenical W (1978) Tetrahedron 34:1661

223. Cafieri F, Fattorusso E, Magno S, Santacroce D, Sica D (1973) Tetrahedron 29:4259
224. Di Blasio G, Fattorusso E, Magno S, Mayol L, Pedone C, Santacroce C, Sica D (1976) Tetrahedron 32:473
225. Ciminiello P, Fattorusso E, Magno S, Mayol L (1985) J Nat Prod 48:64
226. Tada H, Yasuda F (1985) Chem Pharm Bull 33:1941
227. Faulkner DJ (1984) Nat Prod Rep 251
228. Fenical W, Paul VJ (1984) Hydrobiologica 116/117:135
229. Amico V, Oriente G, Piattelli M, Tringali C, Fattorusso E, Magno S, Mayol L (1978) Tetrahedron Lett 3593
230. Paul VJ, Fenical W (1984) Tetrahedron 40:2913
231. Blackman AJ, Wells RJ (1978) Tetrahedron Lett 3063
232. Paul VJ, Fenical W (1984) Tetrahedron 40:3053
233. Paul VJ, Fenical W (1983) Science 221:747
234. Amico V, Oriente G, Piattelli M, Tringali C, Fattorusso E, Magno S, Mayol L (1980) Tetrahedron 36:1409
234a. Tringali C, Oriente G, Piattelli M, Nicolosi G (1984) J Nat Prod 47:615
235. Alvarado AB, Gerwick WH (1985) J Nat Prod 48:132
236. Gerwick WH, Fenical W, Van Engen D, Clardy J (1980) J Am Chem Soc 102:7991
237. Gerwick WH, Fenical W (1983) J Org Chem 48:3325
238. Gerwick WH, Fenical W, Sultanbawa MUS (1981) J Org Chem 46:2233
239. Schmitz FJ, Hollenbeak KH, Carter DC, Hossain MB, van der Helm D (1979) J Org Chem 44:2445
240. Pettit GR, Ode RH, Herald CL, Von Dreele RB, Michel C (1976) J Am Chem Soc 98:4677
241. Cimino G, De Rosa D, De Stefano S, Minale L (1974) Tetrahedron 30:645
242. Gonzalez AG, Estrada DM, Martin JD, Martin VS, Perez C, Perez R (1984) Tetrahedron 40:4109
243. Gonzalez AG, Darias V, Estevez E (1982) Il Farmaco 37:179
244. Schmitz FJ, Chang JS, Hossain MB, van der Helm D (1985) J Org Chem 50:2862
245. Nakano T, Hernandez MI (1983) J Chem Soc Perkin Trans 1:135
246. de Miranda DS, Brendolan G, Imamura PM, Sierra MG, Marsaioli AJ, Ruveda EA (1981) J Org Chem 46:4851
247. Schmitz FJ, Vanderah DJ, Hollenbeak KH, Enwall CEL, Gopichand Y, SenGupta PK, Hossain MB, van der Helm D (1983) J Org Chem 48:3941
247a. Hoffmann HMR, Rabe J (1985) Angew Chem Int Ed Engl 24:94
248. Uchio Y, Toyota J, Nozaki H, Nakayama M, Nishizono Y, Hase T (1981) Tetrahedron Lett 22:4089
249. Weinheimer AJ, Matson JA, Hossain MB, van der Helm D (1977) Tetrahedron Lett 2923
250. Kazlauskas R, Murphy PT, Wells RJ, Schonholzer P, Coll JC (1978) Aust J Chem 31:1817
251. Tursch B, Braekman JC, Daloze D, Herin M, Karlsson E, Losman D (1975) Tetrahedron 31:129
252. Mori K, Suzuki S, Iguchi K, Yamada Y (1983) Chem Lett 1515
253. Weinheimer AJ, Matson JA, van der Helm D, Poling M (1977) Tetrahedron Lett 1295
254. Still WC, Mobilio D (1983) J Org Chem 48:4785
255. Aoki M, Tooyama Y, Uyehara T, Kato T (1983) Tetrahedron Lett 24:2267
256. Schmitz FJ, Gopichand Y, Michaud DP, Prasad RS, Remaley S, Hossain MB, Rahman A, Sengupta PK, van der Helm D (1981) Pure Appl Chem 51:853
257. Herin M, Tursch B (1976) Bull Soc Chim Belg 85:707
258. Herin M, Colin M, Tursch B (1976) Bull Soc Chim Belg 85:801
259. Fenical W (1978) In: Scheuer PJ (ed) Marine natural products, chemical and biological perspectives Vol II, Academic Press, New York, p 173
260. Bandurraga MM, Fenical W, Donovan SF, Clardy J (1982) J Am Chem Soc 104:6463
261. Yasuda F, Tada H (1981) Experientia 37:110
262. Braekman JC, Daloze D, Kaisin M, Moussiaux B (1985) Tetrahedron 41:4603
263. Fusetani N, Kato Y, Matsunaga S, Hashimoto K (1984) Tetrahedron Lett 25:4941
264. Kato Y, Fusetani N, Matsunaga S, Hashimoto K (1985) Chem Lett 1521

265. Kashman Y, Rotem M (1979) Tetrahedron Lett 1707
266. Manes LV, Bakus GJ, Crews P (1984) Tetrahedron Lett 25:931
267. Suzuki T, Suzuki M, Furusaki A, Matsumoto T, Kato A, Imanaka Y, Kurosawa E (1985) Tetrahedron Lett 26:1329
268. Blunt JW, Hartshorn MP, McLennan TJ, Munro MHG, Robinson WT, Yorke SC (1978) Tetrahedron Lett 69
269. Burnell DJ, ApSimon JW (1983) In: Scheuer PJ (ed) Marine natural products, chemical and biological perspectives Vol V, Academic Press, New York, p 287
270. Nigrelli RF (1952) Zoologica (NY Zool Soc) 37:89
270a. Kitagawa I, Nishino T, Kyogoku Y (1979) Tetrahedron Lett 1419
271. Cairns SD, Olmsted CA (1973) Gulf Res Rep 4:205
271a. Kitagawa I, Nishino T, Kobayashi M, Kyogoku Y (1981) Chem Pharm Bull 29:1951
272. Sullivan TD, Ladue KT, Nigrelli RF (1955) Zoologica (NY Zool Soc) 40:49
272a. Kitagawa I, Nishino T, Matsuno T, Akutsu H, Kyogoku Y (1978) Tetrahedron Lett 985
273. Colon RL, Ruggieri GD, Nigrelli RF (1976) In: Webber HH, Ruggieri GD (eds) Food-Drugs from the sea: Proc 1974, Marine Technology Soc., Washington DC, p 366
273a. Kitagawa I, Nishino T, Kobayashi M, Matsuno T, Akutsu H, Kyogoku Y (1981) Chem Pharm Bull 29:1942
274. Ruggieri GD (1965) Toxicon 3:157
275. Anisimov MM, Kuznetsova TA, Shirokov VP, Prokofyeva NG, Elyakov GB (1972) Toxicon 10:187
276. Anisimov MM, Fronert EB, Kuznetsova TA, Elyakov GB (1973) Toxicon 11:109
277. Anisimov MM, Shcheglov VV, Stonik VA, Fronert EB, Elyakov GB (1974) Toxicon 12:327
278. Anisimov MM, Popov AM, Dzizenko CN (1979) Toxicon 17:319
279. Stonik VA, Chumak AO, Isakov VV, Belogovtseva NI, Chirva V Ya, Elyakov GB (1979) Khim Prir Soedin 522
280. Anisimov MM, Prokofieva NG, Korotkikh LY, Kapustina II, Stonik VA (1980) Toxicon 18:221
281. Elyakov GB, Stonik VA, Levina EV, Slanke VP, Kuznetsova TA, Levin VS (1973) Comp Biochem Physiol 44B:325
282. Elyakov GB, Kuznetsova TA, Stonik VA, Levin VS, Albores R (1975) Comp Biochem Physiol 52B:413
283. Kelecom A, Tursch B, Vanhaelen M (1976) Bull Soc Chim Belg 85:277
284. Ruggieri GD, Nigrelli RF (1974) In: Humm H, Lane C (eds) Bioactive compounds from the sea, Dekker, New York, p 183
285. Owellen RJ, Owellen RG, Gorog MA, Klein D (1973) Toxicon 11:319
286. Rio GJ, Ruggieri GD, Stempien Jr MF, Nigrelli RF (1963) Am Zool 3:554
287. Ruggieri GD, Nigrelli RF (1964) Am Zool 4:431
288. Ruggieri GD, Nigrelli RF (1966) Am Zool 6:592
289. Minale L, Pizza C, Riccio R, Zollo F, Pusset J, Laboute P (1985) J Nat Prod 47:558
290. Pizza C, Pezzullo P, Minale L, Breitmaier E, Pusset J, Tirard P (1985) J Chem Res (S), 76; (M), 0969
291. Riccio R, Minale L, Pizza C, Zollo F, Pusset J (1982) Tetrahedron Lett 23:2899
292. Fusetani N, Kato Y, Hashimoto K, Komori T, Itakura Y, Kawasaki T (1984) J Nat Prod 47:997
293. Nakatsu T, Walker RP, Thompson JE, Faulkner DJ (1983) Experientia 39:759
294. Gunasekera SP, Schmitz FJ (1983) J Org Chem 48:885
295. Kobayashi M, Lee NK, Son BW, Yanagi K, Kyogoku Y, Kitagawa I (1984) Tetrahedron Lett 25:5925
296. Higa T, Tanaka J, Tsukitani Y, Kikuchi H (1981) Chem Lett 1647
297. Minale L, Pizza C, Riccio R, Greco OS, Zollo F, Pusset J, Menou JL (1984) J Nat Prod 47:784
298. Guyot M, Davoust D, Belaud C (1982) Tetrahedron Lett 23:1905
299. Guyot M, Morel E, Belaud C (1983) J Chem Res (S) 188; (M) 1823
300. Cheng KP, Bang L, Ourisson G, Beck JP (1979) J Chem Res (S) 84; (M) 1101
301. Matsuno T, Ookubo M, Komori T (1985) J Nat Prod 48:606

302. Pettit GR, Herald CL, Ode RH, Brown P, Gust DJ, Michel C (1980) J Nat Prod 43:752
303. Pietra F (1985) Gazz Chim Italiana 115:443
304. Higa T (1981) In: Scheuer PJ (ed) Marine natural products, chemical and biological perspectives Vol IV, Academic Press, New York, p 93
305. Minale L, Riccio R, Sodano G (1974) Tetrahedron Lett 3401
306. De Rosa S, Minale L, Riccio R, Sodano G (1976) J Chem Soc Perkin Trans 1, 1408
307. Cariello L, Guidici MDN, Zanetti L (1980) Comp Biochem Physiol 65C:37
308. Müller WEG, Zahn RK, Gasic MJ, Dogovic N, Maidhof A, Becker C, Diehl-Seifert B, Eich E (1985) Comp Biochem Physiol 80C:47
309. Schmitz FJ, Lakshmi V, Powell DR, van der Helm D (1984) J Org Chem 49:241
310. Kazlauskas R, Murphy PT, Warren RG, Wells RJ, Blount JF (1978) Aust J Chem 31:2685
311. Morita M, Endo M (1985) J Pharmacobio-Dyn 8:s-74
312. Carté B, Rose CB, Faulkner DJ (1985) J Org Chem 50:2785
313. Jamieson DD, de Rome PJ, Taylor KM (1980) J. Pharm Sci 69:462
314. Luibrand RT, Erdman TR, Vollmer JJ, Scheuer PJ, Finer J, Clardy J (1979) Tetrahedron 35:609
315. Luibrand RT (1986) Unpublished results
316. Fusetani N, Yasukawa K, Matsunaga S, Hashimoto K (1985) Tetrahedron Lett 26:6449
317. Sun HH, Ferrara NM, McDonnell OJ, Fenical W (1980) Tetrahedron Lett 21:3123
318. Francisco C, Banaigs B, Valls R, Codomier L (1985) Tetrahedron Lett 26:2629
319. Tymiak AA, Rinehart Jr KL (1983) J Am Chem Soc 105:7396
320. Endo M, Nakagawa M, Hamamoto Y, Nakanishi T (1983) J Chem Soc Chem Commun. 322, 980
321. Schmitz FJ, Gopichand Y (1978) Tetrahedron Lett 3637
322. Fusetani N, Kato Y, Matsunaga S, Hashimoto K (1983) Tetrahedron Lett 24:2771
323. Patil AD, Rinehart Jr KL (1986) J Org Chem, submitted
324. Schmitz FJ, Prasad RS, Gopichand Y, Hossain MB, van der Helm D (1981) J Am Chem Soc 103:2467
325. Tachibana K, Scheuer PJ, Tsukitani Y, Kikuchi H, Van Engen D, Clardy J, Gopichand Y, Schmitz FJ (1981) J Am Chem Soc 103:2469
326. Shibata S, Ishida Y, Ohizumi Y, Kitano H, Habon J, Tsukitani Y, Kikuchi H (1983) In: Bevan JA (ed) Vascular neuroeffector mechanisms: 4th International Symposium, Raven Press, New York, p 311
327. Murakami Y, Oshima Y, Yasumoto T (1982) Bull Jap Soc Sci Fish 48:69
328. Yasumoto T, Murata M, Oshima Y, Sano M, Matsumoto GK, Clardy J (1985) Tetrahedron 41:1019
329. Uemura D, Takahashi K, Yamamoto T, Katayama C, Tanaka J, Okumura Y, Hirata Y (1985) J Am Chem Soc 107:4796
330. Hirata Y, Uemura D (1985) Abstracts of IUPAC Symposium on Organic Chemistry of Medicinal Natural Products, Shanghai, Nov 10–14
331. Schmitz FJ, Gunasekera SP, Yalamanchili G, Hossain MB, van der Helm D (1984) J Am Chem Soc 106:7251
332. Ksebati MB, Schmitz FJ (1985) J Org Chem 50:5637
333. Ireland CM, Faulkner DJ (1981) Tetrahedron 37, Supplement 1, 233
334. Hochlowski JE, Faulkner DJ (1983) Tetrahedron Lett 24:1917
335. Biskupiak JE, Ireland CM (1985) Tetrahedron Lett 26:4307
336. Bergquist PR, Wells RJ (1983) In: Scheuer PJ (ed) Marine natural products, chemical and biological perspectives, Vol V, Academic Press, New York, p 1
336a. D'Ambrosio M, Guerriero A, Traldi P, Pietra F (1982) Tetrahedron Lett 23:4403
337. Minale L, Cimino G, De Stefano S, Sodano G (1976) Fortschritte der Chem Org Naturst 33:1
338. Kelecom A, Kannengiesser GJ (1979) An Acad Brasil Cienc 51:633
339. Wiedenmayer F (1977) Shallow-water sponges of the western Bahamas, Birkhaeuser Verlag, Basel, Switzerland, p 64
340. Cimino G, De Rosa S, De Stefano S, Self R, Sodano G (1983) Tetrahedron Lett 24:3029
341. Fattorusso E, Minale L, Sodano G (1970) J Chem Soc Chem Commun 751

342. Fattorusso E, Minale L, Sodano G (1972) J Chem Soc Perkin Trans 1:16
343. Stempein Jr MF, Chib JS, Nigrelli RF, Mierzwa RA (1973) In: Worthen LR (ed) Food-Drugs from the sea, proc 3rd, Mar Tech Soc, Washington, DC, p 105
344. Makarieva TN, Stonik VA, Alcolado P, Elyakov YB (1981) Comp Biochem Physiol 68B:481
345. Bergquist PR (1980) NZ J Zool 7:443
346. Chang CWJ, Weinheimer AJ (1977) Tetrahedron Lett 4005
347. Fulmor W, Van Lear GE, Morton GO, Mills RD (1970) Tetrahedron Lett 4551
348. Gorshkov BA, Gorshkova IA, Makarieva TN, Stonik VA (1982) Toxicon 20:1092
349. Gorshkov BA, Gorshkova IA, Makarieva TN (1984) Toxicon 22:441
350. Andersen RJ, Faulkner DJ (1975) J Am Chem Soc 97:936
351. D'Ambrosio M, Guerriero A, De Clauser R, De Stanchina G, Pietra F (1983) Experientia 39:1091
352. Sharma GM, Burkholder PR (1967) J Antibiotics, Ser A 20:200
353. Tymiak AA, Rinehart Jr KL (1981) J Am Chem Soc 103:6763
354. Sharma GM, Burkholder PR (1967) Tetrahedron Lett 4147
355. D'Ambrosio M, Guerriero A, Pietra F (1984) Helv Chim Acta 67:1484
356. Andersen RJ, Faulkner DJ (1973) Tetrahedron Lett 1175
357. Evans DA, Wong RY (1977) J Org Chem 42:350
358. Gopichand Y, Schmitz FJ (1979) Tetrahedron Lett 3921
359. Hollenbeak KH, Schmitz FJ, Kaul PN, Kulkarni SK (1978) In: Kaul PN, Sindermann CJ (eds) Drugs and food from the sea, myth or reality?, University of Oklahoma Press, p 81
360. Sesin DF, Ireland CM (1984) Tetrahedron Lett 25:403
361. Perry NB, Blunt JW, McCombs JD, Munro MHG (1986) J Org Chem 51:5476
362. Tymiak AA, Rinehart Jr KL, Bakus GJ (1985) Tetrahedron 41:1039
363. Carter GT, Rinehart Jr KL, Li HL, Kuentzel SL, Conner JL (1978) Tetrahedron Lett 4479
364. Roll DM, Ireland CM (1985) Tetrahedron Lett 26:4303
365. Heitz S, Durgeat M, Guyot M, Brassy C, Bachet B (1980) Tetrahedron Lett 21:1457
366. Hogan IT, Sainsbury M (1984) Tetrahedron 40:681
367. Kazlauskas R, Murphy PT, Quinn RJ, Wells RJ (1977) Tetrahedron Lett 61
368. Hollenbeak KH, Schmitz FJ (1977) Lloydia 40:479
369. Djura P, Faulkner DJ (1980) J Org Chem 45:735
370. Christophersen C (1983) In Scheuer PJ (ed) Marine natural products, chemical and biological perspectives Vol V, Academic Press, New York, p 259
371. Schmitz FJ, Hollenbeak KH, Prasad RS (1979) Tetrahedron Lett 3387
372. Chantrapromma K, McManis JS, Ganem B (1980) Tetrahedron Lett 21:2605
373. Carté B, Faulkner DJ (1983) J Org Chem 48:2314
374. Frincke JM, Faulkner DJ (1982) J Am Chem Soc. 104:265
375. Arai T, Kubo A (1983) In: The Alkaloids Vol. XXI, Academic Press, New York, p 55
376. Schmitz FJ, Agarwal SK, Gunasekera SP, Schmidt PG, Shoolery JN (1983) J Am Chem Soc 105:4835
377. Harbour GC, Tymiak AA, Rinehart Jr KL, Shaw PD, Hughes Jr RG, Mizak SA, Coats JH, Zurenko GE, Li LH, Kuentzel SL (1981) J Am Chem Soc 103:5604
378. Snider BB, Faith WC (1983) Tetrahedron Lett 24:861
379. Snider BB, Faith WC (1984) J Am Chem Soc 106:1443
380. Walts AE, Roush WR (1985) Tetrahedron 41:3463
381. Cimino G, De Rosa S, De Stefano S, Mazzarella L, Puliti R, Sodano G (1982) Tetrahedron Lett 23:767
382. Manes LV, Naylor S, Crews P, Bakus GJ (1985) J Org Chem 50:284
383. Faulkner DJ (1986) Nat Prod Rep 1
384. Roesener JA, Scheuer PJ (1986) J Am Chem Soc 108:846
385. Matsunaga S, Fusetani N, Hashimoto K, Koseki K, Noma M (1986) J Am Chem Soc 108:847
386. Ireland CM, Scheuer PJ (1980) J Am Chem Soc 102:5691
387. Ireland CM, Durso Jr AR, Newman RA, Hacker MP (1982) J Org Chem 47:1807
388. Biskupiak JE, Ireland CM (1983) J Org Chem 48:2302

389. Wasylyk JM, Biskupiak JE, Costello CE, Ireland CM (1983) J Org Chem 48:4445
390. Schmidt U, Gleich P (1985) Angew Chem Int Ed Engl 24:569
391. Schmidt U, Utz R, Gleich P (1985) Tetrahedron Lett 26:4367
392. Hamada Y, Shibata M, Shioiri T (1985) Tetrahedron Lett 26:5155
393. Hamada Y, Shibata M, Shioiri T (1985) Tetrahedron Lett 26:5159
394. Schmidt U, Griesser H (1986) Tetrahedron Lett 27:163
395. Hamamoto Y, Endo M, Nakagawa M, Nakanishi T, Mizukawa K (1983) J Chem Soc Chem Commun 323
396. Hamada Y, Kato S, Shioiri T (1985) Tetrahedron Lett. 26:3223
397. Suntory Ltd. (1984) Chem Abstr 100:215498n
398. Müller WEG, Maidhof A, Zahn RK, Conrad J, Rose T, Stefanovich P, Muller I, Friese U, Uhlenbruck G (1984) Biol Cell 51:381
399. Carter DC, Moore RE, Mynderse JS, Niemczura WP, Todd JS (1984) J Org Chem 49:236
400. Matsunaga S, Fusetani N, Konosu S (1985) Tetrahedron Lett 26:855
401. Matsunaga S, Fusetani N, Konosu S (1984) Tetrahedron Lett 25:5165
402. Matsunaga S, Fusetani N, Konosu S (1985) J Nat Prod 48:236
403. Ballantine JA, Psaila AF, Pelter A, Murray-Rust P, Ferrito V, Schembri P, Jaccarini V (1980) J Chem Soc Perkin Trans. 1:1080
404. Jaccarini V, Agius L, Schembri PJ, Rizzo M (1983) J Exp Mar Biol Ecol 66:25
405. Pelter A, Ballantine JA, Ferrito V, Jaccarini V, Psaila AF, Schembri PJ (1976) J Chem Soc Chem Commun 999
405a. Pelter A, Abela-Medici A, Ballantine JA, Ferrito V, Ford S, Jaccarini V, Psaila AF (1978) Tetrahedron Lett 2017
406. Cariello L, Guidici MDN, Zanetti L, Prota G (1978) Experientia 34:1427
407. Agius L, Jaccarini V, Ballantine JA, Ferrito V, Pelter A, Psaila AF, Zammit VA (1979) Comp Biochem Physiol 63B:109
408. Kato Y, Fusetani N, Matsunaga S, Hashimoto K (1985) Tetrahedron Lett 26:3483
409. Weinheimer AJ, Chang CWJ, Matson JA, Kaul PN (1978) Lloydia 41:488
410. Komori T, Sakamoto K, Matsuo J, Sanechika Y, Nohara T, Ito Y, Kawasaki T, Schulten HR (1978) 11th Symp Pap IUPAC Int Symp Chem Nat Prod 2:120
411. Komori T, Sanechika Y, Ito Y, Matsuo J, Nohara T, Kawasaki T, Schulten HR (1980) Liebigs Ann Chem 653
412. Baker JT, Murphy V (1981) Handbook of Marine Sciences, Vol II, CRC Press, p 83
413. Dematte N, Guerriero A, De Clauser R, De Stanchina G, Lafargue F, Cuomo V, Pietra F (1985) Comp Biochem Physiol 81B:479
414. Fuhrman FA, Fuhrman GJ, Nachman RJ, Mosher HS (1981) Science 212:557
415. Quinn RJ, Gregson RP, Cook AF, Bartlett RT (1980) Tetrahedron Lett 21:567
416. Cook AF, Bartlett RT, Gregson RP, Quinn RJ (1980) J Org Chem 45:4020
417. Fuhrman FA, Fuhrman GJ, Kim YH, Pavelka LA, Mosher HS (1980) Science 207:193
418. Yong HK, Nachman RJ, Pavelka L, Mosher HS, Fuhrman FA, Fuhrman GJ (1981) J Nat Prod 44:206
419. Kim YH, Nachman RJ, Pavelka L, Mosher HS, Fuhrman FA, Fuhrman GJ (1981) J Nat Prod 44:206
420. Grozinger K, Freter KR, Farina P, Gladczuk A (1983) Eur. J Med Chem-Chim Ther 18:221
421. Bakus GJ (1981) Science 211:497
422. Albericci M, Braekman JC, Daloze D, Tursch B (1982) Tetrahedron 38:1881
423. Endean R, Cameron AM (1983) Toxicon 21:105
424. Bergquist PR, Bedford JJ (1978) Mar Biol 46:215
425. Thompson JE, Walker RP, Faulkner DJ (1985) Mar Biol 88:11
426. Dixon SE (1985) Sea Technology Dec 19
427. Rinehart Jr KL, Armstrong JE, Hughes Jr RG, Theiss WC, Munro MHG, Holt TG, Hummel H, Pomponi SA (1985) Abstracts of the Fifth International Symposium on Marine Natural Products, Paris, September 2–6
428. Blunt JW, Calder VL, Fenwick GD, Lake RJ, McCombs JD, Munro MHG, Perry NB (1987) J Nat Prod 50:290

429. Driscoll JS (1984) Cancer Treatment Rep 68:63
430. Goldin A, Schepartz SA, Venditti JM, DeVita Jr VT (1979) In: DeVita Jr VT, Busch H (eds) Methods in cancer research, Vol. XVI, Academic Press, New York, p 165
431. Frei E (1982) Science 217:600
432. Rockwell S (1980) Br J Cancer 41:118
433. Venditti JM (1981) Seminars in Oncology 8:349
434. Venditti JM (1983) Cancer Treatment Rep 67:767
435. Venditti JM, Wesley RA, Plowman J (1984) Advances in Pharmacology and Chemotherapy 20:1
436. Hamburger AW, Salmon SE (1977) Science 197:461
437. Salmon SE, Hamburger AW, Soehnlen B, Durie BGM, Alberts DS, Moon TE (1978) N Eng J Med 298:1321
438. Elliot J (1979) J Am Med Assoc 242:501
439. Salmon SE, Meyskens Jr FL, Alberts DS, Soehnlen B, Young L (1981) Cancer Treatment Rep 65:1
440. Hamburger AW (1981) J Nat Cancer Inst 66:981
441. Calvert AH, Harrap KR (1983) Developments in Oncology 15:359
442. Shoemaker RH, Wolpert-DeFilippes MK, Kern DH, Lieber MM, Makuch RW, Melnick NR, Miller WT, Salmon SE, Simon RM, Venditti JM, Von Hoff DD (1985) Cancer Research 45:2145
443. Inoue K, Mukaiyama T, Mitsui I, Ogawa M (1985) Cancer Chemother Pharmacol 15:208
444. Hakala MT, Rustum YM (1979) In: DeVita Jr VT, Busch H (eds) Methods in cancer research, Vol XVI, Academic Press, New York, p 247
445. Cragg GM, Suffness M (1985) Abstracts of the Fifth International Symposium on Marine Natural Products, Paris, September 2–6
446. Boyd RB, Shoemaker RH, Cragg GM, Suffness M (1985) SeaPharm Conference on Pharmaceuticals and the Sea, Harbor Branch Oceanographic Institution, Fort Pierce, Florida, USA, October 24–25
447. Cragg GM (1986) Personal communication
448. Nachman RJ, Wong RW, Haddon WF, Lundin RE (1985) J Chem Soc. Perkin Trans 1 1315

Subject Index

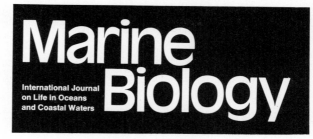

International Journal on Life in Ocean and Coastal Waters

ISSN 0025-3162   Title No. 227

Editor-in-Chief: O. Kinne, Oldendorf/Luhe

Editors: M. Anraku, Tokyo; B. Battaglia, Padova; R. W. Doyle,
Halifax; T. M. Fenchel, Helsingør; J. P. Grassle, Woods Hole;
G. F. Humphrey, Sidney; J. M. Lawrence, Tampa; J. Mauchline,
Oban; J. M. Pérès, Marseille; P. C. Schroeder, Pullman;
M. E. Vinogradov, Moscow

Assistants to the Editors: R. Friedrich, T. E. Galvin

Marine Biology publishes contributions in the following fields:

Plankton Research: Studies on the biology, physiology, biochemistry, ecology, and genetics of plankton organisms under
both laboratory and field conditions.

Experimental Biology: Research on metabolic rates and routes
in microorganisms, plants and animals. Respiration; nutrition;
life cycles.

Biochemistry, Physiology and Behavior: Biochemical research
on marine organisms; photosynthesis; permeability; osmoregu-
lation; ion regulation; active transport; adaptation; analyses of
environmental effects on functions and structures; migrations;
orientation; general behavior.

Biosystem Research: Experimental biosystems and micro-
cosms. Energy budgets. Interspecific interrelationships, food
webs. Dynamics and structures of microbial, plant and animal
populations. Use, management and protection of living marine
resources. Effects of man on marine life, including pollution.

Evolution: Investigations on speciation, population genetics,
and biological history of the oceans.

Theoretical Biology Related to the Marine Environment:
Concepts and models of quantification and mathematical
formulation; system analysis; information theory.

Methods: Apparatus and techniques employed in marine biological research; underwater exploration and experimentation.

Abstracted/Indexed in: Biosis, CAS, Current Contents,
Environment Abstracts, Excerpta Medica, INIS, Technical
Information Center/Energinfo.

Springer-Verlag
Berlin Heidelberg New York
London Paris Tokyo

Springer